ATOMIC AND MOLECULAR DATA AND THEIR APPLICATIONS

ATOMIC AND MOLECULAR DATA AND THEIR APPLICATIONS

ICAMDATA—First International Conference

Gaithersburg, Maryland October 1997

EDITORS
Peter J. Mohr
Wolfgang L. Wiese
National Institute of Standards and Technology,
Gaithersburg, Maryland

American Institute of Physics

AIP CONFERENCE PROCEEDINGS 434

Woodbury, New York

Editors:

Peter J. Mohr
Atomic Physics Division
Building 225, Room B161
National Institute of Standards and Technology
Gaithersburg, MD 20899-0001

Email: mohr@nist.gov

Wolfgang L. Wiese
Atomic Physics Division
Building 221, Room A167
National Institute of Standards and Technology
Gaithersburg, MD 20899-0001

Email: wiese@nist.gov

Authorization to photocopy items for internal or personal use, beyond the free copying permitted under the 1978 U.S. Copyright Law (see statement below), is granted by the American Institute of Physics for users registered with the Copyright Clearance Center (CCC) Transactional Reporting Service, provided that the base fee of $15.00 per copy is paid directly to CCC, 222 Rosewood Drive, Danvers, MA 01923. For those organizations that have been granted a photocopy license by CCC, a separate system of payment has been arranged. The fee code for users of the Transactional Reporting Service is: 1-56396-751-0/ 98 /$15.00.

© 1998 American Institute of Physics

Individual readers of this volume and nonprofit libraries, acting for them, are permitted to make fair use of the material in it, such as copying an article for use in teaching or research. Permission is granted to quote from this volume in scientific work with the customary acknowledgment of the source. To reprint a figure, table, or other excerpt requires the consent of one of the original authors and notification to AIP. Republication or systematic or multiple reproduction of any material in this volume is permitted only under license from AIP. Address inquiries to Office of Rights and Permissions, 500 Sunnyside Boulevard, Woodbury, NY 11797-2999; phone: 516-576-2268; fax: 516-576-2499; e-mail: rights@aip.org.

L.C. Catalog Card No. 98-71563
ISBN 1-56396-751-0
ISSN 0094-243X
DOE CONF- 970960

Printed in the United States of America

Contents

Preface ix
Committees xi
Sponsors xii

GLOBAL TRENDS AFFECTING SCIENTIFIC DATA

Policy and Funding Outlook: U.S. and Global Trends in Science
and Engineering 3
    N. F. Lane
Transnational Exchange of Scientific Data: The "Bits of Power" Report 7
    R. S. Berry

ATOMIC DATA PRODUCTION

The Iron Project: Past, Present, and Future 23
    K. Butler
Ion–Atom Data for Fusion Research 37
    R. Hoekstra, H. Anderson, F. W. Bliek, G. Lubinski, R. Morgenstern,
    R. E. Olson, and H. P. Summers

BENCHMARK DATA

Electron Impact Benchmark Experiments Viewed from Different
Perspectives 57
    G. H. Dunn
Precision Lifetime Measurements on Neutral Atoms—A Sensitive Test
for Atomic-Structure Calculations 67
    U. Volz and H. Schmoranzer

DATA ASSESSMENT AND EVALUATION

X-ray Transition Energies—New Comprehensive Evaluation 89
    R. D. Deslattes, E. G. Kessler, Jr., P. Indelicato, and E. Lindroth
The Critical Assessment of Atomic Transition Probabilities 105
    W. L. Wiese and D. E. Kelleher

Italicized name indicates the author who presented the paper

MAJOR DATA ACTIVITIES AND SERVICES

Atomic Collision Databases and Data Services—A Survey 119
    D. R. Schultz
Atomic and Molecular Data Activities in Japan:
Production, Evaluation, Databases, and Applications 131
    H. Tawara
The Russian Effort in Establishing Large Atomic and Molecular
Databases 147
    L. P. Presnyakov
Atomic and Molecular Databases for Planetary and Terrestrial
Applications 159
    L. R. Brown

ASTROPHYSICAL AND ATMOSPHERIC DATA NEEDS

Atomic and Molecular Data for Cosmology 193
    A. Dalgarno
Atomic Data and Modelling in X-ray Astronomy 203
    T. R. Kallman
A New Perspective on the Sun from SOHO—Challenges for Atomic
Physics 213
    K. Dere and H. Mason
Some Examples of Spectroscopic Data Needs for Optical Remote Sensing
in the Atmosphere 221
    J.-M. Flaud

LABORATORY DATA NEEDS AND APPLICATIONS

The Crucial Role of Atomic and Molecular Processes in the Success
of Controlled Fusion 233
    D. E. Post
The Use of Atomic and Molecular Data in Fusion Plasma Diagnostics 259
    H. P. Summers, H. Anderson, N. R. Badnell, F. W. Bliek, D. C. Griffin,
    M. von Hellermann, R. Hoekstra, A. Howman, L. D. Horton, R. Konig,
    G. M. McCracken, C. F. Maggi, M. G. O'Mullane, M. S. Pindzola,
    R. E. Olson, and M. F. Stamp
The National Ignition Facility and Atomic Data 287
    D. H. Crandall
Atomic Data and Methods for Low-Pressure Discharge Lamps 295
    T. J. Sommerer

Italicized name indicates the author who presented the paper

Understanding the High Intensity Discharge Lamp:
The Need for More Data 315
    H. G. Adler
Industrial Applications of Low Temperature Plasmas 333
    J. N. Bardsley
Molecular Spectroscopy Data Needs 377
    J. T. Hougen

PANEL DISCUSSIONS

International Coordination of A+M Data Efforts 391
    R. Janev (Panel Chair), K. Becker, R. Clark, G. Lister, and K. Niemax
Increasing the Visibility and Publicity for Data Activities and Assuring
the Open Exchange of Data 397
    D. R. Schultz (Panel Chair), R. S. Berry, C. Mendoza, and S. Younger
Standardization of Databases; Data Assessment;
and Uncertainty Statements 401
    P. Smith (Panel Chair), L. Brown, J. Rumble, and H. Tawara
Utilization of the World Wide Web 403
    P. Mohr (Panel Chair), G. Mallard, U. Ralchenko, and D. R. Schultz

Author Index 405

Italicized name indicates the author who presented the paper

Preface

These proceedings contain the written versions of most of the invited talks presented at the first International Conference on Atomic and Molecular Data and Their Applications (ICAMDATA), which was held at the National Institute of Standards and Technology (NIST) from September 29 through October 2, 1997. One hundred seventy-one registered participants from 20 countries attended the conference. Twenty-seven invited talks and about 80 contributed poster papers were presented. Also, representatives of six atomic and molecular data centers from around the world were invited to give short presentations of their databases.

In broad terms, the invited presentations addressed four major areas of importance to atomic and molecular data activities: First, global trends affecting scientific data were addressed. Second, there were several talks on production of collision and spectral radiation data with emphasis on new approaches in generating benchmark data. Third, issues in data assessment and important database and data management activities in several countries were described, and fourth, the data needs of the main user communities, such as the magnetic and inertial fusion research communities, the semiconductor-related plasma processing and the lighting industries, the atmospheric research community and the space astronomy community, were reviewed. Summaries of four panel discussions held at the conference, as provided by the panel chairs and members, are also included in these proceedings. These reflect community views on current issues and problems, and they provide some recommendations for future action.

This conference grew out of a workshop convened at the Harvard Smithsonian Center for Astrophysics on June 14 and 15, 1996. At this workshop, it was emphasized that there is a need for better international coordination in order to promote atomic and molecular data compilation and database activities as a valid enterprise. Needs were also expressed for better coordination of planning efforts among the data centers as well as for longer-range planning with the users and producers on specific needs and priorities. It was generally felt that communication between producers and data centers, including database developers, is good, but the interactions between the user communities and the data producers and centers are weak and need to be increased. The general sentiment among workshop participants was that an international data conference series would be a desirable forum to bring user and producer and database developers closer together and that the meeting could also serve as a focal point for getting messages to the general community on data needs as well as on the general status of data production activities (between meetings, a permanent conference website may serve this purpose). There was interest within NIST in hosting the first data conference in atomic and molecular physics, because NIST has had a long tradition in generating and critically compiling reference data in these fields. An ad hoc international program committee was created in the Summer of 1996 with A. Dalgarno agreeing to serve as the chair and R. Janev as the vice chair. In addition, an international advisory board was created and W. Wiese agreed to serve as the chairman for the first conference at NIST in 1997.

These proceedings are expected to be of interest to both producers and users of data. The contributions to this volume by experts from various user groups should be of importance to atomic and molecular physicists generating new and improved data by providing them with ideas on current data needs of these communities. But they should also be valuable to data users. Anyone who needs to apply atomic and molecular data has to know a good bit about their availability and quality, and these proceedings provide up-to-date surveys on this subject with an appreciable amount of detail.

It is a pleasure to acknowledge the competent assistance of Clifton E. Marlow in the preparation for the conference as well as the editing of these proceedings. Also, we would like to thank many colleagues at NIST, too numerous to mention all by name, for their cheerful help before and during the conference. In particular, this includes the expert assistance of the NIST conference management group. The International Program Committee assisted greatly in the selection of invited speakers and other organizational matters. Finally, we gratefully acknowledge the financial support of the organizations listed below as sponsors and co-sponsors.

Peter J. Mohr
Wolfgang L. Wiese
Gaithersburg, Maryland
March 1998

Conference Committee

Dr. Wolfgang Wiese, NIST-Atomic Physics Division, Conference Chair
Dr. Peter Mohr, NIST-Atomic Physics Division, Conference Secretary

Dr. James Babb, Harvard-Smithsonian Center for Astrophysics
Prof. Kurt Becker, Stevens Institute of Technology
Dr. George Doschek, Naval Research Laboratory
Prof. James Lawler, University of Wisconsin
Dr. David Leckrone, NASA Goddard Space Flight Center
Prof. Michael Pindzola, Auburn University
Prof. Anil Pradhan, Ohio State University
Dr. John Rumble, NIST-Standard Reference Data Program
Dr. David Schultz, Oak Ridge National Laboratory
Dr. Peter Smith, Harvard-Smithsonian Center for Astrophysics

International Program Committee

A. Dalgarno (USA), Chair
R. Janev (Austria), Vice-Chair
D. Schultz (USA), Treasurer
K. Berrington (N. Ireland), Secretary
L. Brown (USA)
M. Capitelli (Italy)
G. Lister (USA)
H. Mason (United Kingdom)
K. Niemax (Germany)
L. Presnyakov (Russia)
E. Roueff (France)
H. Tawara (Japan)
W. Wiese (USA)

International Advisory Board

N. Bardsley (USA)
P. Burke (N. Ireland)
R. Crompton (Australia)
J. Delcroix (France)
G. Dunn (USA)
F. Gianturco (Italy)
R. Hulse (USA)
J. Li (P. R. China)
I. Martinson (Sweden)
C. Mendoza (Venezuela)
J. Rumble (USA)
E. Salzborn (Germany)
M. Seaton (United Kingdom)
I. Sobelman (Russia)
K. Takayanagi (Japan)

Sponsors	Co-Sponsors
U.S. Department of Commerce, Technology Administration,	Electric Power Research Institute
	Office of Fusion Energy Science, Department of Energy
National Institute of Standards and Technology	Institute for Theoretical Atomic and Molecular Physics at the Harvard-Smithsonian Center for Astrophysics
	Lawrence Livermore National Laboratory
Physics Laboratory, Atomic Physics Division, and Standard Reference Data Program	U.S. Naval Research Laboratory
	SEMATECH
	Office of Space Sciences, National Aeronautics and Space Administration

Global Trends Affecting
Scientific Data

Policy and Funding Outlook: U.S. and Global Trends in Science and Engineering

Neal F. Lane

National Science Foundation
Arlington, VA 22230, USA

Abstract. Two "global" trends greatly affect the future of science and engineering research and education. The first is funding, where public investment patterns in the G-7 nations deserve careful examination. The second is an increased emphasis on evaluation and accountablity (which in many ways is a result of the funding environment).

INTRODUCTION

These informal remarks provide a brief look at two trends that impact on the work of the U.S. National Science Foundation (NSF): tightening funding and increasing accountability. These are both "global" trends, for better or worse.

FUNDING TRENDS

Figure 1 shows the total research and development (R&D) spending in major industrialized countries. The U.S. spends roughly the same amount as the other six of the Group of Seven (G-7) nations combined. (The other six countries are Japan, Germany, France, the United Kingdom, Italy, and Canada.) In most of the G-7 countries, the trend of slowing R&D growth is continuing. In the U.S. and Japan, however, a turn-about in this trend may be evident.

Japan has well publicized plans to double government investment in R&D. Since 1995, Japanese government R&D spending has increased appreciably, by 12% in 1996, and a 7% increase is planned for 1997. The doubling plan affects only Japanese government R&D spending, which is roughly 20% of overall Japanese R&D. Whether or not the Japanese achieve the doubling,

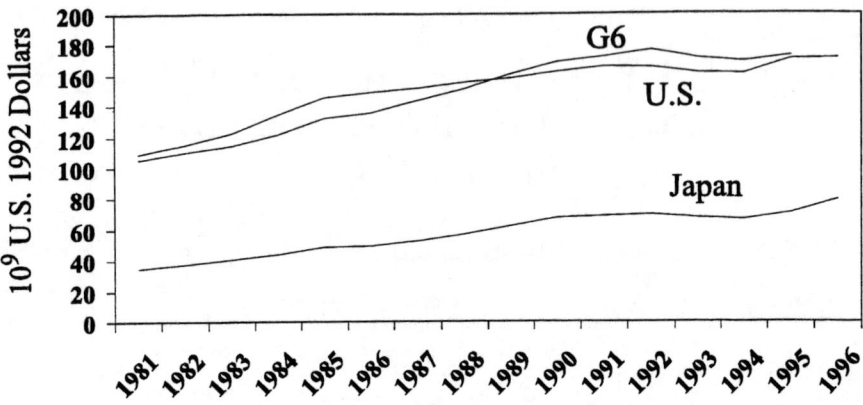

FIGURE 1. U.S. and the other six G-7 countries' R&D spending, in billions of constant 1992 dollars. Source: National Science Foundation, Division of Science Resource Studies.

this commitment by our friends to the East is something other governments would do well to emulate.

Figure 2 shows U.S. R&D Spending. Recent trends hold some promise. In inflation adjusted terms, the total R&D expenditures rose by 4 and 1 percent in 1995 and 1996, respectively, with the 1995 increase being the largest since 1985. But, this growth is driven entirely by investments on the industry side. Industry's projected funding for 1996 represents a 3.5% increase in real terms over the level for 1995.

Federal R&D support in 1996 reached $61.9 billion, a 3.0% decline in real terms from 1995. Since 1987, Federal support has dropped by 23%, after adjusting for inflation. The Federal share of U.S. R&D funds first fell below 50% percent of the national total in 1978. Since then, the Federal share has dropped steadily. The projected value for 1996 of 33.6% is the lowest share since NSF began collecting national R&D data 44 years ago.

RESEARCH ACCOUNTABILITY

The fiscal environment has brought greater competition for funds, and greater scrutiny and accountability. A forthcoming report from the Organisation for Economic cooperation and Development (OECD) notes that this is truly a global trend:

"Research evaluation has emerged as a 'rapid growth industry.' In most OECD countries, there is an increasing emphasis on accountability, as

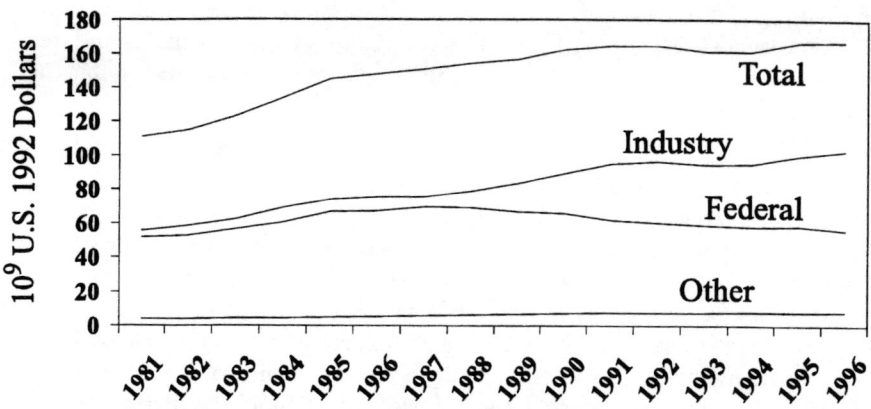

FIGURE 2. U.S. R&D spending in billions of constant (1992) dollars. Source: National Science Foundation, Division of Science Resource Studies; 1996 data are preliminary.

well as on the effectiveness and efficiency of government supported research." [1]

In the U.S., the driving force is the Government Performance and Results Act (GPRA). It is designed to improve the operation of all government programs by establishing a system of program performance goals and a method to measure the results.

Starting with FY 1999, all federal budgets will be "outcome based." What does this mean? It is probably easiest to describe by looking at how NSF is approaching GPRA.

Our approach begins with five outcome goals:

- Discoveries at and across the frontier of science and engineering. – While not every project we support will produce a noted discovery, there should be results from the agency and most parts as a whole.

- Connections between discoveries and their use in service to society. – We will not cause the use of discoveries, but we expect their applications in the service of society to further justify our investment.

- A diverse, globally-oriented workforce of scientists and engineers. – This is the expected outcome of our graduate education and research programs for undergraduates.

- Improved achievement in mathematics and science. – This is our toughest challenge, but we have a mandate from Congress and a presidential directive.

- Meaningful information on the national and international science and engineering enterprise. – This is also related to a Congressional Mandate. The report, *Science and Engineering Indicators*, is one of a number of reports we produce toward this end.

Our strategy for reaching these goals begins with caution. We avoid making bold statements that predict outcomes about NSF programs and their impact on society. We rely on peer review to ensure high standards for all of our investments. All of this requires us to think about the many ways excellent research in science and engineering benefits society–making it one of the best investments taxpayers can make for the future of their country.

Moreover, many researchers choose their fields and projects with societal benefits in mind, even if the research itself is quite fundamental and intellectually challenging. Measuring those benefits, let alone predicting them, rightfully gives us pause. Increasingly, that is what we are going to be asked to do in a balanced budget environment.

REFERENCES

1. Organisation for Economic Cooperation and Development, Working Group on The Science System (April 1997 Working Draft).

Transnational Exchange of Scientific Data: The "Bits of Power" Report

R. Stephen Berry

Department of Chemistry and the James Franck Institute
The University of Chicago
Chicago, Illinois 60637

Abstract. In 1994, the U.S. National Committee for the Committee on Data for Science and Technology (CODATA), organized under the Commission on Physical Sciences, Mathematics and Applications of the National Research Council established the Committee on Issues in the Transborder Flow of Scientific Data. The purpose of this Committee was to examine the current state of global access to scientific data, to identify strengths, problems and challenges confronting scientists now, or likely to arise in the next few years, and to make recommendations on building the strengths and ameliorating or avoiding the problems. The Committee's report appeared as the book *Bits of Power: Issues in Global Access to Scientific Data* (National Academy Press, Washington, D.C., 1997). This presentation is a brief summary of that report, particularly as it pertains to atomic and molecular data. The context is necessarily the evolution toward increasing electronic acquisition, archiving and distribution of scientific data. Thus the central issues were divided into the technological infrastructure, the issues for the sciences and scientists in the various disciplines, the economic aspects and the legal issues. For purposes of this study, the sciences fell naturally into four groups: the laboratory physical sciences, the biological sciences, the earth sciences and the astronomical and planetary sciences. Some of the substantive scientific aspects are specific to particular groups of sciences, but the matters of infrastructure, economic questions and legal issues apply, for the most part, to all the sciences.

INTRODUCTION: BACKGROUND AND ASSUMPTIONS

Science is probably the most international of all human activities today, in the sense that it flourishes by transnational exchanges at all levels, exchanges that are so usual, so important and so productive that national boundaries become virtually transparent during the interactions among scientists. This process goes on in an environment based more and more on electronic exchanges, all the way from formal transmission of finished texts to the intense, personal interchanges by electronic mail that make ideas become results. Consequently, in addition to all the historical issues concerning the exchange of scientific data, scientists now are learning to adapt computers and computer networks to their data acquisition, archiving and distribution. At the same time, because the rest of society, outside the scientific enterprise, uses the same technologies and even sometimes the same data for many purposes irrelevant to the conduct of science, problems arise regarding how the adaptations that scientists want can be kept compatible and consistent with the concerns of others.

As a consequence, the National Research Council (NRC) established the Committee on Issues in the Transborder Flow of Scientific Data in 1994 (henceforth called "the Committee"), at the request of the U.S. National Committee for the Committee on Data for Science and Technology (CODATA). CODATA is the watchdog committee for scientific data of the International Council of Scientific Unions (ICSU), and the U.S. National Committee for CODATA comprises the United States delegation and its advisors. The Committee's charge was fourfold: to outline needs for access to data in the major research areas of current scientific interest falling within the scope of CODATA, i.e., physical, biological, geological and astronomical sciences; to characterize the legal, economic, policy and technical factors that have an influence, positive or negative, on access to data by the scientific community, especially across national boundaries; to identify and analyze barriers to international access to scientific data that may be expected to have adverse impacts on those disciplinary areas, with emphasis on those common to all areas; and finally, to recommend to the sponsors approaches that could help overcome those barriers.

The sponsors of the study were all U.S. Federal agencies supporting scientific research. Hence the study devoted its greatest attentions to publicly-funded research and the data associated with such research. Such research and data are conducted as a public good, a characteristic central to the logic of the broadest recommendations of the report. The emphasis was on basic research, but the Committee did take into account the continuum between fundamental and applied research, between raw data and processed information, and between public and private uses of scientific data. It is no surprise that the most vexing issues the Committee faced were concerned with finding the appropriate balance of diverging interests.

Scientific data have many forms. They may be numeric or symbolic, or may be in the form of images, still or animated sequences. The distinction between "data" and "scientific information" more generally has become diffuse. So has the concept of "primary data," as the output of detectors has gone, more and more, directly into computers for processing before the investigator ever sees any results.

The same electronic devices that process data as they are acquired play more and more important roles in archiving and disseminating scientific data. This makes the exchange of data among scientists ever easier and cheaper. However because many of the data shared by working scientists have value to people outside the scientific community, conflicts arise over the management and distribution of such data. Some of the discussion and recommendations of the Committee deal with just these issues, particularly the economic and legal aspects.

One important assumption the Committee made is a reflection of the attitudes of virtually all scientists in the execution of their research. The most valued goal of scientists is that other scientists should learn of their work and use it. Discovery in a closet is only a game; research requires dissemination of new information. Hence scientists, especially those in basic research, are much more concerned with the accessibility of the work they produce than with immediate financial return from it. This puts the scientist in a role very different from that of the author of a novel or even of a textbook. Consequently scientists who write books and articles for profit find themselves in different niches, according to the task they are carrying out. In their roles as researchers, however, scientists have their common interests best served by as full and open exchange of scientific information as possible, consistent with the preservation of their own capacities to continue their investigations.

THE OVERARCHING PRINCIPLE AND RECOMMENDATION

The act of providing public funds for the support of research and for generation, storage and exchange of scientific data is a statement that these activities are carried out as public goods. The Committee conducted its study with this always in mind, and hence with a principle that informed its conclusions and recommendations, that the full and open exchange of scientific data–the "bits of power" on which the health of the scientific enterprise depends–is vital for advancing the nations progress and for maximizing the social benefits that accrue from science worldwide.

The laws and traditions of the United States have long upheld the tradition of full and open availability of publicly-supported scientific data, on an international basis. The only exceptions have been those concerned with national security. This tradition has become international and explicit in the context of studies of global change, which depend critically on comparative data from all parts of the planet. It has been maintained tacitly in virtually every other sector of the natural sciences.

The two trends that dominate scientific uses of data now are

- The need for scientists to adapt to conducting research with data that come in rapidly increasing quantities, varieties and modes of dissemination, often for purposes far more interdisciplinary than in the past, and

- The worldwide trend toward imposition of increasing economic and legal restrictions on access to scientific data gained from publicly funded research.

The former forces scientists to reexamine how they conduct their calling. The latter impels the scientific community to become more involved in matters of public policies and legislative activities in building a constructive interface between science and the rest of society.

These considerations led the Committee to present the following General Guideline as the first and most general of its Recommendations:

"The value of data lies in their use. Full and open access to scientific data should be adopted as the international norm for the exchange of scientific data derived from publicly funded research. The public-good interests in the full and open access to and use of scientific data need to be balanced against legitimate concerns for the protection of national security, individual privacy and intellectual property."

"Full and open access" is sometimes misinterpreted to mean "access at no cost," but the report states explicitly that this phrase is intended to mean "at lowest feasible cost, ranging from 'free to the user because the costs are paid directly to the supplier of the data,' to 'at the marginal cost of the distribution to the user'."

One concern that prompted the heavy emphasis on this point has been a decision by the European Union (certainly not with any enthusiasm from its scientists) to permit the marketing, for profit, of some of the data generated for global change and meteorological studies. This action was, at least in part, a reaction to a U.S. Court decision that prohibited copyright on a telephone directory, on the basis that copyright is intended to protect works based on intellectual creativity, and not simply "sweat of the brow." This decision appeared to put into jeopardy the protection, by copyright, of intellectual property rights of compilers of scientific data. Subsequent activities in the United States have been directed toward creating new means to protect the works of those compilers, not particularly toward compilations of scientific data. However the consequences of success of these activities would be likely to have significant, even devastating, impact on the exchange of scientific data. Consequently, the National Academy of Sciences, the National Academy of Engineering and the Institute of Medicine, the parent organizations of the National Research Council, have taken steps to make Members of Congress and of the Executive Branch aware of the potential impacts of such activities on the conduct of the sciences.

ISSUES AND RECOMMENDATIONS ON DATA ISSUES IN THE NATURAL SCIENCES

Substantive issues of data in the natural sciences include traditional endemic problems of maintaining consistency among data sets, providing pathways for researchers to find what data compilations are available and providing enough descriptive material (metadata) to enable researchers to use databases with full awareness of the limitations of the data. More recent issues include the vast increase in the quantity of data, particularly in the observational sciences in which automatic methods of data collection and compilation have made virtually qualitative changes in the character of the available information. Data centers have grown up in some fields to distribute and, especially in the physical sciences, to evaluate such bodies of data. The range of kinds of data, extending far beyond simple numerical tables, has also changed enormously for virtually all sciences. One other issue that appears to be more important than in earlier times is a perception of a growing use of data from one discipline by scientists in other areas, and hence the importance of incorporating into databases enough information to allow nonspecialists to use the data in sound ways. Beyond these generalizations, the issues tend to be area-specific, and the natural areas of the laboratory physical sciences, the biological sciences, and the predominantly observational fields of the earth sciences and astronomical and space sciences. The laboratory physical sciences have the fewest and simplest data problems, among all these areas, but this is not to say that this sector is free of data problems, for it is certainly not.

The first recommendation of this group is a general one, simply one to implement the overarching recommendation of the report:

- *Government science agencies and intergovernmental organizations should adopt as a fundamental operating principle the full and open exchange of scientific data. By "full and open exchange," the Committee means that the data and information derived from publicly funded research are made available with as few restrictions as possible, on a nondiscriminatory basis, for no more than the cost of reproduction and distribution.*

The second recommendation arises from the realization of the value of existing international data centers and the desirability of extending and expanding the network of such centers:

- *the International Council of Scientific Unions (ICSU), together with the scientific Specialized Agencies of the United Nations, the Organisation for Economic Co-operation and Development Megascience Forum, and the national science agencies and professional societies of member countries, should consider developing a distributed network of data centers. All publicly funded scientific data activities, including this network, should a) involve both experts from the relevant disciplines and from information management and technology, in the management and preservation of the data; b) develop and maintain up-to-date, comprehensive, on-line directories of data sources and protocols for access; c) provide documentation (metadata) adequate to ensure that each data set can be properly used and understood, with special attention to the needs of individuals outside the core area of each data set, and the CODATA Commission on Standardized Terminology for Access to Biological Data Banks should be enhanced into a true international consultative body* (much as the CODATA Commissions and Special Committees on Nomenclature for the physical sciences function now); *d) incorporate advances in technology to facilitate access to and use of scientific data, while overcoming incompatibilities in formats, media, and other technical attributes, through vigorous coordination and standardization efforts; e) institute effective programs of quality control and peer review of data sets; and f) digitize all key historical data sets and ensure that every important condition for the long-term retention of data be met, including the adoption of appropriate retention and purging criteria and timely transfer of all data sets to new media to prevent their deterioration or obsolescence.*

The remaining recommendations in this area deal with stimulation and recognition of data activities. These are:

- *The ICSU and other professional societies should encourage the study of, and publication of peer-reviewed papers on effective data management and preservation practices, as well as promote the teaching of those practices in institutions of higher learning.*

- *All scientists conducting publicly funded research should make their data available immediately, or following a reasonable period of time for proprietary use. The maximum duration of such a proprietary period should be expressly established by the particular scientific communities, and compliance should be monitored subsequently by the relevant funding agency.*

- *Publicly funded scientific databases should be maintained either directly or under subcontract by the government science agencies with the requisite discipline mission and need. In the United States, the Office of Science and Technology Policy (OSTP) should develop and overall policy for the long-term retention of scientific data, including a contingency plan for protecting those data that may become threatened with the loss of their institutional home.*

- This recommendation complements the one in the next section concerning developing countries. It has several parts:

  a) International development organizations, together with professional societies, should provide training programs for scientists in the use of computers, with emphasis on the management of digital data in specific disciplines; b) foreign aid agencies should make available to individual scientists in developing countries more direct, peer-reviewed grants that include support for access to data and facilitate the involvement of scientists in such nations in their own countries capacity-building initiatives, research policy decisions and national database construction activities; c) scientists in developing countries should be encouraged to organize to promote the policy of full and open access to scientific data in their own countries, as well as to make their data available internationally; d) the ICSU, together with funding agencies and nongovernmental bodies, should strengthen its efforts to assist developing countries in undertaking their own scientific studies, and encourage scientists to take active roles in the international scientific community; e) until affordable and ubiquitous electronic network facilities are available, national and international scientific societies and foreign aid agencies should establish or improve their existing efforts to send extra stocks of scientific publications to libraries and research institutions in developing countries that need them.

- *Finally, the ICSU, together with principal national and international scientific organizations, should convene a series of major international meetings to initiate meaningful action to implement these recommendations.*

ISSUES AND RECOMMENDATIONS IN INFORMATION TECHNOLOGY

The obvious factor dominating information technology is the steadily decreasing cost and increasing capacity for collecting, storing and disseminating scientific data. With this trend have come some problems as well as many benefits. One of the benefits that is perhaps not so obvious but very important and needing emphasis is the potential for electronic communication to bring scientific data at reasonable cost to developing nations. The price of buying and maintaining network and computer facilities is far less than that of buying and maintaining a library. This led to one set of recommendations from the Committee:

- *Foreign aid to developing countries in the form of computers, computer networks, associated software and training and resources to maintain those technologies should be given high priority, on the basis of the high leverage these provide for self- sustaining capacity and long-term socioeconomic returns.*

Another recommendation addresses the growing problem of congestion on the Internet, a problem that has arisen as the predominant users of the network have shifted from scientific and scholarly to commercial and entertainment. This congestion is something every scientist has experienced who has tried to transfer files across an ocean, any time in the past two or three years. The recommendation is:

- *The principal scientific societies and the Internet Engineering Task Force (IETF) should begin a long-term planning effort to assess the carrying capacity and distribution capability of the Internet, and of the coming demand for its use. Scientific societies and government science agencies should consider and evaluate the creation of a dedicated international science network, such as the Internet II now under development.*

Other recommendations regarding information technology addressed technical organization and management, and authentication and security. These are:

- *To improve the technical organization and management of scientific data, the scientific community, through government science agencies, professional societies and actions of individual scientists, a) work with the information and computer science community to increase their involvement in scientific data management; b) support computer science research in database technology, particularly to strengthen standards for metadata, efficient storage of very large data sets, and integration of standards for configuring data sets; c) improve science education and the reward system in the area of scientific data management, and d) encourage the funding of data compilation and evaluation projects, and of data rescue efforts for important data sets in transient or obsolete forms, especially in developing countries.*

- *The U.S. government science agencies, working with their counterparts in other nations, should improve data authentication and apply security safeguards more vigorously.*

- *A consortium of intergovernmental and nongovernmental organizations, including the International Telecommunications Union, the World Bank, the Specialized Agencies of the United Nations, the International Council of Scientific Unions, and other concerned bodies, should mount a global effort to reduce telecommunications tariffs to scientists in developing countries through differential pricing or direct subsidy.*

ISSUES AND RECOMMENDATIONS REGARDING ECONOMIC ASPECTS OF SCIENTIFIC DATA

The discussion of economic issues was pervaded by growing fears that privatization and commercialization of scientific data will price such data beyond the means of publicly supported researchers. While traditional economic analyses may initially lead one to think that privatization would be a healthy move for the scientific data world, a more careful analysis suggests that the appropriate market model for scientific information, including data, is rather different from usual markets. First, the conduct of some scientific research is itself tightly tied to the collection, maintenance and distribution of the data generated by that research. For example, in the observational sciences, with their massive databases, separating the gathering, archiving and maintenance of data from their distribution is likely to be more costly and inefficient than keeping them integrated. Second, the contributors of scientific data, particularly in basic research, are likely to be the consumers of such data, in which case non-monetized exchange of data may be the most efficient method. Third, in many situations, the market for scientific data is not large enough to support more than a single supplier, if that. Finally, most basic research is necessarily publicly funded, so that privatizing the distribution of those data generated by such research would require that funds now channeled to data suppliers would have to be redistributed to individual grantees, making such funds far more vulnerable than they are now.

The issue of potential monopoly control, especially in small markets, is a constant concern here. Only if privatized data distributors were competitive in their value-adding roles could privatizing of publicly supported data be justified. Furthermore the data and the results obtained with the use of such data must be considered public goods–resources that are both nondepletable, indivisible and nonexcludable (once having been supplied to some, cannot easily be denied to others). There is no social cost from repeated use, so price differentiation may be justified in many situations: one price for those conducting publicly supported research and another for those in private, proprietary firms.

The recommendations concerning economic aspects take the form of a set of criteria that the Committee would like to become standard for evaluating the storage and distribution of scientific data generated by publicly funded research:

- *Does the scientific research depend on a substantial public investment one or more facilities that generate the data of interest? If so, the data distribution facilities are most likely to benefit by being vertically integrated with the observational or experimental facilities themselves.*

- *Does the (non-facilities-based) distributed scientific research involve coordination among researchers, possibly in different countries? If so, then data distribution becomes a matter of communication among contributing scientists, and for this community, the price of the data alone should be zero. If the distributor subsequently adds value to the data, then the price should be no higher than the marginal cost of adding value.*

- *Is the community of users roughly the same as the community of contributors? If so, then data distribution should be priced at zero (or at marginal cost, if value is added). If there are many users who are not contributors, such as commercial customers, then some form of price discrimination to ensure zero or low prices to contributing scientific users, with possibly higher prices to others, may be appropriate.*

- *Is the user community large enough to support more than one data distributor? If so, then privatization of data distribution may be a viable policy option. If not, then privatization should occur only if the contractual arrangements are adequately protective of the needs of the scientific community. Necessary, but not sufficient, conditions for privatization to be desirable are these: the distribution of data can be separated easily from their generation; the scientific data set is used by others beyond the research community; it is easy to price-discriminate or product-differentiate between scientific users and others, and it is easy for the government to contractually mandate low prices to scientific users for government-funded data.*

ISSUES AND RECOMMENDATIONS REGARDING LEGAL DEVELOPMENTS AFFECTING ACCESS TO DATA

The dominant subject for the Committee in the legal area is the growing controversy between the ready access of scientists to data generated by public funds and the protection of databases created for commercial purposes. The former aspect would be realized by acceptance of the principle of "full and open access." The latter has become a focus, not only for private firms but for governments, primarily in Europe, that wish to market, for profit, data such as those obtained from earth-observing satellites. At issue is whether a new form of protection of intellectual property, specifically aimed at electronic information including data, should be created. The question came into being when, in 1991, the Supreme Court decided, in *Feist Publications, Inc. v. Rural Telephone Service*, that significant original and creative authorship, at least in the selection and arrangement of contents, is necessary for a work to be eligible for copyright. This was taken by some to threaten the protection of commercial databases. The Commission of the European Communities (CEC) had been looking toward a new framework to protect databases in order to stimulate their commercialization. In 1996, the culmination of this move was the European Directive on Databases, adopted formally by the CEC. Legislation even more protective than the Directive was introduced into the U.S. Congress in 1996, but died. Proposals very similar to the proposed legislation were introduced as treaties at the Diplomatic Conference under the auspices of the World Intellectual Property Organization. At the Conference in December, 1996, a rather different and somewhat more open treaty on copyright was adopted and forwarded to the participating nations; the database treaty proposal was set aside for the time being.

At the time of this writing, several bills stand now in both the Senate and House of the U.S. Congress, with varying degrees of control and protection. Those considered worst from the perspective of the Committee have, of course, been the most protective; some of these would do away with the doctrine of fair use, the doctrine which allows free reproduction of materials for personal use in scholarly, educational or research activities. Another problem for the science community is already incorporated in the European Directive; this, in effect, endows a creator of an electronic database with perpetual property rights, through the automatic restarting of the protection clock whenever any change is made in a database. The result would be monopolistic ownership of databases with no allowance for or consideration of public policy or public goods. Current debate has begun to bring these aspects into the dialogue but the outcome is still very unclear.

The report contains five recommendations concerning legal matters:

The Office of Science and Technology Policy, leaders from the science agencies and professional societies, and all those concerned with sustaining the health of the scientific enterprise should immediately take these actions:

- *Present to all relevant legislative forums the principle of full and open exchange of scientific data resulting from publicly funded research, and clarify the importance of sustaining such exchange to the nation's future whenever these forums consider laws that would apply to exchange of scientific data;*

- *Demand that national and international legislative processes in progress slow to a rational pace, and that the deliberations become more public to allow the scientific and educational communities to present their views and concerns to lawmakers;*

- *Advocate the incorporation of equivalents of "fair use" as part of any regulatory structure applying to databases as such, or to on-line storage and transmission of data and other scientific information. As a corollary, ensure that the public-good aspects of scientific data are preserved and promoted in laws and regulations governing intellectual property on the Internet and in any future electronic networked environments;*

- *Work with Congress and the official U.S. representatives to the World Trade Organization and the World Intellectual Property Organization to ensure that the nations interests in maintaining preeminence in science and technology are not undermined; and*

- *Pursue these issues not only within the United States, but also internationally through international scientific organizations and U.S. foreign-policy channels as they deal with trade and other agreements affecting intellectual property protection.*

SPECIAL CHARACTERISTICS AND ISSUES FOR THE LABORATORY PHYSICAL SCIENCES

Since this report has been prepared in the context of a Conference on Atomic and Molecular Data, it is appropriate that its final section discuss those aspects of the international exchange of scientific data that are specifically relevant in that context. While some atomic and molecular studies are certainly components of earth, astronomical and space sciences, it is rare for atomic and molecular data to arise from these observational sciences. (One example of such an exception was the balloon-based observation of the infrared spectrum of α-Tauri, Aldebaran, by M. Schwartzschild, which turned out to provide the most accurate *experimental* determination of the electron affinity of hydrogen. Others have certainly been the observations and determinations of microwave spectra of various molecular species seen first in the interstellar medium.) An important characteristic that distinguishes the laboratory sciences from the observational sciences is the (at least potential) reproducibility of any result from the former, in contrast to the uniqueness of events recorded by the latter. However, as all working scientists know, when a set of results is compatible with prior knowledge, self-consistent, and based on established methods of a discipline, later workers are unlikely to reproduce the entire data set in order to verify it. They are much more likely to check only a few results by using the same methods to obtain further results or to verify the results in selected cases by measuring the previously-determined quantities by new methods they intend to use for fresh studies.

As in all the natural sciences except sometimes regional studies, the processes and results of the laboratory physical sciences are extremely international. The research itself and its tangible products flow from workers in one country to another as readily as they move from one building to another in the same institution. This transparency to national boundaries is perhaps most striking during the working stages of any project. The substantive results begin as "raw data" (whose "rawness" was discussed above) and pass to processed data, then to normal, refereed publication and, particularly for numerical data in atomic and molecular science, to Data Centers where the results are incorporated into Evaluated Data Bases. The Data Centers, many of them operated within or under auspices of the National Institute of Science and Technology (NIST) and other Federal agencies, have traditionally handled numerical data and the accompanying metadata. As the nature and form of data have broadened, it is natural to expect the scope of material handled by Data Centers to broaden. The symbolic data, images and animations that are now accepted forms of scientific data, will increasingly find their way into evaluated data bases. At present, such data are often not incorporated into normal, refereed manuscripts but are made available--sometimes through the publisher's archiving services, and increasingly through authors' own pages on the World Wide Web.

While much of the data flow in the laboratory sciences lies within their disciplines, broadly defined, it is particularly important in the atomic, molecular and optical sciences, as well as within the chemical, condensed matter and materials sciences, to recognize the two-way flow of data between basic and applied sciences. This has led to a tradition in these fields for commercial databases to be used as frequently as publicly supported sources. One of the most interesting aspects of these commercial databases to follow in the coming years will be their evolution to electronic form. It will be particularly worth watching to see what forms of added value the producers of such databases will introduce. To whatever extent the privatized model applies to scientific data, we might expect some of the most useful innovations in electronic databases to come from these suppliers.

A long-standing topic for the laboratory sciences is the problem of data compatibility. This is endemic, but is perhaps more an issue than in the past because of the expansion of interdisciplinary research. One process now still evolving is the establishment of compatibility and interchangeability of electronic formats, such as TeX, LaTeX, RevTeX and the various commercial word processing programs. The appearance of such programs as one which allows an author to prepare a text that looks, on the computer screen, like a commercial word processor format but is stored in LaTeX form, is a partial step toward such compatibility. Another example of the incompletely resolved compatibility problem in electronic communication is the establishment of methods for compressing data for high-speed transmission. This becomes especially important as the congestion problem grows. Still another is the evolving choice of formats for storage of data. Here, the trend is clearly away from storage of images of text and toward storage of digitized, alphanumeric information, which allows easy updating. Whether free market mechanisms will bring about the compatibility that scientists need, or forums will be necessary to establish common needs, standards and means of achieving those standards, stands as an open issue now.

The databases in the physical sciences are essential nutrients for the health and operation of these sciences. From the report, "Just as funding agencies that support research accept the costs of the equipment needed to carry out this research, they also carry the responsibility to support the data components of the infrastructure that enables the research to go forward. Furthermore, just as basic research support needs to be protected because of the likelihood of long time intervals between the conduct of the research and its eventual applications, so should the development and maintenance of databases be protected from short-term fluctuations in budgets or varying needs for the data in industrial applications. The development of databases includes the compilation and evaluation of data from the variety of sources of the data. Once developed, it is critical that databases be maintained and continuously updated as new, relevant data become available. the dissemination should be via a variety of platforms and should be in user-friendly forms, with cross-referencing to files maintained by other agencies, or available vial other electronic media.

The committee believes that science agencies should maintain responsibility either directly or under subcontract for the development, management, retention, and dissemination of electronic databases that are the product of their research programs. Within the United States, the Office of Science and Technology Policy should develop an overall policy for the long-term retention of scientific data, including a contingency plan for protecting those data that may become threatened with the loss of their institutional home." [1,2]

ACKNOWLEDGMENTS

The author is deeply indebted to all the members of the Committee and staff who worked to carry out the study that produced *Bits of Power*. He would like to acknowledge especially the work of Jerome Reichman in his crucial (and continuing) efforts to educate the rest of the Committee and then to deal with the legal issues of intellectual property protection, and, finally and most of all, the enormous effort by the Study Director, Paul Uhlir, whose skill and labor brought the report into being.

REFERENCES

[1] *Bits of Power: Issues in Global Access to Scientific Data* (National Academy Press, Washington, D.C., 1997), p. 90.

[2] *Preserving Scientific Data on Our Physical Universe: A New Strategy for Archiving the Nations Scientific Information Resources* (National Academy Press, Washington, D.C., 1995).

Atomic Data Production

The Iron Project: Past, Present and Future

Keith Butler

Institut für Astronomie und Astrophysik[1]
Scheinerstr. 1
Munich 81679

Abstract. This paper provides an overview of the Iron Project at the time of writing, what it has achieved, what is currently being done and what we hope to achieve in the near future. Since the methods and initial goals have been summarized by [9], we concentrate here on the more recent results and developments made since the Hummer et al. paper appeared.

INTRODUCTION

The Iron Project is the direct successor to the Opacity Project [15]. The latter aimed to produce high quality atomic data for the calculation of opacities suitable for use in stellar envelope models. This was mostly intended to remove discrepancies in the models concerning Cepheid variables, the cornerstone of distance determinations in astronomy. As a "side" product, the Opacity Project provided a wealth of radiative data, oscillator strengths and photoionization cross sections that can be used in other applications. A recent example is that of Seaton [16] who determines radiative accelerations in stellar envelopes and describes the latest opacities which may be obtained electronically from the Centre de Données astronomiques de Strasbourg (CDS). Codes for interpolation in the tables provided are also available.

For stellar envelope work, collisional data are unnecessary since the densities are so high that LTE pertains. However, the interpretation of other plasmas, such as the solar corona, nebulae and stellar atmospheres collisional data are required since conditions far from LTE can prevail. In the light of this, the members of the Opacity Project decided to apply the methods and experience they had developed in the OP work to the determination of accurate collisional data. Since these methods were collisional by nature, there was no fundamental problem in doing this. However, it was clear at the outset that iron group elements would be of particular importance in an astrophysical context, given their (relatively) high abundance

---

[1] This paper is dedicated to the memory of Maryvonne Le Dourneuf

and the wealth of lines they contribute in all spectral regions. This implied that much larger calculations would be necessary and, equally importantly, that relativistic computations were essential. This latter aspect necessitated much program development as we shall see later. In addition, the radiative data available from the Opacity Project for lowly ionized species of iron were relatively crude. This had not been a problem for envelope work since the temperatures are so high that only highly ionized species contribute significantly to the opacity. For the analysis of other plasmas, it would be important to have more accurate radiative data for these ions.

Given these goals, the calculation of radiative and collisional data for iron group ions, the title of the project was obvious. In the same way that Mike Seaton was (and is) the moving force in the Opacity Project, the Iron Project was very much the brainchild of David Hummer. Although his name only appears on the first of the Iron Project papers he has been very much involved in all aspects of the work and as chairman he has guided the the project from its inception. He has recently stood down as chairman to enjoy his retirement.

The structure of the project is that of a loosely knit group of atomic and astrophysicists scattered throughout the world. Groups from the USA, France, Venezuela, Canada, Germany and the UK are involved. Biannual meetings are the main tool whereby the status and progression of the project are discussed. The results are published in a series of papers appearing the Supplement Series of Astronomy and Astrophysics although occasionally a paper may appear in the main journal. A full list of publications including those in press may be found on the Iron Project home page

- http://www.am.qub.ac.uk/projects/iron

and a list complete up to October 1995 appears in [8]. Here, for convenience, we list the papers that have appeared since that date.

IRON Project Papers: October 1995 – October 1997

XI. The $^2P^o_{3/2}$–$^2P^o_{1/2}$ fine-structure lines of Ar VI, K VII and Ca VIII. H.E. Saraph and P.J. Storey, A&A Suppl. Ser. **115**, 151, 1996

XII. Electron excitation of forbidden transitions in V-like ions Mn III, Fe IV, Co V, and Ni VI. K.A. Berrington and J.C. Pelan, A&A Suppl. Ser. **114**, 367, 1995

XIII. Electron excitation rates and emissivity ratios for forbidden transitions in Ni II and Fe II. M.A. Bautista and A.K. Pradhan, A&A Suppl. Ser. **115**, 551, 1996

XIV. Electron impact excitation for the Fe XIV fine-structure transition $^2P^o_{1/2}$ – $^2P^o_{3/2}$. P.J. Storey, H.E. Mason, and H.E. Saraph, A&A **309**, 677, 1996

XV. Electron excitation of the fine-structure transitions in hydrogen-like ions He II and Fe XXVI. R. Kisielius, K.A. Berrington, and P.H. Norrington, A&A Suppl. Ser. **118**, 157, 1996

XVI. Photoionization cross sections and oscillator strengths for Fe V. Manuel A. Bautista, A&A Suppl. Ser. **119**, 105, 1996

XVII. Radiative transition probabilities for dipole allowed and forbidden transitions in Fe III. Sultana N. Nahar and Anil K. Pradhan, A&A Suppl. Ser. **119**, 509, 1996

XVIII. Electron impact excitation collision strengths and rate coefficients for Fe III. Hong Lin Zhang, A&A Suppl. Ser. **119**, 523, 1996

XIX. Radiative transition probabilities for forbidden lines in Fe II. P. Quinet, M. Le Dourneuf, C.J. Zeippen, A&A Suppl. Ser. **120**, 361, 1996

XX. Photoionization cross sections and oscillator strengths for Fe I. Manuel A. Bautista, A&A Suppl. Ser. **122**, 167, 1997

XXI. Electron excitation of fine-structure transitions involving the $3d^64s^2$ ^5D ground state and the $3d^74s$ ^5F metastable state of Fe I. J. Pelan and K.A. Berrington, A&A Suppl. Ser. **122**, 177, 1997

XXII. Radiative rates for forbidden transitions within the ground state configurations of ions in the carbon and oxygen isoelectronic sequences. M.E. Galavís, C. Mendoza and C.J. Zeippen, A&A Suppl. Ser. **123**, 159, 1997

XXIII. Relativistic excitation rate coefficients for Fe XXII with inclusion of radiation damping. Hong Lin Zhang and Anil K. Pradhan, A&A Suppl. Ser. **123**, 575, 1997

XXIV. Electron excitation of Li-like Fe XXIV between the n=2 and n'=2,3,4 fine-structure levels. K.A. Berrington and J.A. Tully, A&A Suppl. Ser. (in press)

XXV. Electron impact excitation of fine-structure transitions in the ground configuration of Fe XII. A.M. Binello, H.E. Mason and P.J. Storey, A&A Suppl. Ser. (in press)

XXVI. Photoionization cross sections and oscillator strengths for Fe IV. Manuel A. Bautista and Anil K. Pradhan, A&A Suppl. Ser. (in press)

XXVII. Electron impact excitation collision strengths and rate coefficients for Fe IV. Hong Lin Zhang and Anil K. Pradhan, A&A Suppl. Ser. (in press)

XXVIII. Electron excitation of the $^2P_{3/2} \rightarrow {}^2P_{1/2}$ fine structure transitions in fluorine-like ions at higher temperatures. Keith A. Berrington, Hannelore E. Saraph and John A. Tully, A&A Suppl. Ser. (in press)

Although the project is aimed at iron group elements, the imminent launch of the ISO satellite made the computation of data suitable for the analysis of infrared spectra desirable and this was the first priority for the project. The results have been summarized by Butler [8] and will not be discussed further here. The SOHO satellite allows detailed observations, including the dynamics, of the sun to be made. To make full use of the observational data accurate atomic parameters, in particular collisional cross sections for highly ionized ions of iron, were found to be lacking and the provision of such data was made the next goal for the project. This is the work that will be mostly described in this paper. In the following section, a short description of the methods is to be found. Results from recent papers are presented in section 3, while section 4 provides an outlook on future directions and methods.

METHODS

The basis of the calculations up to now has been the close-coupling expansion for the "target" + electron system. If the target has a wavefunction ψ_i and the scattered electron is described by the wavefunction θ_i then each combination $\psi_i \theta_i$ is a free channel. The total wavefunction of the system can be written as an antisymmetrised sum over all channels but, of course, only a finite number of terms, NCHF, can be included. Thus the close-coupling expansion looks like

$$\Psi = \mathcal{A} \sum_{i=1}^{NCHF} \psi_i \theta_i + \sum_j c_j \phi_j \qquad (1)$$

The second summation consists of bound channels which are introduced partly for numerical reasons but also to increase the flexibility of the trial wavefunction. Further details are to be found in [9]. The success of the approximation depends largely on the target description which must be considered with great care, balancing the computational possibilities with the desired accuracy. Much of the work involved in these calculations is devoted to the determination of a "good" target.

The next step is the solution of the Schrödinger equation using this close-coupling wavefunction. Interference between the various channels leads to resonance effects and cross sections that vary rapidly with energy (see section 3). The delineation of the resonances requires that a large number of energy points be obtained. To this end, the R-matrix method has proved to be most efficient. Here, space is divided into two regions: the inner region ($r < a$, where a is the R-matrix boundary), is complicated due to the possibility of exchange between the scattered and target electrons. A set of coupled integro-differential equations must be solved. On the other hand, for $r > a$, exchange is no longer relevant and ordinary coupled differential equations pertain which can be solved rapidly for any energy. A further expansion is made in the inner region, each of the wavefunctions, target and scattered electron, is expanded in terms of a suitably chosen set of basis functions.

The Hamiltonian matrix is formed and diagonalized using this set. This needs to be done only once and provides the R-matrix which is stored. For each energy then, the complete wavefunction is found by matching the radial inner and outer wavefunctions F at the R-matrix boundary via the relations

$$F = RF'. \tag{2}$$

In the outer region F has the form

$$F \sim s + c\mathbf{K} \tag{3}$$

s,c being Coulomb functions and the collisional cross section may be obtained from the $\mathbf{K}$ matrix.

There are two alternative methods for performing relativistic calculations. In the first and more accurate, the Breit-Pauli Hamiltonian

$$H = H^{NR} + H^{mass} + H^{Dar} + H^{so} \tag{4}$$

is used. The mass, Darwin and one-body spin-orbit corrections to the non-relativistic Hamiltonian are included. Note that only the latter is not diagonal in the LS representation so that the mass and Darwin operators can be included in an LS-coupling treatment at almost no extra cost. This is done in all Iron Project calculations. The spin-orbit term on the other hand means that intermediate coupling is required greatly increasing the resources needed. The second possibility is the JAJOM approach in which an algebraic transformation is performed on the LS-coupling $\mathbf{K}$-matrices to give intermediate coupling results. As noted earlier, the mass and Darwin terms are included as are term-coupling coefficients which account for relativistic effects within the target so the main deficiency of the method is the omission of the fine-structure energy splitting. This is only relevant at low energies so a combination of the two methods in which the Breit-Pauli Hamiltonian is used at low energies and the much cheaper JAJOM technique is applied in the high energy region has proved particularly useful.

In either case, the $\mathbf{K}$-matrix for a particular value of J (LS) is obtained. The total cross section is a sum over all J (LS) values. The collision rate is a convolution of the cross section with a Maxwellian distribution and for high temperatures the high energy tail becomes important. Since the R-matrix expansion itself can only provide data up to some limiting energy, there is a need to "top up" the results both in J and in energy. This has been a major obstacle for the Project up to now, particularly as far as allowed transitions are concerned since in this case the convergence in J is very slow. The problem is not new of course but since the Project is producing tens, hundreds or perhaps thousands of collisional rates for each ion there has been a pressing need for some means of automating the process. Recently, Werner Eissner has succeeded in developing a top-up method valid in intermediate coupling which will greatly facilitate future work. Figure 1 demonstrates this. The collision strength for the lowest transition in C-like iron is shown with and without top-up. It is obvious that the points with top-up are converging to the correct high-energy limit while those without are much too small at high energies. The energy scale has been adjusted following [6] to allow the full energy range to be seen.

RESULTS

The results presented here are meant to illustrate the work of the Project of late. The interested reader is referred to the papers cited in the Introduction for details.

Collisional Data

An excellent discussion of the influence of the choice of target is given by Binello et al. [4]. In particular, the role of the second sum in Eq. 1 is described in detail. How important such differences can be is apparent from Figure 2. Two calculations are compared. They differ only in the inclusion of a 4f pseudo-orbital whose bound

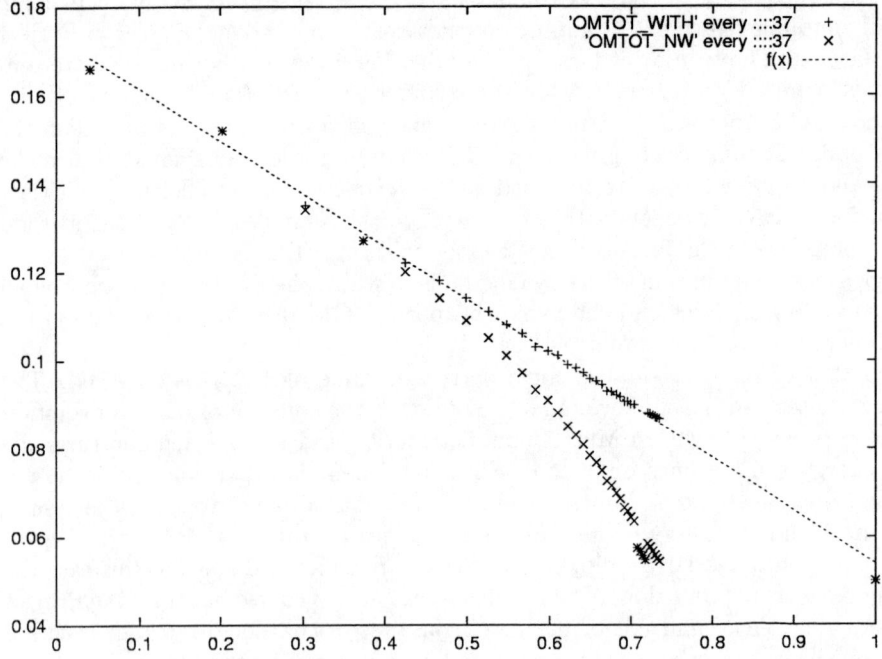

FIGURE 1. Collision strength versus reduced energy for the first transition in C-like iron. The crosses are without top-up, the pluses with. The asterisk is the limit at infinite energy.

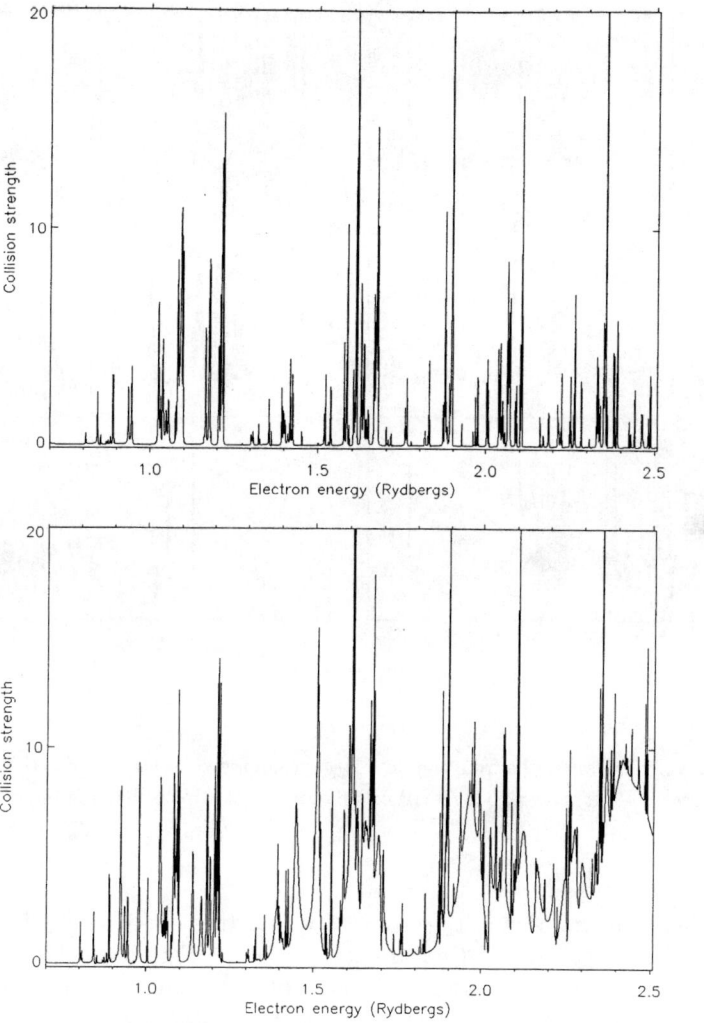

FIGURE 2. Collision strength for the Fe XII $3s^23p^3$ $^2D^0_{3/2}$–$^2D^0_{3/2}$ transition. 7 states including a $\overline{3d}$ orbital (upper panel) and $\overline{3d},\overline{4f}$ orbitals (lower).

channels cause the large increases seen at higher energies. Such orbitals are included to take the channels, particularly the ionization channels, omitted from the close-coupling expansion into account so in some way they do represent real features of the cross sections. On the other hand, what they really mean is a thorny problem and has been recently tackled by the methods touched on in the concluding section. In any case, the differences do give an indication of possible uncertainties in the

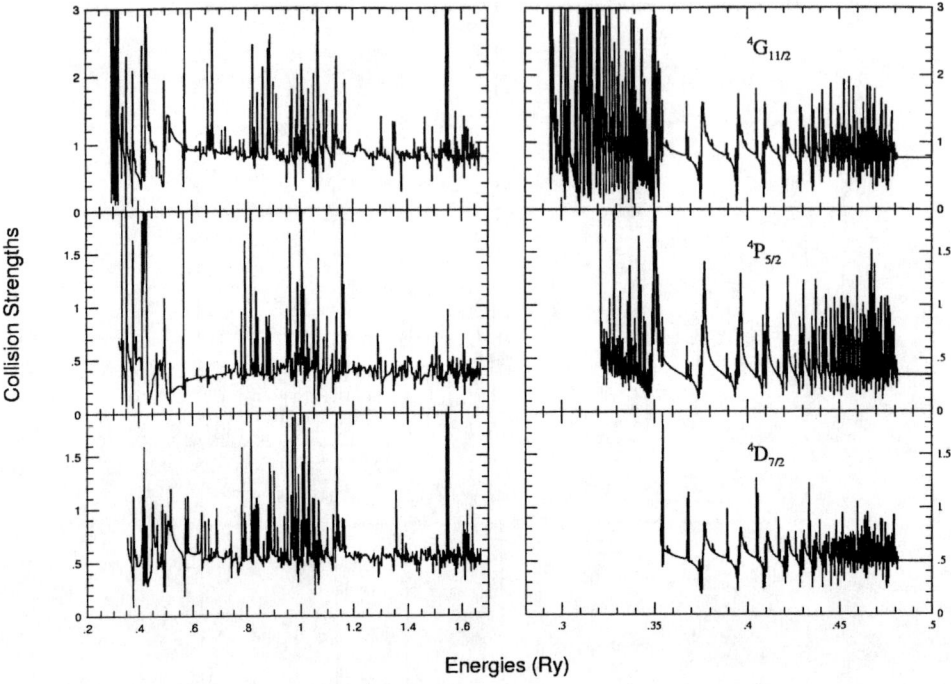

FIGURE 3. Collision strengths from the $3d^5$ $^6S_{3/2}$ ground level to states indicated in a 49 NR state approximation (left) and a 16 level BP calculation (right). Note the differing energy scales.

derived collision rates.

The differences introduced in the cross sections by the use of JAJOM as opposed to the Breit-Pauli type calculation are to be seen in Figure 3. These are cross sections for transitions out of the ground state of Fe IV into various excited states. On the left are the Breit-Pauli results, on the right those given by JAJOM (note the different scales of the plots). The Breit-Pauli collision strengths show much larger resonance structures although much of the detail has been lost as a much coarser energy mesh has been used in this case. However, at the higher energies, where the resonant contributions become less important, the background values are in good agreement thus validating the use of the algebraic transformations in this region.

As a final example we look at the work of Berrington et al. [3] who investigated the $^2P^o_{3/2}$–$^2P^o_{1/2}$ fine structure transition in fluorine-like ions. They found a large discrepancy with the earlier results of Mohan and Le Dourneuf [10] (Figure 4) for Si VI. The curves labelled a and b are the older calculations, curve c is from the Iron Project. The reasons for the differences are two-fold. There is a large

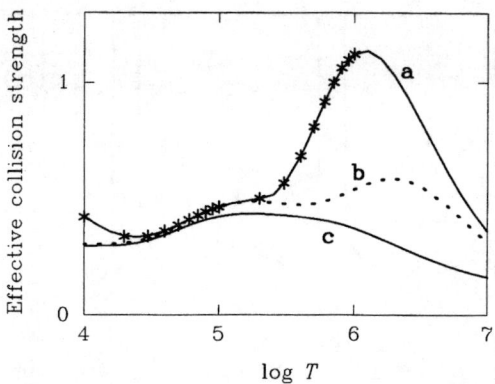

FIGURE 4. Effective collision strengths for Si VI from [10] (a) and (b) and [3], (c). (a) is with trapezoidal integration, (b) integrated correctly.

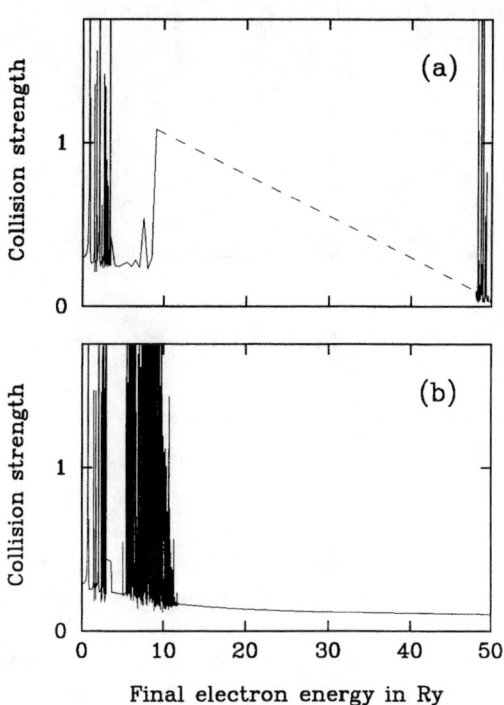

FIGURE 5. The collision strengths corresponding to the data shown in Figure 4. Note the gap and the high energy resonances in (a).

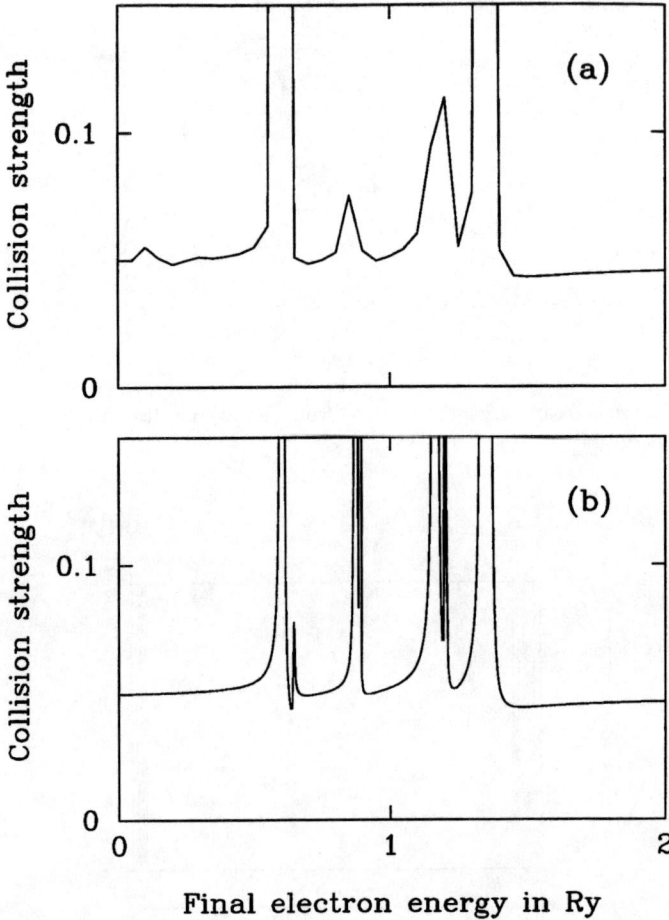

FIGURE 6. Detail in the cross sections showing the coarseness of the mesh in Mohan and Le Dourneuf [10] compared to Berrington et al. [3].

gap in the earlier data between 10 and 48 Ryd (Figure 5) and Mohan and Le Dourneuf integrated over the region using the trapezoidal rule thus overestimating the contribution. Secondly, the resonances at energies above 48 Ryd are spurious. They do not appear in the Iron Project results at all and are responsible for the bump at temperatures about 10^6 K. As a last remark, it may be noted that both Figure 5 and Figure 6 clearly show the coarseness of the energy mesh in the previous work. The codes are now much more computationally efficient making a fine energy grid possible with reasonable use of computer resources.

TABLE 1. Observed [17] and calculated target energies for Fe V (in Ryd).

	Level		E_{obs}	E_{cal}	Level		E_{obs}	E_{cal}
1	$3d^4$	^5D	0.000	0.000	2	$3d^4$	^3P	0.227	0.232
3	$3d^4$	^3H	0.223	0.259	4	$3d^4$	^3F	0.238	0.262
5	$3d^4$	^3G	0.268	0.309	6	$3d^4$	^3D	0.328	0.379
7	$3d^4$	^3P	0.561	0.619	8	$3d^4$	^3F	0.561	0.619
9	$3d^3(^4F)4s$	^5F	1.702	1.696	10	$3d^3(^4F)4s$	^3F	1.780	1.782
11	$3d^3(^4P)4s$	^5P	1.863	1.875	12	$3d^3(^2G)4s$	^3G	1.899	1.910
13	$3d^3(^4P)4s$	^3P	1.936	1.963	14	$3d^3(^2P)4s$	^3P	1.948	1.976
15	$3d^3(^2D)4s$	^3D	1.965	1.985	16	$3d^3(^2H)4s$	^3H	1.970	1.985
17	$3d^3(^2F)4s$	^3F	2.123	2.161	18	$3d^3(^4F)4p$	$^5G^o$	2.333	2.356
19	$3d^3(^4F)4p$	$^5D^o$	2.351	2.386	20	$3d^3(^2D1)4s$	^3D	2.349	2.420
21	$3d^3(^4F)4p$	$^5F^o$	2.366	2.392	22	$3d^3(^4F)4p$	$^3D^o$	2.368	2.397
23	$3d^3(^4F)4p$	$^3G^o$	2.404	2.441	24	$3d^3(^4F)4p$	$^3F^o$	2.429	2.469
25	$3d^3(^4P)4p$	$^5P^o$	2.493	2.534	26	$3d^3(^4P)4p$	$^5D^o$	2.517	2.556
27	$3d^3(^4P)4p$	$^3P^o$	2.508	2.552	28	$3d^3(^2G)4p$	$^3H^o$	2.522	2.559
29	$3d^3(^2G)4p$	$^3G^o$	2.540	2.585	30	$3d^3(^2G)4p$	$^3F^o$	2.550	2.597
31	$3d^3(^2P)4p$	$^3P^o$	2.605	2.623					

Radiative Data

While the main emphasis of the Project has been on the provision of accurate collisional data, radiative data have not been completely ignored. As an example we discuss the work of Bautista and Pradhan [2] on Fe IV. As with the collisional work an adequate target description is needed. There follows a list of the configurations they included in their calculations.

Fe V configurations

Spectroscopic: $3s^23p^63d^4$, $3s^23p^63d^34s$, $3s^23p^63d^34p$

Correlation: $3s^23p^53d^5$, $3s^23p^43d^6$, $3s^23p^63d^24p^2$, $3s3p^63d^44s$, $3s3p^63d^44p$, $3p^63d^6$, $3p^63d^54s$, $3p^63d^44s^2$, $3s^23p^63d^24s4p$, $3s^23p^63d^24s4d$

This extensive target, in which the spectroscopic configurations are considered explicitly in the scattering treatment, leads to excellent agreement between the observed [17] and calculated energies as is indicated Table 1. The calculated bound state energies of Fe IV are also in excellent agreement with experiment and are to be found in the paper cited.

For exact wavefunctions, the length and velocity forms of the oscillator strengths (and the photoionization cross sections) must agree. Their agreement when using approximate wavefunctions is an indication of the quality of the calculations. This may be seen in Figure 7 where the length and velocity f-values of Fe IV have been plotted. The scatter about the line of slope one is small for the vast majority of the data. The ground state photoionization for this ion is particularly interesting as can be seen from Figure 8. The enormous peak and trough between 4 and 5 Rydbergs

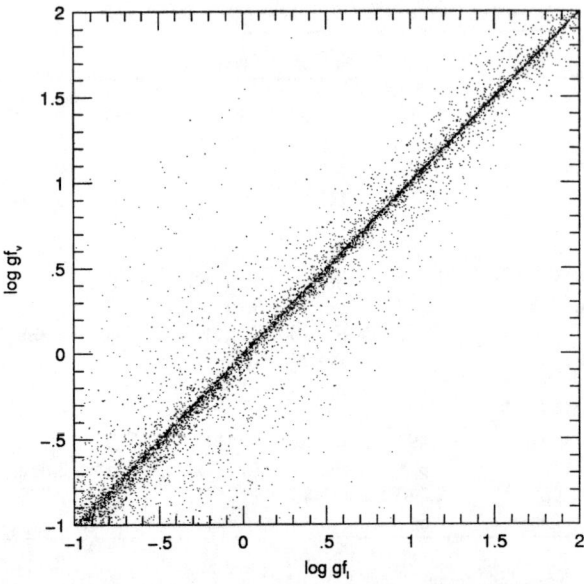

FIGURE 7. Comparison of length and velocity f-values for Fe IV.

in (a) is caused by the $3s^23p^53d^5(^7P^o)3d(^6P^o)$ resonance. This is confirmed by the following two plots. In (b) the $3s^23p^53d^5$ target terms have been excluded from the calculation while in (c) they have been explicitly included. Note that in (a), the resonances only appear in the second summation of the close-coupling expansion Eq. (1). Figure 8a also shows that previous work has underestimated the cross section at low energies.

The Iron Project has now provided improved radiative data for Fe I–V

FUTURE WORK

In the not too distant future, XMM and AXAF are due to be launched opening up new possibilities in X-ray astronomy. The analysis of these observations will require new atomic data, particularly for inner shell processes. This is one of the aspects we shall be looking at. At the same time we shall be making use of recent developments on the theoretical side. The present codes are being extended to make it possible to calculate E2 and M1 radiative transitions. The convergent close-coupling method of Bray and collaborators (e.g. [5]), the intermediate R-matrix [7] and the R-Matrix pseudostate approaches [1] provide practical means of dealing with the intermediate energy range. In particular, the last of these may easily be implemented within the current framework which makes it particularly attractive. The fully-relativistic Dirac scattering code of Norrington [11] promises

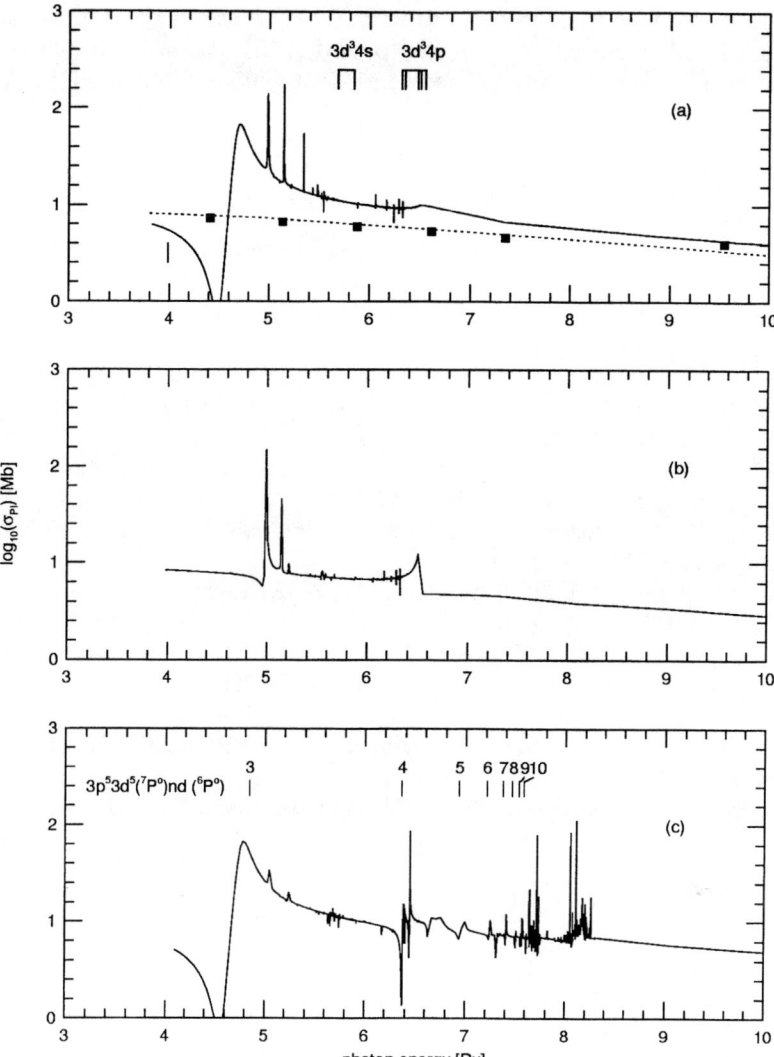

FIGURE 8. Photoionization cross section (σ (Mb)) of the ground state $3d^5(^6S)$ of Fe IV as a function of photon energy (Rydbergs). (a) the cross section obtained with the present 31CC expansion (solid curve); (b) the cross section excluding the $3s^23p^53d^6$ configuration; (c) the cross section with the $3s^23p^53d^5$ target terms of Fe V included explicitly (the Rydberg series $3s^23p^53d^5(^7P^o)nd(^6P^o)$ for $n=3$ to 10 is marked). The dashed curve shows the results of Sawey & Berrington [14] and the filled dots, those of Reilman & Manson [13].

to be a further useful weapon in the arsenal. It seems that less correlation needs to be included in the full treatment compared to the normal Breit-Pauli technique.

Finally, and perhaps most importantly, database access to the data, along the lines of TOPBASE [12], will be provided by TIPBASE which is currently under development. The new Iron Project radiative data will be put into TOPBASE in due course.

REFERENCES

1. Bartschat, K. Hudson, E.T., Scott, M.P., Burke, P.G.,Burke, V.M., J. Phys. B:At. Mol. Phys. **29**, 115, (1996)
2. Bautista, M.A., Pradhan, A.K., A&A Suppl. Ser. (in press).
3. Berrington, K.A, Saraph, H.E., Tully, J.A., A&A Suppl. Ser. (in press).
4. Binello, A.M., Mason, H.E., Storey, P.J., A&A Suppl. Ser. (in press).
5. Bray, I., Stelbovics, A.T., Adv. At. Mol. Phys. **35**, 209, (1995)
6. Burgess, A., Tully, J.A., A&A **254**, 436, (1992).
7. Burke, P.G., Noble, C.J., Scott, M.P., Proc. R Soc. A **410**, 289 (1987)
8. Butler, K., Physica Scripta **T65**, 63, (1996).
9. Hummer, D.G., Berrington, K.A., Eissner, W., Pradhan, A.K., Saraph, H.E., Tully, J.A., A&A **279**, 298, (1993).
10. Mohan, M. Le Dourneuf, M., Phys. Rev A **41**, 2862, (1990).
11. Norrington, P.H., Priv. Comm.
12. Cunto, W., Mendoza, C., Ochsenbein, F. and Zeippen, C.J., A&A, **275**, L5, (1993)
13. Reilman, R.F., Manson, S.T., ApJ Suppl **40**, 815, (1979).
14. Sawey, P.M.J., Berrington, K.A., J. Phys. B:At. Mol. Phys. **25**, 1451, (1992)
15. Seaton, M.J., Yu Yan, Mihalas, D., Pradhan, A.K., MNRAS **266**, 805, (1994).
16. Seaton, M.J., MNRAS **289**, 700, (1997).
17. Sugar, J., Corliss, C., J. Phys. Chem. Ref. Data, **14** Suppl. No. 2 (1985).

Ion - Atom Data for Fusion Research

R. Hoekstra[1], H. Anderson[2], F. W. Bliek[1], G. Lubinski[1],
R. Morgenstern[1], R. E. Olson[3], and H.P. Summers[2]

[1] *KVI, Atomic Physics, RijksUniversiteit Groningen, Zernikelaan 25,
9747 AA Groningen, The Netherlands*
[2] *Dept. of Physics and Applied Physics, University of Strathclyde,
107 Rottenrow, Glasgow G4 0NG, UK*
[3] *Physics Dept., University of Missouri - Rolla, Rolla MO 65401, USA*

Abstract. Charge exchange spectroscopy as a principal tool for fusion-plasma diagnostics relies heavily on the quality of the cross sections for the underlying charge transfer processes. The present status of charge transfer cross sections is discussed both for processes relevant for the spectroscopy of the core plasma and of the edge and/or divertor plasma.

INTRODUCTION

Charge eXchange Spectroscopy (CXS) using neutral (heating) beams as probes has become a principal diagnostic tool on tokamak plasmas [1–4]. At large tokamaks such as JET (Joint European Torus), in particular, CXS based on visible light emission by plasma impurities following electron capture from the neutral beams is employed to determine local fusion-plasma quantities such as ion temperature and density, and plasma rotation. The success of the method is directly linked to the accuracy of the cross sections for the fundamental processes underlying the CXS. The CXS diagnostics is based on charge transfer reactions in collisions between plasma impurity ions, A^{q+}, and neutrals, B, as schematically given by:

$$A^{q+} + B \longrightarrow A^{(q-1)+}(n\ell) + B^+ \qquad (1)$$

which is followed by photon emission

$$A^{(q-1)+}(n\ell) \longrightarrow A^{(q-1)+}(n'\ell') + h\nu \qquad (2)$$

With n and ℓ the principal and angular momentum quantum numbers of the state populated by the charge transfer process. The intensity of the photon emission and the spectral line shapes are measures for the ion abundances and temperatures. While the density determination depends on the absolute photon emission cross sections, the temperature determination is strongly depended on an accurate knowledge of the collision-velocity dependence of the charge transfer cross sections [4,11].

Already for some 10 years, either via EC programmes or EURATOM there exists a close and formal collaboration between the diagnostics and modelling group of JET and the KVI atomic physics group with the objective of producing, gathering and recommending cross sections for the above mentioned ion-atom collision processes. At KVI crossed-beam experiments are performed to determine specific state selective charge transfer cross sections. Photon emission spectroscopy is used to obtain the cross section data, see e.g. Hoekstra et al. [6].

In this paper we review some results of the KVI-JET collaboration with the aim of picturing the present status of the ion-atom data base underlying the CXS diagnostics at large tokamaks and in particular the CXS diagnostics at JET. In the next section of the paper we will first describe the collision systems and energy ranges relevant for CXS. Since neutrals are needed as electron donors, CXS is restricted to the plasma regions were the neutral beams penetrate through the (core) plasma (active CXS) and to the colder, outer regions of the plasma, where there are sufficient neutrals, i.e., the Scrape-Off-Layer and the divertor (passive CXS).

ION-ATOM DATA FOR CXS.

Figure 1 shows a schematic view of the layout of the neutral beam based CXS at JET. As neutral heating beams, deuterium beams and to a lesser extent helium beams are used. The energies of the beams are in the range of 35 to 70 keV amu^{-1}. Note that the deuterium beams have three velocity components, namely components with energies per mass unit of 1, 0.5, and 0.33 of the beam energy. This is due to the fact that in the source not only D^+, but also D_2^+ and D_3^+ ions are produced. For example a 120 keV D beam contains particles with energies of 60 keV amu^{-1}, 30 keV amu^{-1}, and 20 keV amu^{-1}. Together with the fact that ion temperatures in the center of the core plasma can exceed 20 keV, this implies that fundamental data are needed over a broad energy range, say typically 10 to 100 keV amu^{-1}. The most abundant impurity ions in the core plasma are fully stripped ions of the low-Z elements, i.e., He^{2+}, C^{6+}, Be^{4+}, and to a lesser extent O^{8+}. Puffing of N_2 or Ne in the tokamak, to enhance radiative cooling near the walls, leads to the introduction of N^{7+} and Ne^{10+} ions in the core plasma. By boronization of the walls, B^{5+} becomes a prominent impurity ion. So basically the fully stripped ions of all low-Z elements with maybe the exclusion of Li^{3+} and F^{9+} have to be considered.

An important point to note is that the light resulting from the interaction

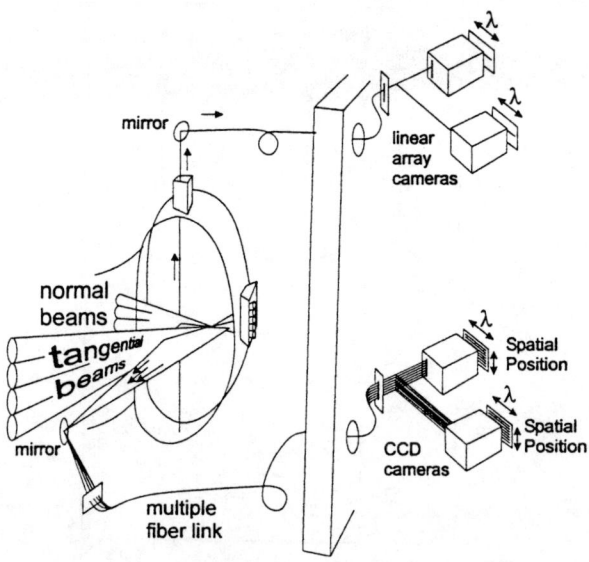

FIGURE 1. Schematic representation of the lay-out of the neutral beam based charge exchange spectroscopy at JET.

of plasma ions with neutral beam atoms is transported to the spectrometers by means of fiber optics. This implies that the wavelength range is limited to approximately the visible spectral range. This wavelength restriction puts severe demands on both the experimental and theoretical studies on state selective charge transfer in ion-atom collisions. For the hydrogen-like ions produced in charge transfer reactions of type (1), the visible light emission originates from transitions between high-lying n shells, which are only weakly populated. For example, for C^{6+} ions colliding on hydrogen the dominantly populated n shell is $n = 4$, while for CXS the most relevant n shell is the $n = 8$ one. The corresponding line emission cross sections differ by orders of magnitude, see figure 2. The line emission used for CXS presents only a very minor fraction of the total photon emission subsequent to electron capture. This not only makes the determination of the relevant cross sections by means of crossed-beam laboratory experiments very difficult, but the theoretical predictions also require extensive computer resources. This can nicely be illustrated by the evolution of the atomic orbital (AO) calculations by Fritsch [8–10]. In the first attempts (AO1) to calculate the capture into the $C^{5+}(8\ell)$ states only the $C^{5+}(4\ell, 5\ell, 8\ell)$ states were included in the basis set. In the AO2 calculations all the $C^{5+}(6\ell, 7\ell)$ states were added to the basis set of AO1. The improved agreement with the experiments (see figure 2) indicates the impor-

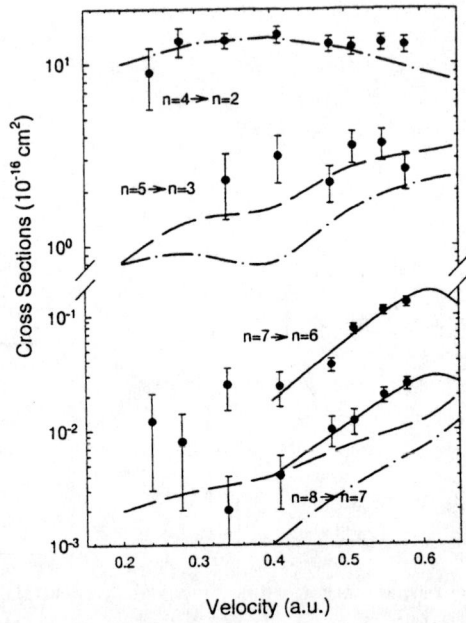

FIGURE 2. Cross sections for line emission following charge transfer in C^{6+} - H collisions. The experimental data are from Hoekstra *et al.* [7]. The theoretical curves represent results of 3 AO calculations with increasing basis set sizes (cf. text). Dashed-dotted line: AO1 (Fritsch [8]), long-dashed line: AO2 (Fritsch [9]) and solid line: AO3 (Fritsch [10]), see text. The dominant line emission channels are also well predicted by other theoretical methods (Ref. [7] and references therein).

tance of stepwise promotion processes in the population of high-n levels. The final step was made (AO3) by also including excited states of the target into the basis set. Basis set sizes (computer memory) are to a large extent still limiting factors in calculating charge transfer into non-dominant high-n levels for heavier elements (for example Ne^{10+}) and for systems with two or more active electrons [11]. For these systems benchmarking by experiments is still of major importance for assessing the accuracy of the cross sections.

Besides using (parasitically) the hydrogen (deuterium) or helium heating beams for CXS, keV amu^{-1} Li beams are used. Such Li beams are dedicated for CXS diagnostics of the edge plasma close to the plasma facing walls [12,13]. Because of the low binding energy of the Li(2s) electron, charge transfer directly populates the high-n levels of the plasma ions, leading to strong photon emission signals in the visible spectral range. In particular, spatially resolved CXS along the Li beam enables the study of the influx into the plasma of material sputtered from the walls. Because of the C and Be (or B) coated walls, C^{q+}, Be^{q+} and B^{q+} ions are the species of prime interest. Charge transfer data in the energy range of several to 10 keV amu^{-1} are needed to cover the energies of Li beams at different tokamaks.

In addition to the above mentioned so-called active CXS methods using injected neutral beams as electron donors, there is the possibility of passive CXS in regions in which neutrals are intrinsically present. Appreciable fractions of neutrals are found only in the colder areas of the plasma, i.e., the edge or scrape-off layer and the divertor. From divertor to edge layer the ion temperatures increase from a few eV up to approximately 1 keV. Because of the low temperatures partially stripped ions of the low-Z elements are the dominant ionic species interacting with neutrals.

In the next section we will review the information available on charge transfer relevant for CXS diagnostics. Throughout the rest of the paper we will speak of hydrogen, and use energies per mass unit, so that the results are directly convertible to deuterium which is most often used as a heating beam.

Data for Active CXS.

Hydrogen

Capture from atomic hydrogen by fully stripped low-Z ions has been one of the main topics in theoretical collision studies, see e.g. [11,14]. Therefore, in particular for the dominant capture channels, a lot of consistent cross sectional data are available. Especially from the comparison of theoretical results with our experimental data for C^{6+} - H (see figure 2) and He^{2+} - H [4,6,10] it now appears that also the predictions for capture in the non-dominantly populated high-n states are accurate enough, so that they can be trusted for other fully stripped ion - atomic hydrogen collision systems, such as Be^{4+} - H. In general, the recommended high-n cross sections are based on AO calculations [10] for energies between 1 and 50 keV amu^{-1}, on classical trajectory Monte Carlo (CTMC) calculations [17] for energies between 50 and 100 keV amu^{-1} and on continuum distorted wave approximations (CDWA) [18] for energies above 100 keV amu^{-1}.

The recommended cross sections are stored in the JET Atomic Data and Analysis Structure package (ADAS [16]). The uncertainties associated with the recommended cross sections are $\leq 25\%$ for capture into the non-dominant high-n levels. The errors in the total charge transfer cross sections, needed for modelling the neutral beam attenuation in the plasma, are believed to be on the order of 10%. This meets the demands of the diagnostics group. Plasma quantities, such as Z_{eff} and neutron rates, derived from the local ion temperatures and densities obtained by CXS are now in agreement with other measurements of these integral plasma quantities. The improvement of the

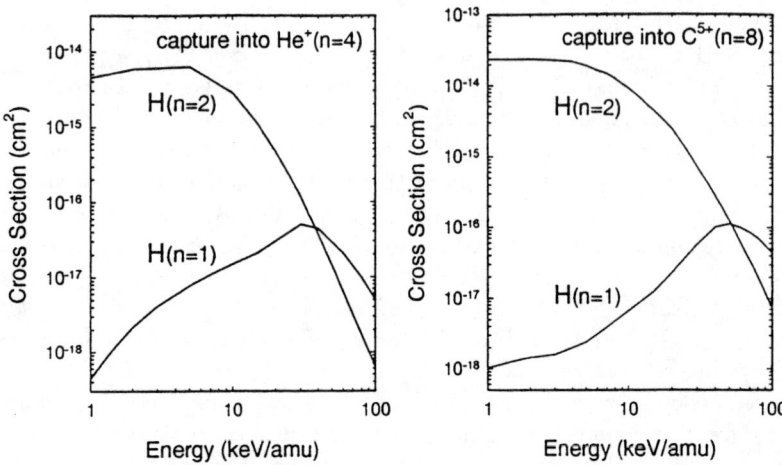

FIGURE 3. Comparison of cross sections for capture into $He^+(n=4)$ and $C^{5+}(n=8)$ in collisions of respectively He^{2+} and C^{6+} on ground state H and $H(n=2)$. The $He^+(n=4)$ and $C^{5+}(n=8)$ states are the upper levels of the transitions used for CXS. The ground state cross sections are the recommended data as present in ADAS. The cross sections for $H(n=2)$ are from our recent CTMC calculations [20].

agreement between CXS and other diagnostics has been strongly linked to the improved quality of the fundamental data for ion - atom collisions.

Although there seems to be a sound basis for CXS at least at JET, discrepancies between different techniques have been encountered at smaller tokamaks, such as TEXTOR (Jülich, Germany), at which neutral H or D beams are used at energies of approximately 25 keV, i.e., beam components of 25, 12.5 and 8.33 keV amu^{-1} for H beams and 12.5, 6.25 and 4.16 keV amu^{-1} for D beams. It was realised that at these energies a small fraction of metastable $H(n=2)$ might have a pronounced effect on the CXS diagnostics [19]. This is caused by the fact that the charge transfer reaction

$$A^{q+} + H(n=2) \longrightarrow A^{(q-1)+}(n\ell) + H^+ \qquad (3)$$

populates directly the high-n levels in the receiver ions from which the visible light emission originates. Because of the resonant nature of charge transfer reactions the cross sections for reaction (3) are much larger than for collisions on ground state ions. This is shown in figure 3. In this figure we compare our recent CTMC results for capture from $H(n=2)$ with the recommended data for capture from ground state H. From the figure it is clear that especially at energies below approximately 25 keV amu^{-1} the cross sections for capture from excited H donors are orders of magnitude larger than for capture from

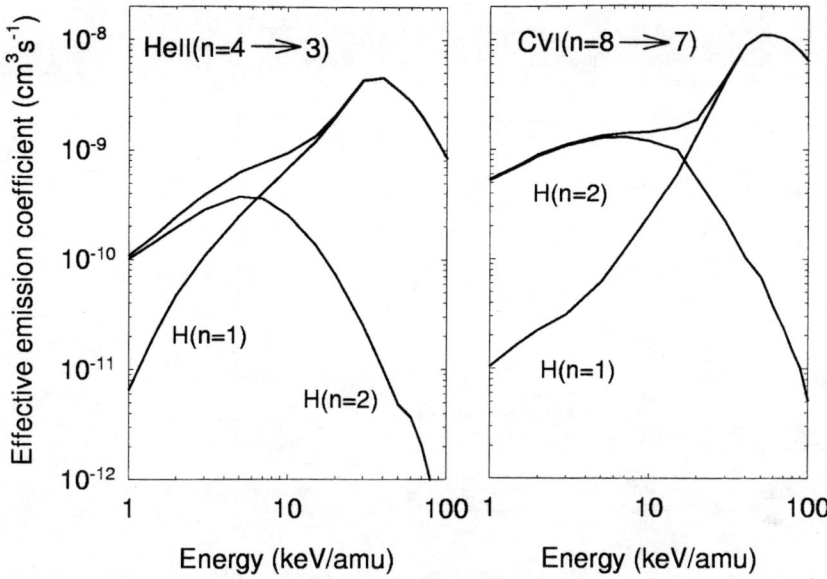

FIGURE 4. Effective line emission coefficients for the HeII($n = 4 \rightarrow n = 3$) and CVI($n = 8 \rightarrow n = 7$) CXS lines resulting from charge transfer from mixed ground state / excited state H beams. The fraction is calculated in a full collisional-radiative model for a deuterium plasma with $Z_{eff} = 2$, ion and electron temperatures of 8 keV and an electron density of 5×10^{13} cm^{-3} [20].

ground state donors. This indicates that the influence of an excited beam component can be considerable even for very small fractions of H($n = 2$).

To get an idea about the influence of excited neutrals it is clear that the fraction must be known. With ADAS we have calculated within a fully collisional-radiative model the excited beam fraction for a deuterium plasma with $Z_{eff} = 2$ and with ion and electron temperatures of 8 keV and with an electron density of 5×10^{13} cm^{-3} [20]. Dependent on the energy of the neutral beam, the fraction is found to range from somewhat below to slightly above 0.1%. As function of the neutral beam energy figure 4 shows the contribution of the ground state and of the small excited H($n = 2$) fraction to the effective line emission coefficients or photon emissivities [2] for the HeII($n = 4 \rightarrow n = 3$) and CVI($n = 8 \rightarrow n = 7$) transitions. It is obvious that notwithstanding the smallness of the fraction of excited beam particles, charge transfer from H($n = 2$) dominates the HeII($n = 4 \rightarrow n = 3$) and CVI($n = 8 \rightarrow n = 7$) line

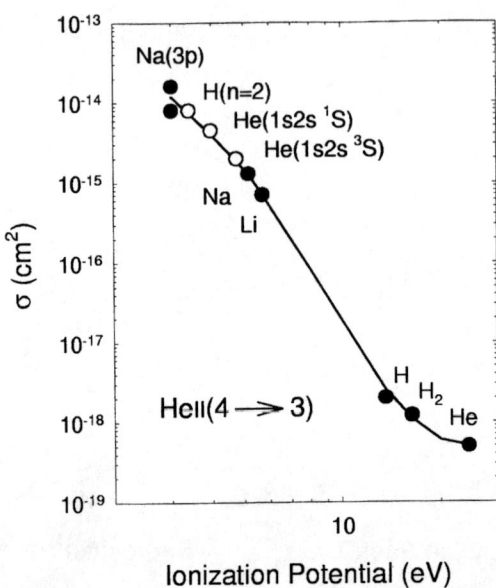

FIGURE 5. HeII($n = 4 \to n = 3$) line emission cross sections resulting from 3 keV amu^{-1} He^{2+} ions colliding on different targets. Closed symbols represent experimental data (He [21], H$_2$ [22], H [6], Li [23], Na [24], and laser excited Na($3p$) [25]) and the open symbols theoretical data (He* [21] and H($n = 2$) [20]).

emission at energies below 10 and 20 keV amu^{-1} respectively. In general it can be stated that the influence of excited H($n = 2$) donors on CXS observations is increasing with increasing Z of the impurity ions.

In contrast to capture from ground state H it is not possible to benchmark the theoretical results for capture from H($n = 2$) directly with experiments, because of experimental problems in producing well-defined, high density H($n = 2$) targets. To get at least indirectly an idea of the validity of the theoretical results figure 5 shows as an example the HeII($n = 4 \to n = 3$) line emission cross sections resulting from 3 keV amu^{-1} He^{2+} ions colliding on different targets. The two cross section entries for laser excited Na($3p$) are to be noted. They represent results from experiments in which the Na($3p$) orbital was aligned either perpendicular or parallel to the velocity of the He^{2+} ions. At this energy of 3 keV amu^{-1} the perpendicular direction is more favourable for capture. The preference for one of the alignment directions depends on the collision energy, see, e.g., Schlatmann *et al.* [25]. By taking the average

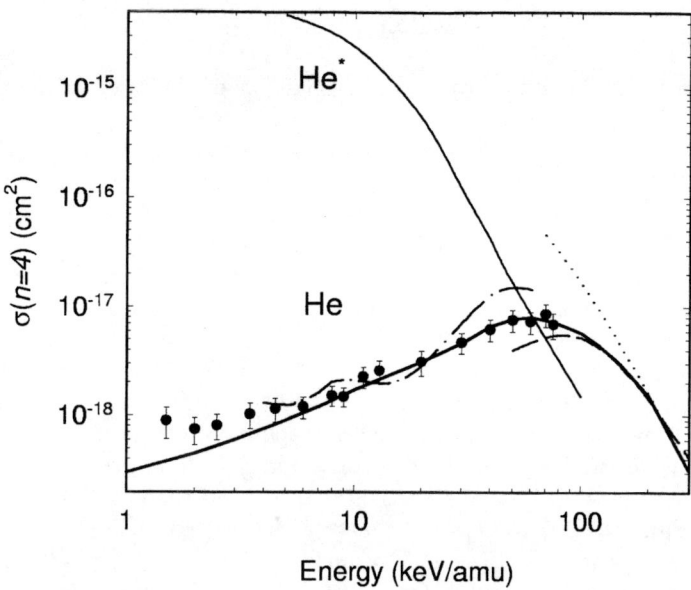

FIGURE 6. Cross sections for capture into $He^+(n = 4)$ in He^{2+} - He and He^{2+} - He^* collisions. Closed symbols represent the experimental data of Folkerts et al. [21] for He^{2+} - He. Theoretical curves: dashed-dotted AO (He) [26], dashed (He) and thin solid (He^*) CTMC [21], dotted CDWA (He) [30]. The thick solid line represents the recommended cross section for capture into $n = 4$ in He^{2+} - He collisions as implemented in ADAS. At the lowest energies the difference between the recommended curve and the experiments arises from cascade corrections, see Folkerts et al. [21].

value of both directions, it is seen that the calculations for H($n = 2$) and also for excited He, He($1s2s\,^1S$) and He($1s2s\,^3S$), are close to values expected from the interpolation between Na($3p$) and Na. In combination with the fact that for keV amu^{-1} collisions of He^{2+} on Na($3p$) and Na there is good agreement between CTMC calculations and experiments, it is believed that the H($n = 2$) CTMC results [20] have an associated uncertainty of approximately 25%. Furthermore, in the 1 - 10 keV amu^{-1} energy range comparable cross sections are predicted by molecular orbital calculations [28,29]. For excited He the CTMC results [21] are confirmed by atomic orbital calculations [26].

In conclusion, for CXS using neutral hydrogen beams there is enough quality data available forming a reliable basis for local ion density and temperature diagnostics.

Helium

For CXS on neutral helium beams, the availability of low-Z ion - He data is limited to total charge transfer cross sections and capture into the dominant, resonant levels. Only the He^{2+} - He system is studied experimentally [21] and theoretically [26] on a level similar to low-Z ion - atomic hydrogen systems. Experimentally, the problem in studying capture into non-dominantly populated high-n states is just counting statistics. Due to the higher ionization potential of He as compared to H the cross sections are much smaller than for H, see figure 5. Only for He^{2+} could we produce beams intense enough to perform the crossed-beam experiments [21], for an ion as C^{6+} only an upper limit for the relevant CVI($n = 8 \rightarrow n = 7$) line emission cross section could be determined [27]. The fact that for He two active electrons have to be considered makes theoretical approaches much more complex [26].

From extensive studies of the He^{2+} - He system it is found that the agreement between theory and experiment is somewhat less than for the He^{2+}- H system. As figure 6 illustrates, unfortunately the largest differences are observed around the energies of the JET beams, i.e., 30 - 60 keV amu^{-1}. In that energy range the recommendation for the cross sections is mainly determined by the experiments, which smoothly bridge the gap between the AO [26] and CTMC [21] results. It is mentioned that very recent calculations [31] show an improved agreement with the experiments.

For the other low-Z ions the visible light CXS relies on cross sections which are either scaled from the hydrogen ones or extrapolated from the cross sections for the dominant channels. More work on these systems is really needed especially for C^{6+} and Be^{4+} which are the dominant impurities at JET. The situation might actually be less critical as compared to the case of CXS based on H beams, because of the role of the (unknown) fraction of metastable He in the beams. As compared to H the effect of excited donors in the He beams is much stronger. This can already be understood from comparing the left panel of figure 3 with figure 6. From figure 3 it is seen that the cross sections for capture into $He^+(n = 4)$ become equal for H and H($n = 2$) donors at a collision energy of approximately 40 keV amu^{-1}. For He^{2+} - He/He* collisions at the same energy of 40 keV amu^{-1} the cross section for capture from He* is still a factor of 8 larger than for capture from ground state He. For C^{6+} ions, preliminary results of CTMC calculations indicate that at a typical He beam energy of 40 keV amu^{-1} the ratio for capture into $C^{5+}(n = 8)$ from metastable He and ground state He is more than a factor of 500 [27]. This indicates that the influence of metastable beam donors is much more important for He than for H beams. In addition, and at least as important, the metastable He-beam component is estimated to be in the percent range [32], i.e., about 10 times the fraction found for H beams. Therefore, for the higher charged low-Z ions the CXS observations may very well be dominated by capture from He($1s2s$). Cross sections for capture from He($1s2s$) can be and actually are

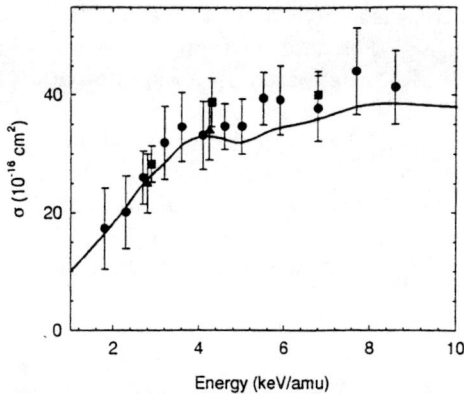

FIGURE 7. Line emission cross sections for AV($n = 7 \to n = 6$) transitions following A^{5+} - Li collisions [34]. The CTMC results are calculated (line) for B^{5+} - Li collisions, the experimental data are for B^{5+} (circles), C^{5+} (triangle) and N^{5+} (squares).

being calculated by the CTMC method. But it is just equally important to obtain accurate knowledge on the fraction of metastables. This fraction is still not known [33]. For He beam-based CXS a better modelling of the fraction of metastables in the He beams is truly needed.

Lithium

For Li beam-based CXS a large collection of data is available for charge transfer into the high-n states of low-Z ions, e.g. refs. [23,34–37] and references therein. The collision energy range studied, typically 1 - 10 keV amu^{-1} encompasses nicely the energies of the Li beams used or planned at different tokamaks. In general, calculations are performed for fully stripped ions, but since Li beams probe the outer, colder regions of the plasma, also cross sections for partially stripped ions must also be considered. From comparisons of line emission cross sections of different ion species with the same charge it seems justified to neglect the effect of the projectile core electrons on the charge transfer processes and to group the ions according to their charge state only. As an example figure 7 summarizes AV ($n = 7 \to n = 6$) line emission cross sections for different 5-fold charged ions ($A^{5+}=B^{5+}$, N^{5+} and C^{5+}). Similar agreements are found for other ions [34,35], indicating that the core effect on charge transfer from Li is marginal at least for the higher charge states ($q \geq 3$) of the low-Z elements. For the lower charge states separate studies are needed just as for heavier elements such as Ar, for which clear core effects are observed [38].

Since ground state Li(2s) is easily excited to Li(2p), also capture from Li(2p) has to be dealt with [37]. For capture from Li(2p) some theoretical work [37,39] has been done on He^{2+}, Be^{4+}, and C^{6+}. Experimentally the information is limited to a translational-energy spectroscopy experiment on collisions of He^{2+} on laser excited Li(2p) [40]. Realizing the difficulties in performing experiments with laser excited targets, the bulk of the data for Li(2p) has to be delivered by theory. In previous sections we have already shown the capacity of theory to produce good cross sections for He* and H*. The same should be possible for Li(2p). A few benchmark experiments would of course be welcome.

Data for Passive CXS

From the previous discussions on the role of excited donors it is clear that in the colder, low-energy regions of the plasma photon emission from high-n states populated by charge transfer is predominantly due to capture from the probably tiny but unknown fraction of excited neutrals. Therefore, the passive CXS information has to be extracted from the (VUV) radiation following capture into the dominant states, which is barely influenced by capture from excited atoms. Already in the mid-eighties, with an astrophysical motivation cross sections have been calculated for many multicharged ion - atom collision systems [41]. At energies well below 1 keV amu^{-1} experimental tests of theory are rather scarce and mainly limited to total charge transfer cross section measurements by the Oak Ridge group using a merged-beam technique (e.g. refs. [42,43]). Already these total charge transfer cross sections showed sometimes large discrepancies with theory, e.g., the case of O^{3+} - H [44]. The first state selective experiments [45-47] on this system even showed that in the energy range of 0.1 - 1 keV amu^{-1} the $O^{2+}(3s\ ^{1,3}L)$ states are dominantly populated. Fully quantal calculations [48,49] predicted capture to proceed mainly into the $O^{2+}(3p\ ^{1,3}L)$ states. More recent calculations [50] reproduce the experimentally observed dominant nature of capture into the $O^{2+}(3s\ ^{1,3}L)$ states.

The most recent and extensive calculations (N^{4+} - H [51-53]) for low energies show good agreement with the merged-beam experiments [43,52]. Nevertheless an experimental test on a state-selective basis would be appropriate to benchmark such modern calculations. For passive CXS the state selective cross sections are the most important ones. Absolute state selective electron capture measurements are basically restricted to energies in the 0.1 - 1 keV amu^{-1} energy domain, e.g., refs. [45,47,54,55]. To make lower collision energies accessible for our photon emission spectroscopy experiments we have constructed a new set-up for low-energy collision studies.

Figure 8 shows schematically the main features of the new set-up. The charge-transfer collisions are studied inside a so-called RF octopole ion guide.

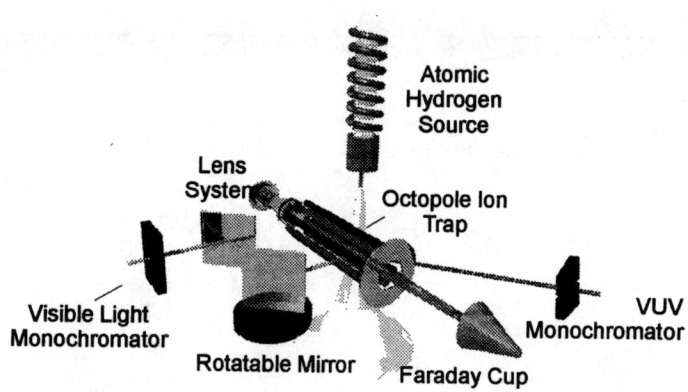

FIGURE 8. Schematic representation of the octopole ion-guiding set-up at KVI developed for photon emission spectroscopy experiments at energies well below 1 keV amu^{-1}.

By floating the complete octopole on a voltage V_{oct} the collisions take place with an energy of $q(V_{ecr} - V_{oct})$, with the q the charged state of the ion and V_{ecr} the potential on the ion source. Radial trapping of the divergent decelerated ion beam is achieved by applying an RF voltage to the rods of the octopole. The voltage between neighboring rods has a phase difference of 180°. Typical RF frequencies are several to tens of MHz. For sufficiently high frequencies the ions experience a potential proportional to r^6, which confines the radial motion of the ions. This technique of RF multipole ion beam guiding was pioneered by Teloy and Gerlich [56]. For a review see Gerlich [57]. This technique has been introduced in the field of multicharged ion - atom collisions by Okuna et al. [58] with the objective of measuring total charge changing cross sections [59]. The geometry of our ion guide is such that it is possible to retain a crossed-beam configuration and a maximum observation of the collision center. Ion beams originally having an energy of approximately 1 keV amu^{-1} can be decelerated down to energies of approximately 5 eV amu^{-1} without appreciable loss of beam intensity [60]. The lower limit is fully defined by the intrinsic energy spread of the ion beams extracted from our ECR-type ion source. For He-like low-Z ion beams the energy spread is of the order of 2 eV amu^{-1}. Therefore, certainly at energies below 5 eV amu^{-1} the collision energy becomes too ill-defined to perform conclusive measurements.

The first system we studied was C^{4+} on H_2, which is one of the few systems for which at least in the 0.1 - 1 keV amu^{-1} energy range relative data

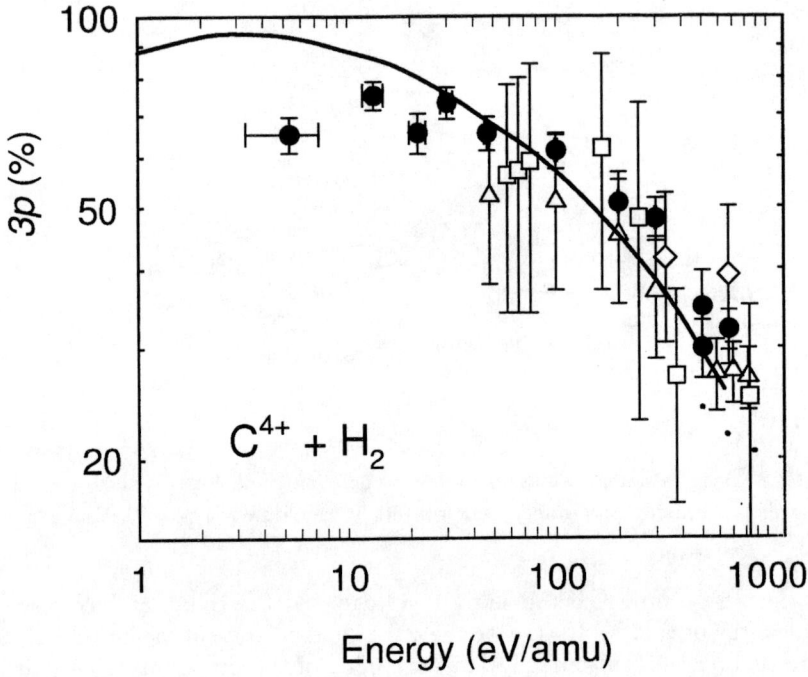

FIGURE 9. Relative cross sections for capture into $C^{3+}(3p)$ in C^{4+} - H_2 collisions. Circles: present measurements within the octopole ion guide, triangles: Hoekstra et al. [54], squares: Baptist et al. [61], diamonds: McLaughlin et al. [55], solid line: MO calculation [63], and dotted line: AO calculation [62].

are available for comparison. Charge transfer in C^{4+} - H_2 collisions populates the $C^{3+}(3\ell)$ states only [6]. In our case the subsequent $C^{3+}(3\ell \to 2\ell')$ photon emission can be detected in one spectrum, because the spectrometer is equipped with a position sensitive micro-channel plate detector enabling the detection of photons in a spectral range of about 20 nm. The $C^{3+}(3\ell \to 2\ell')$ transitions have wavelengths of 31.2, 38.4, and 42.0 nm for $3p$, $3d$, and $3s$ respectively. Therefore the relative capture cross sections $\sigma_{3\ell}$ can be determined accurately. All effects due to target fluctuations and beam overlaps cancel. Since the spectrometer is positioned under the double magic angle there is no influence of possible polarization effects on the measured photon emission yields [6]. Finally, the photon emission yields are directly related to the capture cross sections because the branching ratios for the $C^{3+}(3\ell \to 2\ell')$

transitions are 1.

Figure 9 summarizes the results for the relative cross section for capture into $C^{3+}(3p)$ together with existing data. It is seen that the present measurements merge nicely with previous ones. The uncertainties in the present measurements are much smaller than in the other experiments. Concerning the simplicity of the MO calculations (only three states included) [63] there is a rather good agreement between experiment and theory, except for the energies below 50 eV amu^{-1}.

It should be mentioned that in the first experiments we were not able to determine absolute state selective cross sections. The outcome of the experiment depended strongly on certain experimental settings, where factor of 2 differences were observed. After some modifications we are now able to perform absolute measurements down to 50 eV amu^{-1} with an absolute error of approximately 20%. Towards lower energies the error increases to about 40 - 50 % at 5 eV amu^{-1}. This will be improved in the near future. So we have reached experimentally the point that absolute state selective cross sections can be measured at energies well below 1 keV amu^{-1}.

CONCLUSIONS

The success of charge exchange spectroscopy as a principal diagnostics for fusion plasmas is strongly interlinked with an accurate knowledge of the underlying charge transfer processes. In this paper we review the present status of charge transfer cross sections relevant, on one hand, for the spectroscopy of the core plasma and, on the other hand, for the edge and/or divertor plasma. This is done with emphasis on the data relevant for JET. For CXS based on deuterium or hydrogen beams there is a sound cross sectional basis, for CXS based on helium beams further improvement is needed, especially the role of metastables needs to included. For passive CXS we have reached the point that for the first time theoretical predictions can be tested at energies well below 1 keV amu^{-1}.

ACKNOWLEDGMENTS

The authors like to acknowledge the support of the 'Stichting voor Fundamenteel Onderzoek der Materie' (FOM), financially supported by the 'Nederlandse Organisatie voor Wetenschappelijk Onderzoek'(NWO) and EURATOM. The work has also received support from the US Department of Energy - Office of Fusion Energy.

REFERENCES

1. Isler, R. C., *Plasma Phys. Contr. Fusion* **36**, 171 (1994).
2. Boileau, A., von Hellermann, M., Horton, L. D., and Summers, H. P., *Plasma Phys. Contr. Fusion* **31**, 779 (1989).
3. Wolfrum, E., Aumayr, F., Wutte, D., Winter, HP., Hintz, E., Rusbüldt, D., and Schorn, P., *Rev. Sci. Instrum.* **64**, 2285 (1993).
4. von Hellermann, M., Mandl, W., Summers, H.P., Boileau, A., Hoekstra, R., de Heer, F. J., and Frieling, G. J., *Plasma Phys. Contr. Fusion* **33**, 779 (1991).
5. Frieling, G. J., Hoekstra, R., Morgenstern, R., and de Heer, F. J., *Z. Phys. D* **21**, S163 (1991)
6. Hoekstra, R., Morgenstern, R., and de Heer, F. J., *J. Phys. B: At. Mol. Opt. Phys.* **24**, 4025 (1991).
7. Hoekstra, R., Ćirić, D., Zinoviev, A. N., Gordeev, Yu. S., de Heer, F. J., and Morgenstern, R., *Z. Phys. D* **8**, 57 (1988)
8. Fritsch, W., *Phys. Rev. A* **30**, 3324 (1984) and Fritsch, W., and Lin, C. D., *Phys. Rev. A* **29**, 3039 (1984)
9. Fritsch, W., *J. Physique. Colloq.* **50**, CI87 (1988)
10. Fritsch, W., *VIth International Conference on the Physics of Highly Charged Ions*, New York: AIP, Conference Proceedings **274**, 1992, pp. 24-33
11. Fritsch, W., and Lin, C. D., *Phys. Rep.* **202**, 1 (1991).
12. Schorn, R. P., Wolfrum, E., Aumayr, F., Hintz, E., Rusbüldt, D., and Winter, HP., *Nucl. Fusion* **32**, 352 (1992).
13. Winter, HP., *Comm. At. Mol. Phys.* **12**, 165 (1982).
14. Janev, R. K., and Winter, HP., *Phys. Rep.* **117**, 265 (1985).
15. Frieling, G.J., Hoekstra, R., Smulders, E., Dickson, W. J., Zinoviev, A. N., Kuppens, S. J., and de Heer, F. J., *J. Phys. B: At. Mol. Opt. Phys.* **25**, 1245 (1992).
16. Summers, H. P., *Atomic Data and Analysis Structure* JET Joint Undertaking Report JET-IR(49)-06, http://patiala.phys.strath.ac.uk/adas/.
17. Schultz, D. R., Meng, L., Reinhold, C. O., Olson, R. E., *Phys. Scr. T* **37**, 89 (1991)
18. Belkić, D., Gayet, R., and Salin, A., *At. Nucl. Data Tabl.* **51**, 59 (1992).
19. Isler, R. C., and Olson, R. E., *Phys. Rev. A* **37**, 3399 (1988).
20. Hoekstra, R., Anderson, H., Bliek, F. W., von Hellermann, M., Maggi, C. F., Olson, R. E., and Summers, H. P., *Plasma Phys. Contr. Fusion* submitted
21. Folkerts, H. O., Bliek, F. W., Meng, L., Olson, R. E., Morgenstern, R., von Hellermann, M., Summers, H. P., and Hoekstra, R, *J. Phys. B: At. Mol. Opt. Phys.* **27**, 3475 (1994).
22. Hoekstra, R., Folkerts, H. O., Beijers, J. P. M., Morgenstern, R., de Heer, F. J. *J. Phys. B: At. Mol. Opt. Phys.* **27**, 2021 (1994).
23. Hoekstra, R., Wolfrum, E., Beijers, J. P. M., de Heer, F. J., Winter, HP., and Morgenstern, R., *J. Phys. B: At. Mol. Opt. Phys.* **25**, 2587 (1992).
24. Schlatmann, A. R., Hoekstra, R., Folkerts, H. O., and Morgenstern, R., *J. Phys. B: at. Mol. Opt. Phys.* **25**, 3155 (1992).

25. Schlatmann, A. R., Hoekstra, R., Morgenstern, R., Olson, R. E., and Pascale, J., *Phys. Rev. Lett.* **71**, 513 (1993).
26. Fritsch, W., *J. Phys. B: At. Mol. Opt. Phys.* **27**, 3461 (1994).
27. Bliek, F. W., Folkerts, H. O., Morgenstern, R., Hoekstra, R., Meng, L., Olson, R. E., Fritsch, W., von Hellermann, M., and Summers, H. P., *Nucl. Instrum. Meth. B.* **98**, 195 (1995).
28. Blanco, S. A., Falcón, C. A., Reinhold, C. O., Casaubón, J. I., and Piacentini, R. D., *J. Phys. B: At. Mol. Phys.* **20**, 6295 (1987).
29. Jouin, H., and Harel, C., *J. Phys. B: At. Mol. Opt. Phys.* **24**, 3219 (1991).
30. Gayet, R., Belkić, D., and Salin, A., private communication
31. Busnengo, H. F., Martínez, A. E., Rivarola, R. D., and Tawara, H., *J. Phys. B: At. Mol. Opt. Phys.* **30**, L805 (1998).
32. Summers, H. P., von Hellermann, M., Breger, P., Frieling, G. J., Horton, L. D., König, R., Mandl, W., Morsi, H., Wolf, R., de Heer, F. J., and Hoekstra, R., *Atomic Processes in Plasmas*, New York: AIP conference proceedings **257**, 111 (1992).
33. Maas, A., thesis, JET Joint Undertaking and RU Utrecht, 1995
34. Hoekstra, R., Olson, R. E., Folkerts, H. O., Wolfrum, E., Pascale, J., de Heer, F. J., Morgenstern, R., and winter, HP., *J. Phys. B: At. Mol. Opt. Phys.* **26**, 2029 (1993).
35. Wolfrum, E., Hoekstra, R., de Heer, F. J., Morgenstern, R., and Winter, HP., *J. Phys. B: At. Mol. Opt. Phys.* **25**, 2597 (1992).
36. Olson, R. E., Pascale, J., and Hoekstra, R., *J. Phys. B: At. Mol. Opt. Phys.* **25**, 4241 (1992).
37. Schweinzer, J., Wutte, D., and Winter, HP., *J. Phys. B: At. Mol. Opt. Phys.* **27**, 137 (1994).
38. Laulhé, C., Jacquet, E., Boduch, P., Chantepie, M., Ghérardi, N., Husson, X., Lecler, D., and Pascale, J., *J. Phys. B: At. Mol. Opt. Phys.* **30**, 2899 (1997).
39. Olson, R. E., and Hoekstra, R., *Nucl. Instrum. Meth. B.* **98**, 214 (1995).
40. Gieler M, Aumayr, F., Weber, M., Winter, HP., and Schweinzer, J., *J. Phys. B: At. Mol. Opt. Phys.* **26**, 2153 (1993).
41. Heil, T. G., *Nucl. Instrum. Meth. B* **23**, 222 (1987).
42. Havener, C. C., Huq, M. S., Krause, H. F., Schulz, P. A., and Phaneuf, R. A., *Phys. Rev. A* **39**, 1725 (1989).
43. Huq, M. S., Havener, C. C., and Phaneuf, R. A., *Phys. Rev. A* **40**, 1811 (1989).
44. Havener, C. C., Meyer, F., W., and Phaneuf, R. A., it XVIIth Conference on the Physics of Electronic and Atomic Collisions, Bristol: Adam Hilger, 1992, pp. 381-390.
45. Wilson, S. M., McCullough, R. W., and Gilbody, H. B., *J. Phys. B: At. Mol. Opt. Phys* **21**, 1027 (1988).
46. Hoekstra, R., Boorsma, K., de Heer, F. J., and Morgenstern, R., *J. Physique Colloq.* **50**, C349 (1988).
47. Beijers, J. P. M., Hoekstra, R., and Morgenstern, R., *J. Phys. B: At. Mol. Opt. Phys.* **29**, 1397 (1996).
48. Heil, T. G., Butler, S. E., and Dalgarno, A., *Phys. Rev. A* **27**, 2365 (1983).

49. Bienstock, S., Heil, T. G., and Dalgarno, A., *Phys. Rev. A* **27**, 2741 (1983)
50. Gargaud, M., McCarroll, R., and Opradolce, L., *Astron. Astrophys.* **208**, 251 (1989).
51. Shimakura, N., Itoh, M., and Kimura, M., *Phys. Rev. A*, **45**, 267 (1992).
52. Folkerts, L., Havener, C. C., Shimakura, N., and Kimura, M., *Phys. Rev. A*, **51**, 3685 (1995).
53. Stancil, P., Zygelman, B., Clarke, N., and Cooper, D., *J. Phys. B: At. Mol. Opt. Phys.* **30**, 1013 (1997).
54. Hoekstra, R., Beijers, J. P. M., Schlatmann, A.R., Morgenstern, R., and de Heer, F. J.,*Phys. Rev. A*, **41**, 4800 (1990).
55. McLaughlin, T. K., McCullough, R. W., and Gilbody, H. B., *J. Phys. B: At. Mol. Opt. Phys.* **25**, 1257 (1992).
56. Teloy, E., and Gerlich, D., *Chem. Phys.* **4**, 417 (1974).
57. Gerlich, D., *Adv. Chem. Phys.* **LXXXII**, 1 (1992).
58. Okuno, K., Soelima, K., and Kaneko, Y., *Nucl. Instrum. Meth. B* **53**, 387 (1991).
59. Kravis, S., Saitoh, H., Okuno, K., Soejima, K., Kimura, M., Shimamura, I., Awaya, Y., Kaneko, Y., Oura, M., and Shimakura, N., *Phys. Rev. A* **52**, 1206 (1995).
60. Bliek, F. W., Hoekstra, R., and Morgenstern, R., *Hyperf. Int.* **99**, 193 (1996).
61. Baptist, R., Bonnet, J. J., *Nucl. Instrum. Meth. B* **23**, 123 (1987).
62. Fritsch, W., *Phys. Rev. A* **46**, 3910 (1992).
63. Gargaud, M., and McCarroll, R., *J. Phys. B: At. Mol. Phys.* **18**, 463 (1985).

Benchmark Data

Electron Impact Benchmark Experiments Viewed from Different Perspectives

Gordon H. Dunn

JILA, University of Colorado, Boulder, CO 80309-0440

Abstract. With the wide use of modern computing to generate vast amounts of atomic and molecular data, it is particularly important that there be experimental benchmarks to which computed data may be compared. Benchmark experiments also serve the purpose of anchoring other experiments, and revealing new and unanticipated phenomena is a further benefit of such experiments. In this paper we mention and/or illustrate a few examples of benchmarks in the area of electron scattering which serve each of these purposes - and sometimes combinations of them. Careless presentation of uncertainties and mindless or offhand interpretations of quoted uncertainties are a problem in communication of experimental data and the manner in which they relate to computed data and other experimental results.

BENCHMARKS

The term 'benchmark' has been defined (1) as "a mark made in some durable object as a wall or other landmark of known position and elevation, for use as a reference point in surveys or tidal observations." Though one can in principle make some identification with this definition for the purposes of this discussion, perhaps a more suitable definition (1) is "a standard or reference by which others can be measured or judged."

For the discourse here, there are a variety of perspectives from which one can view benchmarks. 1) *Benchmarks for experiments.* Sometimes in experimental work, the tedium and difficulty of performing all the experimental calibrations and ties to basic standards are so great that it is better to originate some benchmarks for later experiments to anchor to rather than try to tie to each of the basic standards over and over again at the expense of time and effort. 2) *Benchmarks for theory.* Also, since so much data is obtained by computation these days, some experimental benchmarks must be relied upon for judgement of these computed data. 3) *Benchmarks establishing new phenomena.* Furthermore, often it is only through experiments that one realizes that physical mechanisms are operative in a given process, i.e. new phenomena are discovered through experimental benchmarks.

In this brief presentation we will examine a few examples of benchmarks which enter from each of the perspectives mentioned. For convenience, examples are chosen primarily (though not exclusively) from work in which the author has been involved without any implication that the examples chosen are particularly more suitable than others which might have been selected. All examples are chosen from the area of electron scattering.

Benchmarks for Experiments

Emission Cross Sections in the Visible Wavelength Region

Conceptually, the measurements of cross sections for electron-impact excitation of atoms are arguably as simple as one can imagine. One merely must take a gas sample at a suitable intermediate pressure, bombard it with electrons, and observe photons resulting from the impacts. By defining geometry, measuring pressure, beam current, and optical sensitivity, there is enough information to deduce a cross section. However, as elaborated by Van Zyl *et al* (2) at the time of their writing there were over 200 papers dealing with the experimental study of electron-impact excitation of He atoms, and the disparity between results was an embarrassment from the naive perspective of the simplicity of such experiments as painted above. In fact, absolute radiometry, absolute density determination, determination of geometry, and even the collection and measurement of electron beam current demand more care and effort to tie them to absolute standards than was typical of the numerous papers noted. Thus, Van Zyl *et al* (2) set out to generate benchmarks which could be used by other experimentalists to serve as a secondary standard for their measurements. This work provides benchmark cross sections in the visible range 417 nm to 728 nm with uncertainties of about 3% at the 98% confidence level (referred to as high confidence level or HCL, corresponding to approximately 1.2% standard deviation).

To insure accurate density of helium, a unique pressure generator (3) was built which would give the helium densities with uncertainties of less than 1.2% (HCL). An electron beam traversed a gas cell, and the beam's current and geometry were carefully measured in the cell as well as after having passed through the cell. A thoroughly characterized spectroradiometer viewed the collision cell through a quartz window whose transmission was carefully measured, and the spectroradiometer was rotated to view two different radiance standards: one was a tungsten strip lamp calibrated by the (then) National Bureau of Standards, the other was a copper melting point (4) black body. Corrections were made for size of source effect and for lifetime of the excited states. Comparisons were made in the paper with previous experiments and theoretical calculations. There was generally good agreement ($\leq$ 5%) with Born approximation calculations (5) at the highest energy measured (2 keV), and agreement (average within 6%) with previous measurements of Moustaffa Moussa *et al* (6). Distorted Wave Polarized Orbital calculations at low energies are within reasonable agreement (~ 10%). Other comparisons show differences ranging to a factor of two or more, and illustrate the need for a benchmark of this nature.

Since the lines measured are not affected by complications of polarization and are not very sensitive to problems of radiation trapping, the results of these experiments serve as accurate benchmarks for other emission cross sections of atoms and molecules where radiation in the visible region is being investigated. This benchmark role has been used not only for electron-impact excitation, but for atom- and ion-impact excitation as well.

Atomic excitation in the visible range has been reviewed by Heddle and Gallagher (7).

Emission Cross Sections in the Ultraviolet Wavelength Region

Benchmark cross sections in the ultraviolet wavelength regions have been strongly sought after so that electron-impact sources can be used as reliable radiometric standards. Though this was a goal in the optical region, there are - after all - reasonable other secondary standards in that wavelength range. In the EUV region, the only primary radiometric source is a well-characterized electron storage ring (synchrotron). We describe briefly here two very careful efforts to relate line cross sections in the VUV and EUV regions to the primary radiation standard of synchrotrons and to accurate density determinations.

In the 1980's J.S. Risley and his collaborators published a series of papers describing (8) methods and measurements relating electron-impact excitation of gases to the well-characterized SURF synchrotron source at NIST. They proposed (9) a radiometric secondary standard based on these benchmarks. A very comprehensive review (10) of the cross sections for hundreds of EUV lines was published by them in 1989. The review reports not only the many measurements of such cross sections, but has a thorough discussion of the normalization of an overwhelming number of the measurements to theory in order to obtain "absolute values". This review is an excellent starting point for anyone addressing EUV emission benchmarks. The disparity of results from among the many workers emphasizes the need for benchmarks.

More recent benchmark work (11) was done in the UV range utilizing the Berlin electron storage ring (BESSY) in a collaboration between Physikalische-Technische Bundesanstalt (PTB) and Giessen University. The work was strongly oriented toward providing secondary radiometric standards in this wavelength range. These measurements (11) led to 18 cross section benchmarks in the spectral range 46 to 100 nm. Considering an example, a benchmark cross section for the Ar II 91.98 emission line at 2 keV is approximately a factor of two less uncertain than that of McPherson et al, (8),(10) the uncertainty being 4.4% at a high confidence level (1.73 σ). The actual cross section *values* for this benchmark found by these two groups differ by less than 1%.

Both of these benchmark works were exemplified by very careful and systematic characterization of the sources, the spectroradiometers, gas density measurement systems and geometries. Polarization of the radiation of both the standard sources as well as that of the line emission were important considerations in these measurements, as were the polarization responses of the spectroradiometers.

Benchmarks to Test Theory/Computation

As mentioned in the introduction, the author has chosen to select an example with which he has been associated. There is an almost unlimited number of other examples which could be put forth, since most experiments are done with some element in mind of "testing

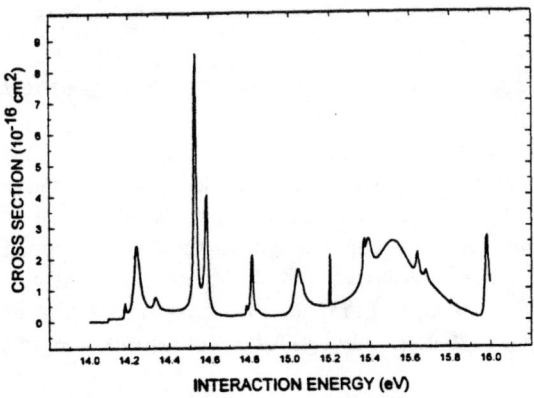

Figure 1. Theoretical electron-impact excitation cross section versus center-of-mass interaction energy for the $3s^2\ {}^1S \rightarrow 3s3p\ {}^3P^o$ transition in Ar^{6+}. The curve is adapted (16) from 8 state CCR results of ref. (12).

the theory".

One of the features stunningly impressive in results of R-matrix calculations of electron-impact excitation of atomic ions is the very strong and copious role of resonances on the cross sections for such excitation. The attractive Coulomb field is, of course, characterized by infinite numbers of series of infinite numbers of resonances; so it is perhaps not surprising that resonances show up in cross section calculations for excitation in such a striking way. The resonances often play a more dominant role in excitation of optically forbidden transitions, such as, for example, intercombination transitions or non-dipole transitions.

An example of an 8 state R-matrix close-coupling calculation by Griffin *et al* (12) is shown in Fig. 1 for the $3s^2\ {}^1S \rightarrow 3s3p\ {}^3P^o$ intercombination transition in Mg-like Ar^{6+}. The figure illustrates numerous resonances in only a 2 eV energy span above threshold, and also demonstrates the theoretical strong dominance of resonances, since the direct excitation is shown by the tiny 'rectangular' feature near 14.1 eV. It should be noted that this is not an extraordinary example. Such resonances occur commonly, and often dominate as here. It becomes extremely important to have experimental benchmarks to test the large numbers of calculations being produced in the various extensive computational data projects.

Most of the work involving electron-ion excitation was initially done using crossed beams and fluorescence detection. For purposes of measuring cross sections for intercombination or otherwise forbidden transitions, this method is not very adaptable. An exception is when the excited state decays to a different excited state, making available detectable fluorescence. An example of this is the work by Rogers et al (13) on excitation of the 2^3P level of He-like Li^+, where the 548 nm fluorescence from the decay to the 2^3S state was observed and the cross section measured. Serving the purpose of a benchmark to theory, the work led to a substantial improvement of the theory (14). However, a more

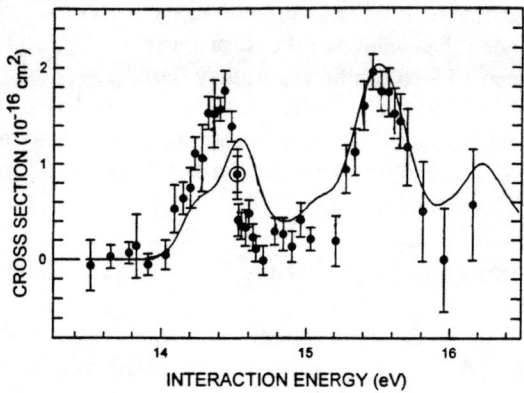

Figure 2. Electron-impact excitation cross section versus center-of-mass energy for the $3s^2\ ^1S \rightarrow 3s3p\ ^3P^o$ transition in Ar^{6+}. Points are measured (16) using MEIBEL and bars represent uncertainties at a high confidence level; bars on point with open circle around it represent expanded combined absolute uncertainty. The solid curve is a convolution of a Gaussian of width 0.24 eV FWHM with CCR results from ref. (12) as shown in Fig. 1. Resonance interference is inferred (16).

global approach to the investigation of non-allowed transitions was needed.

A new approach looking at inelastically scattered electrons with full collection of the electrons made possible the absolute cross section measurements for non-allowed transitions. In this technique, referred to as the MEIBEL technique, described in detail by Bell *et al* (15), electrons and ions are merged using crossed **E** and **B** fields. They then interact over a common path of some distance, and are separated in a second region of crossed **E** and **B** fields. Electrons having lost energy to an inelastic collision are separated from the main electron beam and led to impact upon a position-sensitive detector. Careful measurement of the beam currents, the beam geometry and overlaps of the beams, and the signal counts yields absolute cross sections for the inelastic process.

Results (16) using this technique for the intercombination transition $3s^2\ ^1S \rightarrow 3s3p\ ^3P^o$ Mg-like Ar^{6+} are shown in Fig. 2. This shows the same energy range and the same transition for which the theory is depicted in Fig. 1. The bars in the figure represent uncertainties at a high confidence level (~ 90% CL, 1.7 σ). The experiment makes use of an electron beam with energy width of about 0.24 eV; so many of the detailed features of Fig. 1 are smeared out. However, Fig. 2 also shows the convolution of the theoretical results with an 0.24 eV Gaussian energy distribution. It was concluded (16) that there is substantial interference of the resonances between 14 and 15 eV, causing them to change relative magnitudes and leading to substantial disagreement of the theory with the experiment in this region. This is in some contrast to the energy range 15 - 16 eV; though even here there seems to be some difference between the experiment and the convoluted theory.

Griffin *et al* (17) had made the point that there is extreme sensitivity of some resonances to the exact energies of the resonances and caution should be exercised in using theoretical results with such resonances. In an effort to rationalize the theory and experiment, Griffin et al (18) found that moving the location of the resonances in Ar^{6+} had little effect on the convoluted result between 15 - 16 eV; however, they found a very large effect when the resonances between 14 -15 eV were moved. Previous experimental results (19) using MEIBEL for Kr^{6+} showed similar interference and strong sensitivity to moving (20) of the resonances. There is a continuing need for further benchmark measurements on resonances, especially where they dominate.

Benchmarks Establishing New Phenomena

The earliest direct measurements of dielectronic recombination were partly rationalized with theory by attributing differences to enhancement of DR by ambient electric fields. A systematic experimental study (21) of DR versus field strength was then carried out for Mg^+ and reasonable agreement (10 - 20%) with theory was found. Despite nagging inconsistencies between theory and some experiments at ORNL (22), little further follow-up was done because the field strengths were not well-established in the ORNL experiments and the Mg^+ results seemed to indicate that all was well. However, later experiments (23) on C^{3+} with a well-known field indicated a similar inconsistency as the ORNL experiments, though the limited amount of data and large uncertainties made the results inconclusive. Further and more detailed experimental investigation was thus called for. In the intervening decade between the Mg^+ measurements and the undertaking of the new experiments, the use of heavy ion storage rings came to be the "method of choice" for experimental measurements of DR.

The heavy ion storage ring CRYRING at Stockholm University was used to store Li-like Si^{11+}. In one side of the ring is a solenoidally confined electron beam, this device being called "the cooler" of the ring. By changing the energy of the electrons in the cooler, the relative energy of electrons and ions can be varied. The recombination product Si^{10+} was detected with a surface barrier detector downstream of the cooler section. Ion current, electron density and signal combined to give a cross section for DR.

Normally, the ion beam lines up along the B field of the solenoid for an interaction path of about 85 cm, and since the electrons follow the magnetic field, the ion and electron trajectories are co-linear. By applying a small transverse magnetic field to the solenoid, the ions 'see' a Lorentz electric field given by $E_L = v \times B_\perp$. By varying $B_\perp$, and measuring the cross section as a function of relative energy at each $B_\perp$, one can thus measure cross sections for DR in an electric field (DRF) as a function of energy and E_L. Results showing cross section versus relative energy (24) are reproduced in Fig. 3 for nine different values of the electric field thus imposed. The results very dramatically illustrate the field effect for high Rydberg levels corresponding to formation at energies 20 eV and up.

The integrated collision strength from 20 eV to 25 eV versus electric field was compared (24) with theory of the type used to describe the Mg^+ results (25), (26), and though the general magnitude of the field enhancement is in rough agreement, there is fairly large disagreement in the functional behavior.

This continued disparity between some DRF measurements and the standard theory for

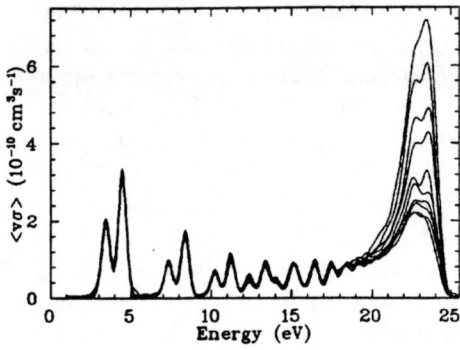

Figure 3. Reaction rate versus relative energy of electron and Si^{11+} for DRF. Cross sections are given for nine different fields: 0.0, 9.2, 18.4, 32.0, 46.0, 68.8, 91.5, 137.5, and 183.1 V/cm. Data are from ref. 24.

electric field mixing of states led Robicheaux and Pindzola (27) to hypothesize a new mechanism wherein the magnetic field also plays a significant role. They made model calculations which suggested that the disparity between theory and experiment for some of the earlier experiments as well as for Si^{11+} may be due to not accounting for the role of the magnetic field. However, the detailed calculations for Si^{11+} would be overwhelmingly large even with today's computers; so an in-depth comparison could not be made. Furthermore, the **B** field in the experiment was not systematically varied with fixed **E** field with the intent of looking for this effect.

Thus, a benchmark experiment has served the purpose of bringing to light a possible new phenomenon. Further benchmark experiments along with theory developed to make calculations computationally tractable will be necessary to firmly establish the new mechanism.

A NOTE ON UNCERTAINTIES

The notion of benchmark carries with it a concept of quality. It carries the feeling that one can refer back to it with some level of confidence in order to judge other objects of a similar nature. In scientific terms, of course, this comes down to specification of uncertainties. Thus, the usefulness of benchmarks - for whatever of the purposes discussed here (and others not discussed) - is to a large extent determined by the quality of the statements of uncertainty in the benchmark. *This means that not only must the uncertainty be given, but that an explanation of what the uncertainty means must accompany the measurement.*

It is very seldom that an uncertainty accompanies theoretical calculations. Presumably this is because it is difficult to assess this - one can only compare with other theories, look for some sort of convergence, compare with experimental benchmarks, etc.

However this is not true of experimental measurements. The assessment of experimental uncertainties has been treated numerous times (28) and is the focus of careers and the subject of international agreements (29). A treatment of the subject is well beyond the scope of these paragraphs here. The author here instead makes a plea that more informed and responsible attention be given uncertainties by authors reporting experimental results. It is all too frequent that one finds uncertainties expressed only as ±XX - with no definition of whether this is a statistical uncertainty only or whether it is a combination of statistical and systematic uncertainties. If statistical, there is often no indication of whether the uncertainty is for the mean or for a measurement. There is often no indication of whether the uncertainty is at the level of a standard deviation or is at the 90% confidence level, 98% confidence level, or ...whatever. Sometimes no uncertainties are even given!

In the same breath, the author makes a plea to those comparing data: experiment/theory; experiment/experiment; or whatever, that the person(s) making such comparisons pay attention to whether the uncertainty statements have the same meanings, state the meanings, and when appropriate try to put the uncertainties on the same 'footing' (e.g. same confidence level). Since NIST is in the business of benchmarks/standards, there has been great attention paid by that institution to the issue of uncertainties. One example of a publication from them has already been referenced. The author recommends a recent short publication (30) as a very readable and very valuable guide to reporting data. Though not everyone is reporting NIST results, a more informed and responsible approach to reporting experimental results is needed and perhaps such a focus will help.

ACKNOWLEDGEMENTS

The author's work in the area of electron scattering has for some time been supported in part by the Office of Fusion Energy Sciences, U.S. Department of Energy under Contract No. DE-A105-86ER53237 with the National Institute of Standards and Technology.

REFERENCES

1. Random House Websters College Dictionary, 1995.
2. Van Zyl, B., Dunn, G.H., Chamberlain,G. , and Heddle, D.W.O. , Phys. Rev. A **22**, 1916 (1980).
3. Van Zyl, B., Chamberlain,G.E., and Dunn,G.H., J. Vac. Sci. Technol. **13**, 721 (1976).
4. NBS-certified freezing-point standard reference material. Freezing temperature of sample = 1358.0 ±0.5 K.
5. Inokuti, M., and Kim,Y.-K., Phys. Rev. **186**, 100 (1969); Kim,Y.-K., and Inokuti, M., Phys. Rev. A **3**, 665 (1971).
6. Moustaffa Moussa,H.R., DeHeer,F.J., and Schutten,J., Physica **40**, 517 (1969).
7. Heddle, D.W.O.,and Gallagher, J.W., Rev. Mod. Phys. **61**, 221 (1989).
8. McPherson, A., Rouze, N.,Westerfeld, W.B., and Risley, J.S., Applied Optics **25**, 298 (1986); Kendrick, R.L., McPherson, A., Rouze, N., Westerfeld,W.B. and Risley, J.S., Applied Optics **26**, 2029

(1987); For particular details see especially the thesis of McPherson, A., North Carolina State University, 1984.

9. Risley, J.S., and Westerfeld, W.B., Applied Optics **28**, 389 (1989).

10. van der Burgt, J., Westerfeld, W.B. and Risley, J.S., J. of Phys. and Chem. Ref. Data **18**, 1757 (1989).

11. Jans,W., Möbus, B., Kühne, M., Ulm, G., Werner, A., Schartner, K-H., Applied Optics **34**, 3671 (1995); Jans, W., Möbus, B. Kühne, M., Ulm, G., Werner, A., Schartner, K-H., Phys. Rev. A **55**, 1890 (1997).

12. Griffin, D.C. Pindzola, M.S., and Badnell, N.R., Phys. Rev. A **47**, 2871 (1993).

13. Rogers, W.T., Olsen, J.Ø., and Dunn, G.H., Phys. Rev. A **18**, 1353 (1978).

14. Christensen, R.B., and Norcross, D.W., Phys. Rev. A **31**, 142 (1985).

15. Bell, E.W., Guo, X.Q., Forand, J.L., Rinn, K., Swenson, D.R., Thompson, J.S., Dunn, G.H., Bannister, M.E., Gregory, D.C., Phaneuf, R.A., Smith, A.C.H., Müller, A., Timmer, C.A., Wåhlin, E..K., DePaola, B.D., and Belić, D.S., Phys. Rev. A **49**, 4585 (1994).

16. Chung, Y-S., Djurić, N., Wallbank, B., Dunn, G.H., Bannister, M.E., and Smith, A.C.H., Phys. Rev. A **55**, 2044 (1997).

17. Griffin, D.C., Pindzola, M.S., Robicheaux, F., Gorczyca, T.W., and Badnell, N.R., Phys. Rev. Lett. **72**, 3491 (1994).

18. Griffin, D.C., Pindzola, M.S., Shaw, J.A., and Badnell, N.R., (Private communication, 1996).

19. Bannister, M.E., Guo, X.Q., Kojima, T.M., and Dunn, G.H., Phys. Rev. Lett. **72**, 3336 (1994).

20. Gorczyca, T., Pindzola, M.S.,. Badnell, N.R, and Griffin, D.C., Phys. Rev. A **51**, 488 (1994).

21. Müller, A. , Belić, D.S., De Paola, B.D., Djurić, N., Dunn, G.H., Mueller, D.W., Timmer, C., Phys. Rev. A **36**, 599 (1987); Phys. Rev. Lett. **56**, 127 (1986).

22. Dittner, P.F., Datz, S., Miller, P.D., Pepmiller, P.L., and Fou, C.M., Phys. Rev. A **35**, 3668 (1987).

23. Young, A.R., Gardner, L.D., Savin, D.W., Lafyatis, G.P., Chutjian, A., Bliman, S., and Kohl, J.L., Phys. Rev. A **49**, 357 (1994). Savin, D.W., Gardner, L. D., Reisenfeld, D.B.,Young, A.R., and Kohl, J. L., Phys. Rev. A **53**, 280 (1996).

24. Bartsch, T., Müller, A., Spies, W., Linkemann, J., Danared, H., DeWitt, D.R., Gao, H., Zong, W., Schuch, R., Wolf, A., Dunn, G.H., Pindzola, M.S., and Griffin, D.C., Phys. Rev. Lett. **79**, 2233 (1997).

25. Bottcher, C. , Griffin, D.C., Pindzola, M.S., Phys. Rev. A **34**, 860 (1986).

26. La Gattuta, K., Nasser, I., Hahn, Y., Phys. Rev. A **33**, 2782 (1986).

27. Robicheaux, F., and Pindzola, M., Phys. Rev. Lett. **79**, 2237 (1997).

28. For example: Bevington, P.R., *Data Reduction and Error Analysis for the Physical Sciences*, New York, McGraw Hill, 1969; Ku, H.H., editor, *Precision Measurement and Calibration: Statistical Concepts and Procedures*, NBS Special Publication 300, Vol.I, Washington, D.C., U.S. Government Printing Office, 1969.

29. For example: ISO, *Guide to the Expression of Uncertainty in Measurement*, prepared by ISO Technical Advisory Group 4.

30. Taylor, B.N., and Kuyatt, C., *Guidelines for Evaluating and Expressing the Uncertainty of NIST Measurement Results*, NIST Technical Note 1297, 1994 Edition, Washington D.C., U.S. Government Printing Office, 1994.

Precision Lifetime Measurements on Neutral Atoms - A Sensitive Test for Atomic-Structure Calculations

U. Volz[1,2] and H. Schmoranzer

Fachbereich Physik, Universität Kaiserslautern
D-67653 Kaiserslautern, Germany

Abstract. A number of systematic comparisons between experimental and theoretical line strengths and lifetimes for neutral atoms are given which provide some insights into the achievements, but also the short-comings of atomic-structure calculations. The first example deals with the resonance transitions in the neutral alkalis. The line strengths of the resonance transitions in all alkalis are known very accurately from recent lifetime measurements. A comprehensive overview of recent experimental and theoretical work is given. The second example deals with the optical and near-infrared transitions in the noble gases. The $4s$–$4p$ transition array in argon is chosen as an example. The comparison of theoretical results with recently measured accurate lifetimes for the Ar $4p$ states reveals the deficiencies of the existing atomic-structure calculations on noble gases.

INTRODUCTION

The talks and posters presented at the ICAMDATA conference have documented the large and still growing need for atomic data from applications in plasma physics and related areas. The chances that all required oscillator strengths will ever be determined at a reasonable level of accuracy by experimental means alone are small, given the large range of wavelengths and charge states that has to be covered. Theoretical calculations, on the other hand, are not limited in their range of application as experiments are and will therefore play an increasing role in the generation of atomic data in the future. A substantial amount of data has already been produced by international theoretical collaborations known as the OPACITY and IRON projects. The determination of reliable uncertainty limits for calculated atomic properties, however, especially for oscillator strengths, has remained a very difficult problem as of now. Experimental benchmark values of excited atomic state lifetimes support the further development of theoretical models and computational

---

[1] Also at: National Institute of Standards and Technology, Gaithersburg, MD 20899, USA
[2] Electronic mail: agschmor@rhrk.uni-kl.de

methods in several respects. They allow for an assessment of the uncertainties in line strength calculations, and they help to decide between competing theoretical approaches which have led to different and incompatible results.

The objective of this paper is to show that a deeper insight can be obtained if comparisons between experiment and theory are done in a systematic way. Firstly, we state that the primary theoretical measure for the strength of a transition $i \to f$ is its line strength S_{if}. Other quantities like oscillator strengths, transition probabilities or excited state lifetimes depend also on the transition energy and should be converted to line strengths before being compared. In this way the question whether experimental or theoretical energies should be used in the calculation, *e.g.*, of oscillator strengths is avoided. Secondly, comparisons between experiment and theory should be done for line strengths S_{if} and the corresponding transition energies E_{if} simultaneously. This might reveal correlations between the errors made in the calculation of these quantities. It might be that certain combinations of S_{if} and E_{if} are less sensitive to the approximations made in the calculation than S_{if} and E_{if} itself. This paper will show examples for this. Finally, the comparisons should be extended to the same type of transition in similar atoms. The obvious approach to this from the theoretical viewpoint is the isoelectronic sequence. However, since the most accurate experimental data available on transition line strengths and excited state lifetimes is for neutral atoms we have chosen a different approach here. We consider corresponding transitions in the chemically similar atoms in a column of the periodic system.

The first example given in this paper deals with the resonance transitions in the alkali atoms. The alkalis as effective one-electron systems are not only a natural testing ground for atomic-structure calculations, but a lot of new very accurate measurements of lifetimes and line strengths performed during the last few years also have made the alkali resonance transitions the best investigated transitions in the whole periodic system. Experimental line strengths with typical uncertainties of ±0.2% are available now for each transition ranging from $2s$–$2p$ in lithium up to $7s$–$7p$ in the radioactive element francium. The second example deals with the prominent orange and red transitions in the heavy noble gases (*e.g.* Ne $3s$–$3p$, Ar $4s$–$4p$, *etc.*). The similarity of the excited states of heavy noble gases to the alkali states manifests itself in similar lifetimes and transition energies. The p^5 core of the noble gases, however, adds additional complexity to the spectra. For instance, the excited np doublets of the alkalis are replaced by sets of 10 excited np states in the noble gases. This opens up another opportunity to compare the ratios of measured lifetimes with theoretical predictions. In the present paper the $4s$–$4p$ transition array is chosen as an example.

THE RESONANCE TRANSITIONS OF THE ALKALIS

New experimental data

Until very recently only few lifetime measurements on the first excited states of alkalis had been done with an accuracy of better than one percent. The lifetime measurements with the smallest quoted uncertainties (±0.16%) were performed by Gaupp et al. [1] on the Li $2p$ $^2P_{1/2}$ and Na $3p$ $^2P_{1/2}$ states. Their lifetimes, however, were about 0.7% longer than the majority of the theoretical predictions, sparking a controversial discussion about the accuracy of *ab initio* calculations. Within the past three years the situation has improved quite dramatically. More than 20 new lifetime measurements with uncertainties below one percent have been reported which cover the upper levels of all alkali resonance transitions from Li $2s$–$2p$ to Fr $7s$–$7p$ (see Table 1). The new measurements on the Li $2p$ and Na $3p$ states resulted in lifetimes consistently about 0.7% shorter than the earlier results of Gaupp et al., thus resolving the prior discrepancy between experiment and *ab initio* calculations in favor of theory. With the exception of the radioactive element francium each of the alkalis has been treated by at least two, in the case of sodium even four, different experimental methods, and the results are in a remarkably good agreement. In the following section we give a short overview of the various methods used and consider their specific advantages and problems.

Beam-Gas-Laser-Spectroscopy

This method utilizes a fast beam of ions with energies ranging from 30 keV to 200 keV which is neutralized by charge exchange collisions in a gas cell. The atoms are then excited by a laser perpendicular to the atom beam and the decay fluorescence is detected as a function of the distance from the laser excitation. An accurate determination of the atomic beam velocity is required to convert the measured spatial decay constant into a lifetime. The basic advantages of this method include a high statistical quality of the decay curves and a precise time scale. The low particle density in the beam makes radiation trapping negligible, and quenching effects due to collisions with residual gas particles are small. A specific problem of the method is its sensitivity to the adjustment of the detection system. The parallelity between the movement of the fluorescence detector and the atomic beam has to be ensured within fractions of 1 mrad. Another difficulty arises from the divergence of the atom beam. A detailed knowledge of the beam characteristics and of the spatial detection efficiency of the fluorescence detector is necessary to estimate the magnitude of this effect.

Volz et al. [4,8] carried out a comprehensive set of lifetime measurements on alkalis by means of beam-gas-laser-spectroscopy. Their measurements cover the Li $2p$, Na $3p$, K $4p$, and Rb $5p$ fine-structure states with uncertainties between ± 0.13% and ± 0.26%. Recently, a new measurement on Li $2p$ with an improved accuracy of ± 0.15% was performed in the same group by Schmitt et al. [7]. A

TABLE 1. New experimental data on upper state lifetimes and line strengths for the resonance transitions in alkali metals (uncertainties are given in parentheses).

Authors	Ref.	Method[a]	Lifetime τ_i (ns) $J=1/2$	Lifetime τ_i (ns) $J=3/2$	Line strength S_{if} (a.u.) Sum	Line strength S_{if} (a.u.) Ratio
Li 2s–2p						
McAlexander et al.	[2]	PAS	26.99(16)[b]		33.14(20)	—
Linton et al.	[3]	MS	27.09(8)[b]		33.02(9)	—
Volz et al.	[4]	BGLS	27.11(6)[b]		33.00(7)	—
McAlexander et al.	[5]	MS+PAS	27.101(9)[b]		33.008(11)	—
Martin et al.	[6]	MS+PAS	27.134(12)[b]		32.968(14)	—
Schmitt et al.	[7]	BGLS	27.09(4)[b]		33.02(5)	—
Na 3s–3p						
Volz et al.	[8]	BGLS	16.299(21)	16.254(22)	37.26(5)	0.50014(44)
Oates et al.	[9]	NLW	16.286(35)[c]	16.237(35)	37.30(8)	—
Jones et al.	[10]	PAS	16.279(16)[d]	16.230(16)[d]	37.31(4)[d]	—
Tiemann et al.	[11]	MS	16.270(53)[d]	16.221(53)[d]	37.33(12)[d]	—
K 4s–4p						
Volz et al.	[4]	BGLS	26.79(7)	26.45(7)	50.47(13)	0.5003(11)
Wang et al.	[12]	PAS	26.72(5)[d]	26.37(5)[d]	50.62(8)[d]	—
Rb 5s–5p						
Volz et al.	[4]	BGLS	27.70(4)	26.24(4)	53.64(8)	0.5011(7)
Freeland et al.	[13]	PAS	27.74(6)[d]	26.23(6)[d]	53.63(12)[d]	—
Simsarian et al.	[14]	PL/MOT	27.64(4)	26.20(9)	53.73(15)	0.5013(19)
Cs 6s–6p						
Young et al.	[15]	PL	34.75(7)	30.41(10)	60.54(17)	0.5059(20)
Rafac et al.	[16]	BGLS	34.93(10)	30.50(7)	60.31(15)	0.5047(18)
Rafac et al.	[17]	ABS	—	—	—	0.50482(23)
Fr 7s–7p						
Simsarian et al.	[14]	PL/MOT	29.45(11)	21.02(11)	53.08(25)	0.526(3)

[a] BGLS: Lifetime / beam-gas-laser-spectroscopy.
PL: Lifetime / pulsed laser spectroscopy, time-correlated single photon counting.
PL/MOT: as PL but using a sample of cold atoms in a MOT.
NLW: Natural line width / laser spectroscopy on cold atoms in a MOT.
MS: Molec. long-range coeff. C_3 / conventional molecular spectroscopy (bound-bound).
PAS: Molec. long-range coeff. C_3 / photoassociative molecular spectroscopy (free-bound).
ABS: Line strength ratio / absorption spectroscopy.
[b] Weighted average for ^2P. The difference between the lifetimes for $^2P_{1/2}$ and $^2P_{3/2}$ of 0.002 ns is insignificant in view of the experimental uncertainties.
[c] Only the lifetime of the $^2P_{3/2}$ level was measured. The total line strength S_{if} and the lifetime of the $^2P_{1/2}$ level are calculated assuming a D_1/D_2 line strength ratio of 0.5.
[d] A molecular long-range interaction coefficient C_3 was measured. The total line strength S_{if} and the lifetimes of the $^2P_{1/2}$ and $^2P_{3/2}$ levels are calculated from the measured C_3 coefficient assuming a D_1/D_2 line strength ratio of 0.5. This assumption might be problematic in the case of Rb.

similar technique was used by Rafac et al. [16] to measure the lifetimes of the Cs 6p fine-structure levels. Their results have uncertainties of about ± 0.25%. Rafac and Tanner also provided an independent very accurate measurement of the line strength ratio of the D lines in Cs by absorption spectroscopy [17].

Pulsed-Laser Spectroscopy

The most intuitive way to measure the lifetime of an atomic state is to excite atoms in a low density sample with a short laser pulse and record the decay fluorescence as a function of time. Two techniques for the recording of the decay curves are available. In the 'analog' variant the current from the photodetector, usually a photomultiplier, is amplified and sampled by a box-car averager. The accuracy is thus limited by nonlinearities in the detection system. The 'digital' variant, termed time-correlated single-photon counting, overcomes this problem by detecting the time intervals between the incoming laser pulses and the resulting fluorescence photons. The average number of detected fluorescence photons per laser pulse has to be kept below 0.01 in order to prevent preferential counting of early events. Decay curves with very good statistical quality can therefore only be obtained if lasers with high pulse repetition rates are used.

Young et al. [15] determined the lifetimes of the Cs 6p fine-structure levels by time-correlated single-photon counting with uncertainties of ± 0.20% ($^2P_{1/2}$) and ± 0.33% ($^2P_{3/2}$), respectively. They used a femtosecond-pulse laser with a 76 MHz repetition rate and a collimated thermal cesium atomic beam. Simsarian et al. [14] demonstrated recently that atoms captured in a magneto-optical trap can also be used for lifetime measurements by the single-photon counting technique. The fact that within a trap a comparatively small number of atoms can be used very efficiently made it possible to investigate the radioactive element francium for which no stable isotope exists. Simsarian et al. carried out measurements on the Fr 7p and on the Rb 5p fine-structure levels with repetition rates of 100 kHz and obtained results with uncertainties between ± 0.15% and ± 0.52%.

Natural Line Width

The only experimental determination of a natural line width with an accuracy better than one percent so far has been carried out by Oates et al. [9] on the D_2 line of sodium. The measurement was done with magneto-optically trapped sodium atoms which were further cooled down to 50 µK in an optical molasses. The Doppler-broadening was essentially removed by using only velocity-selected atoms in the measurement. An effective two-level system of two single magnetic hyperfine sublevels was created by optical pumping. This removed the line broadening due to the Zeeman splitting. The effect of saturation broadening was corrected for by an extrapolation to zero power. The final result for the Na D_2 natural line width has an uncertainty of ± 0.22%.

Molecular Spectroscopy

Another way to determine atomic line strengths that is completely opposed to the lifetime measurements discussed above is to analyse the long-range behavior of interaction potentials in homonuclear diatomic molecules. If two like atoms in states i and f, respectively, form a molecule, and the states i and f are connected by an electric dipole transition, the leading term of the potential at long range is the resonant dipole-dipole interaction potential C_3/R^3 [18]. The coefficient C_3 is proportional to the line strength $S_{if} = |\langle i ||\mathbf{D}|| f \rangle|^2$ of the atomic transition $i \rightarrow f$. C_3 can be determined from an analysis of the rovibrational energy levels of the molecular potential close to the dissociation limit.

The evaluation starts with a construction of the short-range part of the potential curve by means of a Rydberg-Klein-Rees (RKR) analysis of the measured energies of deeper bound rovibrational levels. The asymptotic long-range part of the potential is expressed in terms of a number of adjustable parameters and joined smoothly to the short-range part at intermediate distances. The parameters in the long-range part, including the C_3 coefficient, are then varied in order to reproduce the measured energy levels close to dissociation.

An obvious difficulty in this method is that a comprehensive theoretical model for the molecular potential at long-range is needed for the evaluation. The long-range form C_3/R^3 which is assumed only in the limit of Hund's case a is modified by the transition to Hund's case c at long range caused by the spin-orbit interaction [19]. Hyperfine-structure [20] and the retardation of the electromagnetic interaction [21] add even more complexity to the problem. The accuracy of the determined C_3 coefficient also depends strongly on the quality of the short-range part of the potential used in the evaluation.

Conventional bound-bound molecular spectroscopy. Since it is impossible to access vibrational levels close to the dissociation limit directly from the molecular ground state due to unfavorable Franck-Condon factors several intermediate steps are necessary.

In the measurements of Linton et al. [3] on the $A\ ^1\Sigma_u^+$ state of Li_2 the molecules were excited to the interacting molecular states $E\ ^1\Sigma_g^+$ and $F\ ^1\Sigma_g^+$ by two-step laser excitation. The subsequent fluorescence on transitions to rovibrational states of the $A\ ^1\Sigma_u^+$ potential was detected with a high-resolution Fourier-transform spectrometer. The C_3 coefficient, which in case of the A state equals 1/3 of the total line strength of the atomic $2s$–$2p$ transitions, was obtained with an uncertainty of $\pm$ 0.28%. The accuracy of the C_3 determination was limited by the fact that the highest vibrational levels closer than 2.5 cm^{-1} to the dissociation limit were not observed.

In the measurements of Tiemann et al. [11] on the $A\ ^1\Sigma_u^+$ state of Na_2 the molecules were first optically pumped to a higher vibrational state of the ground state potential and then excited to the $A\ ^1\Sigma_u^+$ state with a second laser. All the vibrational levels of the A state up to the dissociation limit were observed. The final result for the C_3 coefficient has an uncertainty of $\pm$ 0.33%.

Photoassociation spectroscopy. The formation of a bound molecule during a collision between two free atoms, induced by the absorption of a photon, is called photoassociation. In a sample of ultracold trapped atoms the spread in collision energies is reduced to a few MHz which makes photoassociation of ultracold atoms a high resolution spectroscopic tool. The occurence of photoassociation is detected either by monitoring the corresponding loss of atoms from the trap or by photoionizing the molecules with a second laser. Photoassociation spectroscopy is very useful for probing the asymptotic behavior of molecular potentials at long range because the Franck-Condon factors favor transitions to highly excited rovibrational levels close to the dissociation limit.

McAlexander et al. [2] applied photoassociation spectroscopy to ultracold lithium atoms captured in a magneto-optical trap. Two rovibrational series belonging to the $1\ ^3\Sigma_g^+$ and the $A\ ^1\Sigma_u^+$ potentials of Li_2 were observed. The first evaluation of the longer $1\ ^3\Sigma_g^+$ series had to rely on *ab initio* potentials at short range which limited the accuracy of the C_3 determination to $\pm\ 0.6\%$ [2]. In the recent evaluation of the $A\ ^1\Sigma_u^+$ series the RKR potential provided by Linton et al. [3] could be used at short-range [5]. This made it possible to determine the C_3 coefficient of the A state with an extremely small uncertainty of $\pm\ 0.033\%$ only. An independent analysis of the same data by Martin et al. [6] came to a slightly lower result for the line strength. The difference between the two results of 0.040 a.u. is mostly due to an adiabatic correction of 0.027 a.u. applied by McAlexander et al. By contrast Martin et al. found adiabatic corrections to be negligible within their error bars.

Photoassociation spectroscopy on purely long-range states. In a number of measurements the difficulties associated with the short-range part of the potential curve were avoided by using a so-called "purely long-range" molecular state [22]. Such a state is generated by a spin-orbit avoided crossing between two potential curves of the same symmetry at long range. A purely long-range state has a shallow potential well outside of the region where exchange interaction or the overlap of the wave functions of the two atoms becomes important. The attractive part as well as the repulsive part of its potential curve depend only on the long-range forces and on the atomic fine-structure splitting. The most popular example is the 0_g^- purely long-range state which exists in all alkali dimers.

Jones et al. [10], Wang et al. [12], and Freeland et al. [13] used photoassociation spectroscopy to investigate the 0_g^- purely long-range states in Na_2, K_2, and Rb_2, respectively. With an uncertainty of $\pm\ 0.1\%$ the measurement of the Na $3s$–$3p$ line strength by Jones et al. is the most accurate one despite the fact that only the seven lowest vibrational levels of the 0_g^- potential were used in the evaluation. A specific problem in the measurement on Rb_2 is that exchange interaction is no longer negligible because the vibrational wave functions in the 0_g^- potential of Rb_2 extend further towards smaller internuclear distances than in Na_2 or K_2. Another problem is that the C_3 coefficient obtained from the 0_g^- state does not correspond to the line strength of either the D_1 or the D_2 transition alone. The assumption of a D_1/D_2 line strength ratio of 0.5 that is used in the evaluation of the spectra is no longer exactly valid in the case of Rb.

Ab initio calculations

Most of the *ab initio* calculations on alkali atoms carried out so far cover only the resonance transitions of the lightest alkalis lithium and sodium. The most accurate nonrelativistic results for the lithium atom have been obtained from variational calculations using explicitly correlated (Hylleraas-type) wave functions [23–25]. This method, however, is restricted to two- and three-electron systems. Very accurate results have also been calculated with the multi-configuration Hartree-Fock (MCHF) method for lithium and sodium [26–28]. The increase of the number of possible excited-state configurations with the number of electrons in the atom puts a practical limit on the feasability of MCHF calculations. As of now no large-scale MCHF calculations on heavier alkali atoms have been reported.

The only *ab initio* method which has been applied consistently to all alkali atoms is the relativistic many-body perturbation theory (RMBPT) [29–35]. In the second part of this section we will give a number of systematic comparisons, following the guidelines outlined in the introduction, of results from RMBPT with experimental results.

New calculations on lithium and sodium

For the lithium atom the most accurate results were obtained from variational calculations using explicitly correlated (Hylleraas-type) wave functions [23,24]. In these calculations the interelectronic distances r_{ij} appear explicitly in the trial wave functions. One of the main obstacles in this method, the very slow convergence of integrals with several interelectronic distances in the integrand, has been removed recently by a novel asymptotic expansion method discovered by Drake and Yan [25]. As a consequence the accuracy of calculations on lithium could be improved by several orders of magnitude. Yan and Drake determined nonrelativistic energies for the $2s$, $2p$, and $3d$ states in lithium with uncertainties of less than two parts in 10^{10} [24] and nonrelativistic line strengths for the transitions between these states with uncertainties of a few parts in 10^6 [23]. Effects associated with the finite nuclear mass are very important at this level of precision and were included in the calculation. Relativistic and QED effects were taken into account by corrections. The high accuracy of the nonrelativistic results, however, could not be fully maintained for the relativistic results. The final result for the $2s$–$2p$ line strength still has a very low uncertainty of less than ± 0.003% and agrees with the most accurate experimental result obtained by McAlexander *et al.* [5].

Considerable improvements in accuracy have also been achieved in calculations using the multi-configuration Hartree-Fock (MCHF) method [26–28]. In this method the wave functions are constructed as superpositions of single-configuration state functions. The wave functions are determined variationally so as to leave the expectation value of the nonrelativistic energy stationary with respect to variations in the radial functions and in the expansion coefficients of the superposition. Once radial functions have been determined a configuration interaction calculation can

TABLE 2. Recent large-scale line strength calculations on Li $2s$–$2p$ and Na $3s$–$3p$ (uncertainties are given in parentheses).

Authors	Ref.	Method[a]		Line strength S_{if} (a.u.)
Li $2s$–$2p$				
Yan and Drake	[24]	Hylleraas	NR, $M = \infty$	33.00068(5)
			NR, $M < \infty$	33.00558(5)
			R, $M < \infty$	33.0036(8)[b]
Godefroid et al.	[26]	MCHF	NR, $M < \infty$	33.0031
Froese Fischer et al.	[28]	MCHF+BP	R, $M < \infty$	33.0024
Na $3s$–$3p$				
Guet et al.	[32]	MBPT, all orders	R, $M = \infty$	37.38(11)
Jönsson et al.	[27]	MCHF	NR, $M = \infty$	37.35
			R, $M = \infty$	37.26

[a] NR: nonrelativistic, R: relativistic
$M = \infty$: infinite nuclear mass, $M < \infty$: finite nuclear mass
[b] Calculated from the lifetimes 27.1055(10) ns ($^2P_{1/2}$) and 27.1025(5) ns ($^2P_{3/2}$) quoted in Ref. [5] as private communication.

be performed. Relativistic effects and finite nuclear mass effects can be included in the calculation at this point by using the Breit-Pauli Hamiltonian and the specific mass shift operator for the evaluation of the interaction matrix [28]. With modifications of the codes for parallel execution on clusters of workstations MCHF calculations an order of magnitude larger than a few years ago are possible now.

Large-scale MCHF calculations have been carried out recently for lithium and sodium. The MCHF + Breit-Pauli calculations by Froese Fischer et al. [28] on lithium account more precisely for relativistic effects than those of Yan and Drake but they are less efficient in the treatment of electron correlation. The MCHF + BP result for the $2s$–$2p$ line strength agrees very closely with Yan and Drake's result but stays slightly outside the error bars of the latter one.

The MCHF calculation of Jönsson et al. [27] on the $3s$–$3p$ transition in sodium is one of the most extensive MCHF calculations as of now. The configuration expansions in this calculation included about 66,000 configurations for the $3s$ state and even 135,000 configurations for the $3p$ state. Relativistic effects were roughly accounted for by multiplying the nonrelativistic result with the ratio between the line strengths obtained from the Dirac-Fock and Hartree-Fock methods. The line strength calculated by Jönsson et al. differs only by 0.3% from an earlier result by Guet et al. [32] obtained by an all-order many-body perturbation theory. All experimental values for the Na $3s$–$3p$ line strength lie between these two theoretical results.

Relativistic many-body perturbation theory

The starting point of the many-body perturbation theory (MBPT) is the frozen-core Dirac-Hartree-Fock approximation (DHF). The residual Coulomb interaction between the electrons not included in the DHF approximation is treated as a perturbation. The algebraic expressions for the various higher-order corrections can be represented graphically by Brueckner-Goldstone diagrams. A powerful feature of MBPT is the possibility to sum up the contributions from classes of higher-order diagrams by means of all-order methods. Third-order MBPT is usually sufficient for the calculation of line strengths for highly charged ions because of the rapid convergence of MBPT for such systems. Higher-order terms, however, are important for neutral atoms.

Third-order MBPT. Johnson et al. [29,30] applied third-order MBPT to the resonance transitions of all alkali atoms including francium. Fig. 1 displays the relative deviations of the calculated line strengths and transition energies from the corresponding experimental values. The line strengths in the first order of MBPT are obtained directly from the frozen-core DHF wave functions. As Fig. 1 shows, the agreement with experiment is poor for the first-order line strengths and for the corresponding zero-order transition energies. The error of the calculated line strengths exceeds +30% for Cs $6s$–$6p$ and Fr $7s$–$7p$. It is noteworthy that the relative deviations of the line strengths and of the transition energies from experiment have opposite sign and differ by a factor of about -1.6.

In the next step the random phase approximation (RPA) is applied. This removes the differences between length-form and velocity-form results observed in first order (up to +12% for Fr $7s$–$7p$). The RPA correction accounts for the po-

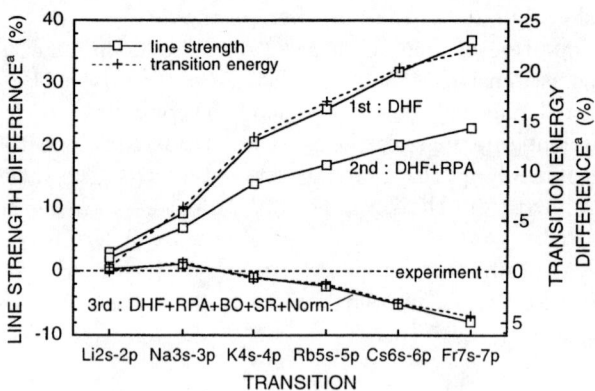

FIGURE 1. Percentage differences[a] between results from up to third-order MBPT [30] and experiment for the line strengths and the transition energies of the alkali resonance transitions. Note that the scale for the transition energy differences (right side) is inverted and expanded. ([a] calculated as $100 \times \log(S^{\text{th}}/S^{\text{exp}})$ and $100 \times \log(E^{\text{th}}/E^{\text{exp}})$, respectively.)

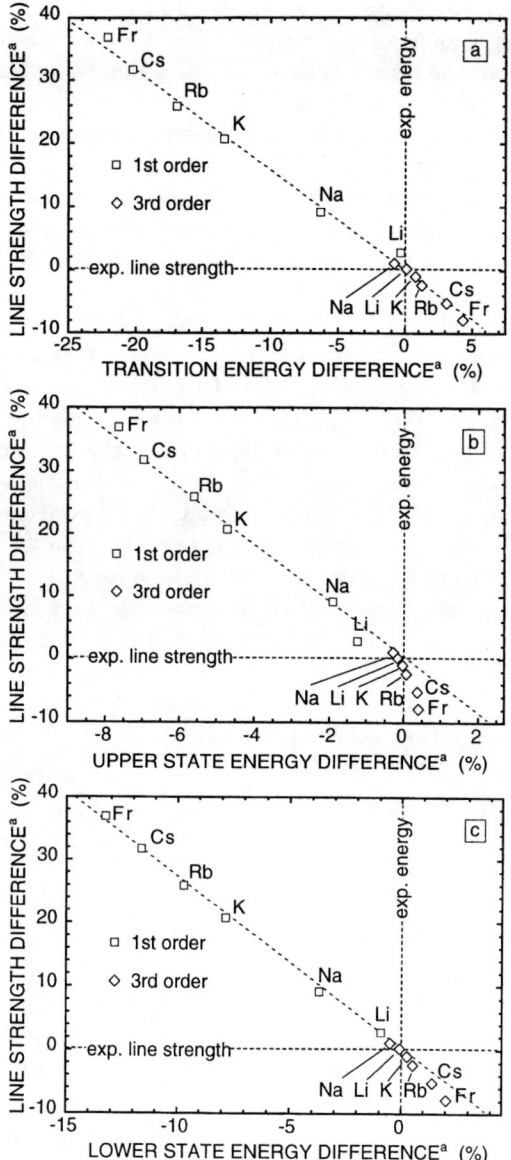

FIGURE 2. Percentage differences[a] between line strengths from up to third-order MBPT [30] and experiment for the alkali resonance transitions plotted as a function of the corresponding percentage differences[a] in the transition energies (top), and in the upper state (center) and lower state ionization energies (bottom).
([a] calculated as $100 \times \log(S^{\text{th}}/S^{\text{exp}})$ and $100 \times \log(E^{\text{th}}/E^{\text{exp}})$, respectively.)

larization of the core by the valence electron. It includes the contributions from the second order of MBPT as leading terms. As Fig. 1 shows the RPA correction removes only about 1/3 of the errors in the first-order line strengths.

The most important contributions in the third order of MBPT are the Brueckner-orbital corrections. The remaining third-order contributions arise from structural radiation and normalization diagrams. These corrections are much smaller than the BO corrections and tend to cancel each other partially for the alkali resonance transitions. The third order of MBPT gives satisfactory results only for the lighter alkalis. Substantial deviations from experiment (up to -8% for Fr $7s$–$7p$) remain for the heavy alkalis. A notable fact is that again the relative deviations of the third-order line strengths and of the corresponding second-order transition energies from experiment have opposite sign and differ by a factor of about -1.6.

The relation between the errors of calculated line strengths and the errors of calculated transition energies that is indicated by Fig. 1 certainly deserves some further exploration. In Fig. 2 the relative deviations of the first-order and third-order MBPT line strengths from experiment are plotted as functions of the relative deviations of the corresponding transition energies (top), upper state ionization energies (center), and lower state ionization energies (bottom). Obviously the first-order results (Dirac-Hartree-Fock) form straight lines in all three diagrams, however, only in the top diagram the results from first-order and third-order MBPT occupy the same line. It is a peculiar and unexpected fact that the experimental data point lies on the connection line between the first-order and the third-order MBPT data points if the line strengths are plotted as a function of the transition energies. The empirical relation between line strengths and transitions energies suggested by Fig. 2a can be written in the form

$$\frac{S_{if}^{(n)}}{S_{if}^{(\text{exp})}} = \left(\frac{E_{if}^{(n-1)}}{E_{if}^{(\text{exp})}}\right)^{-\gamma} \iff S_{if}^{(n)}\left(E_{if}^{(n-1)}\right)^{\gamma} = S_{if}^{(\text{exp})}\left(E_{if}^{(\text{exp})}\right)^{\gamma} \quad (1)$$

where the exponent $-\gamma$ is the slope of the line in Fig. 2a and $n = 0$ or 2 is the MBPT order (note that Fig. 2a is a log-log plot, *i.e.* a line corresponds to a power law). The numerical value of γ is about 1.6. Eq. 1 may be interpreted in such a manner that the product of the line strength and the γth power of the transition energy is insensitive to the higher-order terms neglected in the calculation.

TABLE 3. All-order MBPT calculation of line strengths for the alkali resonance transitions (length-form results unless otherwise noted).

	Ref.	Line strength sum D_1+D_2 (a.u.)					
		Li $2s$–$2p$	Na $3s$–$3p$	K $4s$–$4p$	Rb $5s$–$5p$	Cs $6s$–$6p$	Fr $7s$–$7p$
mod. 3rd order	[30]	33.01	37.32	50.23	52.88	58.68	51.40
all-orders	[34,35]	—	—	—	—	60.34	53.65
all-orders	[31–33]	32.99	37.38	—	—	61.05	—
all-orders[a]	[31,33]	33.02	—	—	—	60.22	—

[a] Velocity form.

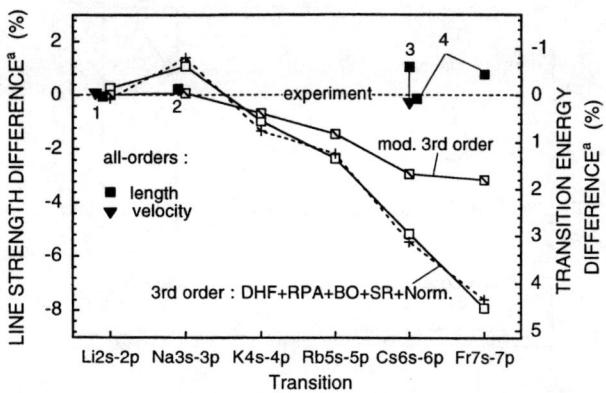

FIGURE 3. Percentage differences[a] between line strengths from all-order MBPT calculations and experiment for the alkali resonance transitions (1: [31]; 2: [32]; 3: [33]; 4: [34,35].) Third-order results [30] and modified third-order results [30] are included for comparison. ([a] calculated as $100 \times \log(S^{\mathrm{th}}/S^{\mathrm{exp}})$.)

All-order MBPT. In many-body perturbation theory it is possible to include complete families of higher-order contributions by means of all-order methods. Calculations of this type have been performed by Blundell *et al.* for lithium [31] and cesium [33], by Guet *et al.* [32] for sodium, and by Dzuba *et al.* for cesium [34] and francium [35] (see Table 3). These calculations have reproduced the experimental line strengths as well as the experimental transition energies to within 1% for Cs $6s$–$6p$ and Fr $7s$–$7p$ and to within 0.1% for Li $2s$–$2p$ (see Fig. 3).

All-order MBPT calculations are very time-consuming, thus higher-order correlation effects are sometimes included in a semi-empirical way. Dzuba *et al.* [34,35] included higher-order correlations by introducing empirical screening factors for the Coulomb interaction in their all-order calculations. Johnson *et al.* [30] proposed an approximate treatment of higher-order corrections by a phenomenological modification of the third-order Brueckner-orbital corrections. This approach is less complete than a full all-order calculation and gives less accurate results for the heavier alkalis (see Fig. 3).

Line strength ratios. The ratio between the line strengths of the two fine-structure components D_1 and D_2 of the resonance doublets is another interesting point. The nonrelativistic prediction of 0.5 is valid only for the lighter alkalis. For the heavier alkalis Cs and Fr a substantial increase of the line strength ratio has been predicted by theory [33–35]. The experimental results confirm this increase as Fig. 4 shows. The results from all-order MBPT calculations for Cs and Fr agree closely with experimental data. Surprisingly, there is little difference between the first-order and the all-order results, although there are substantial second-order and third-order corrections to the line strength ratios.

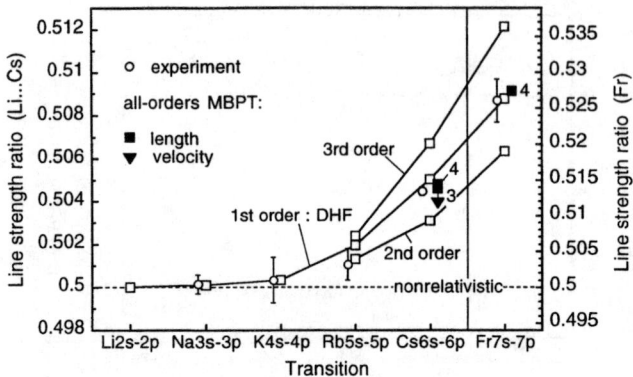

FIGURE 4. Experimental (○) and theoretical line strength ratios (D_1/D_2) for the alkali resonance transitions. (up to third-order MBPT: [30]; all-order MBPT 3: [33]; 4: [34,35].)

THE OPTICAL TRANSITIONS OF THE HEAVY NOBLE GASES

The orange and near-infrared transitions of the heavy noble gases (Ne $3s$–$3p$, Ar $4s$–$4p$, etc.) resemble the resonance transitions of the alkalis in some respects. For instance, the average wavelengths and upper state lifetimes of the Ar $4s$–$4p$ and Kr $5s$–$5p$ transition arrays are around 800 nm and 27 ns, respectively, which is quite close to those of the resonance transitions in K and Rb. The wavelengths and upper state lifetimes of the Ne $3s$–$3p$ transition array are shorter, around 600 nm and 18 ns, respectively, the same is true for Na $3s$–$3p$. This similarity is of course due to the fact that the valence electron makes the same ns–np transition. The difference lies in the core which is a closed p^6 shell for the alkalis, but an open p^5 shell for the noble gases. The hole in the p^5 core of the noble gases acts like a spin-1/2 particle with orbital angular momentum 1. The addition of the angular momenta of core and valence electron results in 10 energy levels in the excited p^5p configurations and 4 levels in the excited p^5s configurations. In Paschen notation these states are conveniently named in order of descending energy. In argon, for instance, the $3p^54p$ and $3p^54s$ levels are termed $2p_1\ldots 2p_{10}$ and $1s_2\ldots 1s_5$, respectively. The levels $1s_3$ and $1s_5$ are metastable and can be used as starting points for lifetime measurements on the $2p_i$ levels. Since there are no other E1 decay channels, the lifetimes of the $2p_i$ levels are related solely to the transitions to the $1s_f$ levels.

The angular momentum coupling scheme which is closest to the real situation is Racah or jK coupling. In this scheme the total angular momentum j_c of the core and the orbital angular momentum l_v of the valence electron are added first to give an intermediate quantum number K, before finally the valence spin is added. The actual coupling scheme, however, is between LS coupling and jK coupling, and an intermediate coupling calculation is necessary to get a correct description. The purpose of the following section is to show that conventional single-configuration intermediate coupling calculations are not sufficient to predict the ratios between the lifetimes of the $2p_i$ levels correctly. It is necessary to include configuration mixing in order to obtain a more accurate prediction of the lifetime ratios. We will use the Ar $4s$–$4p$ transition array as an example. However, it should be noted that the Ne $3s$–$3p$ and Kr $5s$–$5p$ transition arrays show a similar behavior.

Experimental lifetimes for the Ar $4p$ states

The most accurate experimental lifetimes for the Ar $4p$ states so far have been determined in our group by means of the beam-gas-laser-spectroscopy (BGLS) method (Volz et al. [36]). Our measurements cover the levels $2p_2$ to $2p_9$ with uncertainties of a few 0.1%.

In Table 4 our BGLS lifetime results for the Ar $4p$ levels are compared with the most recent experimental data from the literature. Quadfasel and Helbig [37] used the pulsed-laser technique and obtained results for all ten levels with uncertainties of about ±4%. With the exception of the $2p_9$ level their results agree with the data from BGLS within their error bars. Wiese et al. [38] derived a unified set of recommended transition probabilites for the transition arrays $4s$–$4p$ and $4s$–$5p$ in argon from a critical analysis of experimental data. The recommended transition probabilities for the $4s$–$4p$ transitions are based on relative transition probabilities

TABLE 4. Lifetime measurements on the ArI $4p$ levels (uncertainties are given in parentheses).

Level[a]		Lifetimes (ns)		
		Quadfasel and Helbig [37] PL[b]	Wiese et al. [38] CC[c]	Volz et al. [36] BGLS[d]
$2p_1$	$4p'[1/2]_0$	21.7(1.1)	22.4(1.1)	—
$2p_2$	$4p'[1/2]_1$	28.3(1.0)	28.4(1.0)	27.85(7)
$2p_3$	$4p'[3/2]_2$	29.0(1.0)	28.9(1.0)	29.01(7)
$2p_4$	$4p'[3/2]_1$	29.3(1.3)	30.2(1.3)	29.83(8)
$2p_5$	$4p[1/2]_0$	24.4(1.4)	24.9(1.4)	24.84(18)
$2p_6$	$4p[3/2]_2$	29.4(1.3)	29.0(1.3)	28.52(7)
$2p_7$	$4p[3/2]_1$	30.2(1.2)	29.7(1.2)	29.62(7)
$2p_8$	$4p[5/2]_2$	30.6(1.0)	31.0(1.0)	31.17(7)
$2p_9$	$4p[5/2]_3$	30.7(1.4)	30.2(1.4)	29.00(7)
$2p_{10}$	$4p[1/2]_1$	40.5(2.2)	39.2(2.2)	—

[a] In jK coupling notation $4p'$ and $4p$ denote the two possible core angular momenta $j_c = 1/2$ and $j_c = 3/2$, respectively. The intermediate angular momentum K and the total angular momentum J are given as $[K]_J$.
[b] Pulsed laser.
[c] Critical compilation. We quote here the error bars of the lifetimes of Quadfasel and Helbig because these lifetimes were used by Wiese et al. to normalize their recommended transition probabilities.
[d] Beam-gas-laser-spectroscopy.

and branching ratios from several emission experiments and on the pulsed-laser lifetime measurements of Quadfasel and Helbig. Wiese et al. quoted uncertainties of ±5% and ±8% for the transition probabilities of the stronger transitions. However, the lifetimes calculated from these transition probabilities should be about as accurate as the lifetimes that were used for the normalization of the data from the emission experiments. In fact, the r.m.s. difference between the lifetimes obtained from Wiese et al.'s recommended transition probabilities and the more accurate BGLS results is 1.8% only. We use the results of Wiese et al. for the levels $2p_1$ and $2p_{10}$ which have not been measured by BGLS.

Calculations for the Ar 4s–4p transition array

The calculations available for the Ar 4s–4p transition array (see Table 5) are generally far less advanced than the calculations on alkalis described in the previous section. The calculations by Lilly [39] and by Aymar et al. [40] used semi-empirical methods. In the more advanced calculation by Aymar et al. electron correlation was taken into account by means of second-order perturbation theory. The only *ab initio* calculation is a small-scale configuration interaction calculation carried out by Gruzdev and Loginov [41].

The simple approach used by Lilly [39] combined a semi-empirical intermediate coupling algorithm with the Coulomb approximation for the determination of the

TABLE 5. Theoretical lifetimes for the ArI 4p levels.

Level		Lifetimes (ns)					
		Lilly[a] [39]	Aymar et al.[b] [40]			Gruzdev and Loginov[d] [41]	
			1st	2nd	2nd,adj.[c]	sc	mc
$2p_1$	$4p'[1/2]_0$	22.6	19.4	18.5	21.66	21.1	19.5
$2p_2$	$4p'[1/2]_1$	25.7	22.0	24.8	27.87	25.3	28.1
$2p_3$	$4p'[3/2]_2$	27.6	23.7	25.5	29.05	27.3	27.5
$2p_4$	$4p'[3/2]_1$	29.0	24.8	26.0	29.82	28.2	28.9
$2p_5$	$4p\,[1/2]_0$	22.8	19.5	22.0	24.75	22.3	19.4
$2p_6$	$4p\,[3/2]_2$	26.9	23.0	25.3	28.71	26.0	27.7
$2p_7$	$4p\,[3/2]_1$	28.7	24.6	25.8	29.53	28.0	29.9
$2p_8$	$4p\,[5/2]_2$	31.4	26.8	27.0	31.20	30.8	29.7
$2p_9$	$4p\,[5/2]_3$	28.6	24.5	25.2	28.99	28.0	27.5
$2p_{10}$	$4p\,[1/2]_1$	43.3	37.0	32.2	38.30	39.2	34.4

[a] Coulomb approximation, intermediate coupling.
[b] Parametrized central field, first-order (1st) and second-order (2nd) perturbation theory (length form).
[c] Adjusted second-order results. The transition probabilities of Aymar et al. in first order and their second-order corrections have been reduced by empirical factors $\alpha = 0.864$ and $\beta = 0.676$ so as to get best agreement with the BGLS lifetimes of Volz et al. [36].
[d] Configuration interaction, single configuration (sc) and multi-configuration (mc) result (average of length and velocity form).

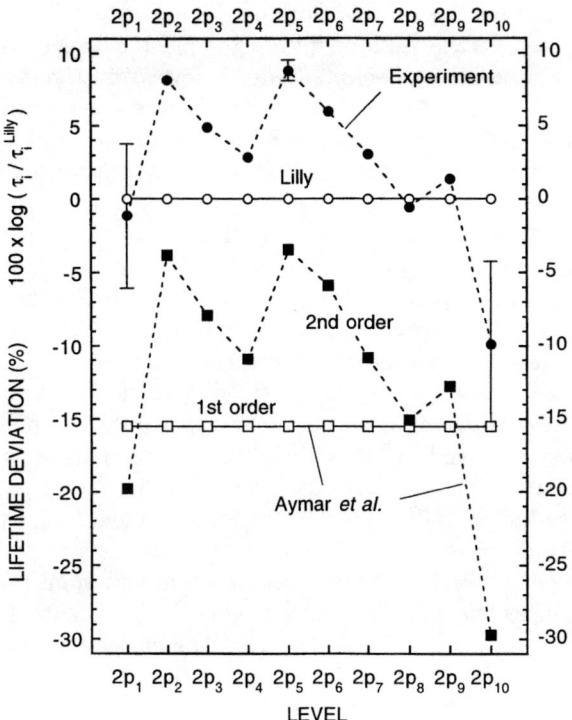

FIGURE 5. Experimental and theoretical lifetimes for the Ar $4p$ states. The diagram shows the percentage deviations[a] of the experimental lifetimes and of the lifetimes calculated by Aymar et al. [40] with respect to the lifetimes calculated by Lilly [39]. The experimental lifetimes were taken from Volz et al. [36] for $2p_2\ldots 2p_9$, and from Wiese et al. [38] for $2p_1$ and $2p_{10}$. The error bars of the experimental lifetimes are too small to be visible except for $2p_1$, $2p_5$, and $2p_{10}$. ([a] calculated as $100 \times \log(\tau_i/\tau_i^{\text{Lilly}})$).

radial transition moment σ^2. The average deviation of Lilly's results from experiment is 2.3% only which shows that the obtained σ^2 of 9.25 a.u. is a quite good approximation. The individual deviations, however, range from -9% in the case of $2p_5$ to $+10\%$ in the case of $2p_{10}$, i.e. the theory fails to predict the lifetime ratios correctly.

Fig. 5 shows the deviations between experiment and theory more clearly. The results of Lilly have been chosen as reference values in Fig. 5 and thus appear on the zero line. The clearly visible pattern in the deviations of the experimental lifetimes from Lilly's results may be interpreted as a signature of the breakdown of the single configuration approximation. Any theory that relies on a single transition moment σ^2 for the description of the whole transition array will produce this pattern in the differences between experimental and theoretical lifetimes.

The first step in the calculations of Aymar et al. [40] was the construction of a semi-empirical potential for the valence electron. The parameters of the potential were adjusted so as to get the best possible agreement between calculated and observed energies for a number of selected energy levels. In the second step the angular coupling coefficients of the wave functions were obtained from a semi-empirical intermediate coupling calculation identical to the one performed by Lilly. In the third step second-order perturbation theory was employed to account for electron correlation.

With average deviations from the experimental lifetimes of about -18% in first order and still -14.5% in second order the quality of Aymar et al.'s results seems to be poor. The transition moment σ^2 of 10.77 a.u. obtained in first order is far too large. A look at Fig. 5, however, reveals an interesting feature of the second-order results. Aymar et al.'s second-order corrections display the same pattern as the differences between the experimental lifetimes and the results of Lilly. This indicates that, despite of the poor agreement with experiment on an absolute scale, the lifetime ratios are improved by the second-order corrections.

The poor agreement of the second-order results of Aymar et al. with experiment is most likely due to the neglect of core-excited configurations in the second-order corrections. Core polarization effects were thus excluded from the calculation. For the corresponding $4s$–$4p$ transitions in potassium the inclusion of core polarization reduces the calculated line strengths by 8% [30]. If the analogy between noble gases and alkalis holds in this case the neglect of core polarization may well be responsible for the major part of the 14.5% difference between experimental and second-order theoretical results.

The second-order results of Aymar et al. can be brought into perfect agreement with the experimental results by reducing the first-order results and the second-order corrections for the transition probabilities by empirical factors α and β, respectively. A least squares-fit to the eight BGLS lifetimes yields reduction factors $\alpha = 0.864$ and $\beta = 0.676$. The reduction factors account in an empirical way for correlation effects like core polarization that were not included in the original calculation. The r.m.s. deviation of the adjusted second-order results (see Table 5) from the eight BGLS lifetimes is 0.3% only which is comparable to the uncertainties in the lifetime measurements. This is quite remarkable as only two adjustable parameters were used.

The configuration interaction (CI) calculations of Gruzdev and Loginov [41] were based on term-independent frozen-core Hartree-Fock wave functions. The number of configurations included in this calculation was quite small. Only 7 and 4 configurations of the type $3p^5\ nl$ were taken into account in the CI expansions for the $4s$ and $4p$ states, respectively. Core-excited configurations were not used, i.e. the core polarization effect was ignored.

The over-all agreement with the experimental lifetimes is surprisingly better for the single-configuration results of Gruzdev and Loginov. For the three levels $2p_1$, $2p_5$, and $2p_{10}$ the CI results are much further away from the measured lifetimes than the single-configuration results. This indicates that the number of configurations included in this CI calculation was far too small to allow for good results.

CONCLUSIONS

Two contrasting examples have been given in which accurate lifetime measurements allowed for a rigorous check of atomic-structure calculations.

In the case of the resonance transitions for the alkali atoms very accurate line strengths have become available from recent precision lifetime measurements. The line strengths obtained from large-scale *ab initio* calculations show a high degree of agreement with the experimental data which underlines the high quality of *ab initio* atomic-structure calculations for alkali atoms.

The theoretical methods applied so far to the heavy noble gases are less advanced, and as a consequence the agreement of theoretical results with experimental data is often poor. The high accuracy of the lifetimes obtained from beam-gas-laser-spectroscopy enabled us to uniquely observe configuration mixing effects which was impossible so far due to the larger experimental uncertainties of earlier measurements. New theoretical efforts will be needed in this case which account more precisely for electron correlation, particularly for the core polarization effect.

ACKNOWLEDGMENTS

The authors would like to thank the co-workers at the beam-gas-laser-spectroscopy group at Kaiserslautern, Germany who have participated in the BGLS lifetime measurements presented in this paper, namely D. Marger, J. P. Hermann, H. Liebel, M. Majerus, H. Roth, and A. Schmitt. We are also grateful to W. R. Johnson and A. P. Derevianko for helpful discussions and to C. Froese Fischer, D. J. Heinzen, J. E. Simsarian, and C. E. Tanner for making their results available to us prior to publication.

REFERENCES

1. Gaupp, A., Kuske, P., and Andrä, H. J., *Phys. Rev. A* **26**, 3351 (1982).
2. McAlexander, W. I., Abraham, E. R. I., Ritchie, N. W. M., Williams, C. J., Stoof, H. T. C., and Hulet, R. G., *Phys. Rev. A* **51**, R871 (1995).
3. Linton, C., Martin, F., Russier, I., Ross, A. J., Crozet, P., Churassy, S., and Bacis, R., *J. Mol. Spectrosc.* **175**, 340 (1996).
4. Volz, U., and Schmoranzer, H., *Phys. Scr.* **T65**, 48 (1996).
5. McAlexander, W. I., Abraham, E. R. I., and Hulet, R. G., *Phys. Rev. A* **54**, R5 (1996).
6. Martin, F., Aubert-Frécon, M., Bacis, R., Crozet, P., Linton, C., Magnier, S., Ross, A. J., and Russier, I., *Phys. Rev. A*, **55**, 3458 (1997).
7. Schmitt, A., Volz, U., Liebel, H., Kohl, M., Henkel, R., and Schmoranzer, H., in preparation (1998); Schmitt, A., Volz, U., and Schmoranzer, H., NIST (U.S.) Spec. Publ. (U.S. Government Printing Office, Washington, D.C., 1998), to be published.
8. Volz, U., Majerus, M., Liebel, H., Schmitt, A., and Schmoranzer, H., *Phys. Rev. Lett.* **76**, 2862 (1996).

9. Oates, C. W., Vogel, K. R., and Hall, J. L., *Phys. Rev. Lett* **76**, 2866 (1996).
10. Jones, K. M., Julienne, P. S., Lett, P. D., Phillips, W. D., Tiesinga, E., and Williams, C. J., *Europhys. Lett.* **35**, 85 (1996).
11. Tiemann, E., Knöckel, H., and Richling, H., *Z. Phys. D* **37**, 323 (1996).
12. Wang, H., Gould, P. L., and Stwalley, W. C., *J. Chem. Phys.* **106**, 7899 (1997).
13. Freeland, R. S., Tsai, C. C., Marinescu, M., Cline, R. A., Miller, J. D., Williams, C. J., Dalgarno, A., and Heinzen, D. J., in preparation (1998).
14. Simsarian, J. E., Orozco, L. A., Sprouse, G. D., and Zhao, W. Z., *Phys. Rev. A*, submitted (1997).
15. Young, L., Hill III, W. T., Sibener, S. J., Price, S. D., Tanner, C. E., Wieman, C. E., and Leone, S. R., *Phys. Rev. A* **50**, 2174 (1994).
16. Rafac, R. J., Tanner, C. E., Livingston, A. E., Kukla, K. W., Berry, H. G., and Kurtz, C. A., *Phys. Rev. A* **50**, R1976 (1994).
17. Rafac, R. J., and Tanner, C. E., *Phys. Rev. A*, submitted (1997).
18. King, G. W. and Van Vleck, J. H., *Phys. Rev.* **55**, 1165 (1939).
19. Movre, M. and Pichler, G., *J. Phys. B* **10**, 2631 (1977).
20. Williams, C. J. and Julienne, P. S., *J. Chem. Phys.* **101**, 2634 (1994).
21. McLone, R. and Power, E., *Mathematika* **11**, 91 (1964); Meath, W. J., *J. Chem. Phys.* **48**, 227 (1968).
22. Stwalley, W. C., Uang, Y.-H., and Pichler, G., *Phys. Rev. Lett.* **41**, 1164 (1978).
23. Yan, Z.-C. and Drake, G. W. F., *Phys. Rev. A* **52**, R4316, (1995).
24. Yan, Z.-C. and Drake, G. W. F., *Phys. Rev. A* **52**, 3711 (1995).
25. Drake, G. W. F. and Yan, Z.-C., *Phys. Rev. A* **52**, 3681 (1995).
26. Godefroid, M. R., Froese Fischer, C., and Jönsson, P., *Phys. Scr.* **T65**, 70 (1996).
27. Jönsson, P., Ynnerman, A., Froese Fischer, C., Godefroid, M. R., and Olsen, J., *Phys. Rev. A* **53**, 4021 (1996).
28. Froese Fischer, C., Saparov, M., Gaigalas, G., and Godefroid, M. R., *At. Data Nucl. Data Tables*, submitted (1997).
29. Johnson, W. R., Idrees, M., and Sapirstein, J., *Phys. Rev. A* **35**, 3218 (1987).
30. Johnson, W. R., Liu, Z. W., and Sapirstein, J., *At. Data Nucl. Data Tables* **64**, 279 (1996).
31. Blundell, S. A., Johnson, W. R., Liu, Z. W., and Sapirstein, J., *Phys. Rev. A* **40**, 2233 (1989).
32. Guet, C., Blundell, S. A., and Johnson, W. R., *Phys. Lett. A* **143**, 384 (1990).
33. Blundell, S. A., Johnson, W. R., and Sapirstein, J., *Phys. Rev. A* **43**, 3407 (1991).
34. Dzuba, V. A., Flambaum, V. V., and Sushkov, O. P., *Phys. Lett. A* **142**, 337 (1989).
35. Dzuba, V. A., Flambaum, V. V., and Sushkov, O. P., *Phys. Rev. A* **51**, 3454 (1995).
36. Volz, U., Marger, D., Roth, H., Liebel, H., and Schmoranzer, H., in preparation (1998).
37. Quadfasel, U. and Helbig, V., unpublished (1983), results are quoted in [38].
38. Wiese, W. L., Brault, J. W., Danzmann, K., Helbig, V., and Kock, M., *Phys. Rev. A* **39**, 2461 (1989).
39. Lilly, R. A., *J. Opt. Soc. Am.* **66**, 245 (1976).
40. Aymar, M., Feneuille, S., and Klapisch, M., *Nucl. Instr. Meth.* **90**, 137 (1970).
41. Gruzdev, P. F. and Loginov, A. V., *Opt. Spectrosc.* **38**, 234 (1975).

Data Assessment and Evaluation

X-ray Transition Energies-
New Comprehensive Evaluation

R. D. Deslattes,* E. G. Kessler, Jr.,*
Paul Indelicato,[†] and Eva Lindroth[‡]

*Physics Laboratory, Atomic Physics Div., NIST, Gaithersburg, MD 20899, USA
[†]Laboratoire Kastler-Brossell, Ecole Normale Supérieure et Univ. P. et M. Curie (Unité associée au CNRS #18), 4 Place Jussieu, F-75252Paris Cedex 05, France
[‡]Dept. of Atomic Physics, Stockholm University, Frescativägen 24 S-104 05 Stockholm, Sweden

Abstract. In this brief report we describe some of the background for our continuing efforts toward a new comprehensive database for x-ray transitions, and indicate the present status of this effort, one that is still a work in progress. In contrast to previous, entirely empirical compilations, ours uses advanced theoretical calculations combined with a sparse network of accurate experimental measurements. The theoretical results closely reproduce high quality experimental data, where they are available, for atoms from Z=22 to Z=90. This suggests that the theoretical estimates are potentially useful in cases where adequate experimental data are absent. Above and below this range, problems emerge which need further examination both theoretically and experimentally.

INTRODUCTION

This report summarizes some efforts directed toward an improved database for x-ray transitions that were presented verbally by one of the authors (RDD) at the first ICAMDATA conference. A more complete version of the tabular output is scheduled to appear in a forthcoming revised edition of the International Tables for Crystallography (1). Further documentation of the work is intended for publication in a more easily accessed archival journal (2).

The contents of a very few x-ray tabulations are widely disseminated in handbooks, texts, and recently on the web. For simplicity, we consider only two of these which, in spite of their being both entirely empirically based, differ considerably in their treatment of primary data. One of these, prepared under the guidance of Y. Cauchois, lists available measurements in wavelength order with uncertainties as assigned by the original authors (3, 4). The second, by J. A. Bearden, employs a moderate filtration of input data, reassesses assigned uncertainties, and attempts to improve the scale linkage to macroscopic dimensional quantities (5). In addition to the many paper representations of these datasets appearing in handbooks and textbooks, the same information is now also widely available on the internet.

There are significant practical questions as to how to improve this situation and meet the evident needs of the principal user communities. On the one hand, while it would be technically possible to establish a program of measurements to improve the purely empirical approach, it is unrealistic to imagine that such work could be supported or staffed in today's world. The alternative of undertaking a massive theoretical effort to escape the limitations of previous tabulations, and the experimental database, would have difficulties of its own. Specifically, such a "theoretical" tabulation would also involve a large effort and correspondingly high cost. In the end, such an effort could neither document its own validity, nor approach the metrological accuracy now attainable.

Other practical questions arise from what might be termed "proliferation" problems. For example, there is potentially intractable complexity associated with the effects of chemistry and the state of atomic aggregation, a circumstance that makes it desirable to focus on spectra resembling those of free atoms. Even in this limited subspace, multi-vacancy processes multiply beyond practical limits of computational effort, and potential utility. The problem has generally been contained by including in tabulations only the principal allowed transitions, the so-called diagram lines, the strongest "forbidden" transitions, along with some of the more prominent satellites.

AN IMPROVED SPARSE EXPERIMENTAL DATASET

Following the first measurement of x-ray reference wavelengths linked to visible standards by combined x-ray and optical interferometry, the number of such measurements slowly expanded over a period of 20 years (6, 7). Most of this expansion came about in response to needs for reference markers for experiments on exotic atoms where the relatively efficient curved crystal spectrometers needed are not generally capable of a direct wavelength determination (8). In addition to the work at NBS/NIST, there have been important contributions based on the more accurate crystal lattice spacing measurements made at the PTB (9, 10). Indeed, the best documented and most effective measurement campaign described to date has been carried out by E. Förster and his collaborators in Jena (11, 12).

In several cases it has been possible to expand this core measurement set by considering earlier work in which several elemental spectra were obtained using a common instrument platform and protocol, and at least one member of the collection of lines reported belongs to our core dataset of directly measured lines. Examples of this situation include the work of Bearden and collaborators on K- and L-series lines (5, 13) in the region 40<Z<50.

Absorption edge locations measured with the new methodology are few in number. The earliest group were determined to provide edge filters for Ly α radiation from H-like ions or to serve as discriminants for mesic x-ray transitions such as those used for determining the mass of the muon(14) and the mass of the kaon (15). In early 1996, this severe data shortage was considerably relieved by an excellent and well-documented set of measurements carried out by the PTB at HASYLAB Synchrotron radiation facility (16).

To further address the remaining shortage of high quality direct edge measurements, we introduced an alternative estimation procedure. This procedure parallels that used by Bearden and Burr in their global adjustment of x-ray energy levels (17), but is of more limited scope and is directed toward a slightly different goal. Our approach is based upon combining emission line energies with photo-electron binding energies to obtain estimates of the binding energies for deeply bound electrons. There are, of course, important symmetry restrictions and a clear need for care in comparing the threshold location for physically distinct processes, as has been described by Powell (18). Nonetheless, where good experimental data for metallic elements are available, the empirical comparisons described below suggest that the procedure is reasonably reliable.

EARLY ENCOUNTERS OF THEORY AND EXPERIMENT

The first conveniently accessible all-Z calculations were carried out using a relativistic Hartree-Fock-Slater approach(19); the calculated values were compared with energy levels derived from experimental data by Bearden and Burr (17). Such "experimental" energy levels have uncertainties significantly larger than those associated with the highest quality emission line measurements. Our view was that extracting transition energies, i.e., level differences, from the theoretical database and comparing these with the most reliable transition energy measurements would offer the possibility of more refined and informative testing of the theory. This expectation was quickly realized and a very clear pattern of discrepancy increasing as Z appeared (20, 21). This discrepancy was largely removed by use of an improved expression for the Z dependence of nuclear radii and inclusion of the diffuse nuclear surface (7, 22).

Following this early work, there was a clear interest in going beyond the DHS approach, and several individual calculations were made in response to our requests, primarily in the MCDF approach emerging from the work of Desclaux. To obtain comparable results but in a package allowing simpler application by experimentalists, we turned to the suite of programs developed by Grant (23). These yielded improved concordance with experiment, discrepancies considerably reduced from the initial DHS comparisons, but with the improving experimental database two new systematic trends emerged (24). The first was the observation of nearly constant offsets between theory and experiment for Kα and Kβ, a component of which was recognized as arising from uncalculated correlation in the K shell. There was, however, the additional and initially puzzling fact that the Kβ offset was considerably greater than that of Kα. Since attention had previously been focused on the QED and relativistic aspects that dominate wavefunction behavior near the nucleus, an inner region common to the two line groups, this puzzle required an increased focus on the outer hole states. Qualitatively, the additional shift for the 3p hole states was soon recognized as due to its coupling to Auger decay continua. Although semi-quantitative results were obtained, it was clear that real and difficult problems had to be confronted, as has now happened in the theoretical developments which are briefly described next.

DEVELOPMENT OF NEW THEORETICAL PROCEDURES

In this section, we present a schematic description of the theoretical procedures that evolved over the course of this undertaking. The interested reader will find a more comprehensive description elsewhere as well as in previous overviews that document various stages in the development of this toolset (25-29).

Realistic calculations of transition energies between hole states for $10<Z<100$, require consideration of contributions arising from nuclear size and relativistic effects, Coulomb and Breit correlation, radiative (QED) corrections (1-e and 2-e Lamb shift, etc.) and Auger shifts. All of these need to be treated in a relativistic framework, while their relative sizes change strongly with Z. In the introduction, we have indicated the reasons for restricting consideration to the case of free atoms. The applicability of this large approximation for at least metallic solids may be seen in the comparisons given below where rare gas data fall within the range of metallic results. With respect to other solid elements or solid compounds, our restriction to free atoms is a significant limitation.

A second limitation is that it has not been possible to include the coupling between the hole and open outer shells. Coupling between a $j=1/2$, $j=3/2$, or $j=5/2$ hole and an external 3d or 4f shell can generate hundreds of levels, with splittings that can reach an eV. Including radiative and Auger processes among hundreds of excited and final states would lead to an even larger proliferation of final states, so we have chosen to compute the weighted average energy for each hole state, and ignore possible distortion of the line profile due to the coupling between inner vacancies and outer shells.

The need to cover the range 10<Z<100 including the effects of relativity non-pertubatively led us to carry out the calculation exact to all orders (in $Z\alpha$) by using the Dirac equation. We thus have used many-body methods, based on the Dirac equation, in which the main contributions to the transition energy are evaluated using the Dirac-Fock method. We use the Breit operator for the electron-electron interaction, to include magnetic spin-spin, spin-other orbit and orbit-orbit interactions in the lower orders in $Z\alpha$ and $(v/c)^2$ retardation effects. Higher order retardation effects (30) are also included. Many-body effects are calculated by using relativistic Many-Body Perturbation Theory (RMBPT). Since inner vacancy levels are auto-ionizing, one must include shifts in their energy due to the coupling between the discrete levels and Auger decay continuua.

Uncorrelated Dirac-Fock Energies.

The first step in the calculation minimizes the energy (with relativistic corrections) in the independent electron approximation, for each hole state (25-27, 29, 31). This provides a suitable starting point for adding many-body and QED contributions. We follow the Dirac-Fock method in the implementation of Desclaux which allows treatment of arbitrary atoms with arbitrary structure (32, 33). We have included full exchange and relaxation (to account for inactive orbital rearrangment due to the hole presence). The electron-electron interaction used in this program contains all magnetic and retardation effects, which is very important for large Z. The magnetic interaction is treated on an equal footing with the Coulomb interaction, to account for higher order effects in the wavefunction (that are also useful for evaluating radiative corrections to the electron-electron interaction). All these calculations must be done with proper nuclear charge models, to account for finite-nuclear-size corrections to all contributions (32, 33). Nuclear deformations must be included for heavy nuclei (34, 35). For all elements for which experiments have been performed, we used experimental nuclear charge radii. For the others we used a formula from Johnson & Soff, corrected for nuclear deformations for Z>90 (36). Contribution of deformation to the rms radius (the only parameter of importance to the atomic calculation) is roughly constant (0.11 fm) for Z>90. There is an unknown region, between Bi and Th (83<Z<90) where deformation effects start to be important, but for which they are not known. When experiments are available for a particular isotope, we calculated separately the energies for each isotope.

As mentioned in the introduction, there are special difficulties involved when dealing with atoms with open outer shells (obviously this is the most common case). Computing all energies E_J for total angular momentum J would be both impossible and useless. The Dirac-Fock method circumvents this difficulty. One can evaluate directly an average energy that corresponds to the barycenter of all E_J with weight (2J+1). There are still a few cases for which the average calculation cannot converge (when the open shells have identical symmetry). In that case the outer electrons have been rearranged in an identical fashion for all hole states of the atom, to minimize possible shifts due to this procedure.

Correlation and Auger Shifts.

Once the Dirac-Fock energy is obtained, many-body effects beyond Dirac-Fock relaxation must be taken into account. These include relaxation beyond the spherical average, correlation (due to both Coulomb and magnetic interaction), and corrections due to the autoionizing nature of hole states (Auger shift). Since the many body generalization of the Dirac-Fock method, the so-called MCDF (Multiconfiguration Dirac-Fock), is very inefficient for hole states, we turned to RMBPT to evaluate those quantities. These many-body effects contribute very significantly to the final value. Coulomb correlation is mostly constant along the periodic table (at the level of a few eV). Magnetic correlations are very strong at high Z. Auger shift is very important for p states. The interested reader will find more details of these complicated calculations in the original references (25-27, 29). As these calculations are very time consuming, they are performed only for selected Z and interpolated. Since the Auger shifts do not always have a smooth Z-dependence, care has been taken to evaluate them at as many different Z's as practical to ensure a good reproduction of irregularities.

QED Corrections

The QED corrections originate in the quantum nature of both the electromagnetic and electron fields. They can be divided into two categories, radiative and non-radiative. The first one includes self-energy and vacuum polarization, which are the main contributions to the Lamb shift in one-electron atoms. These corrections scale as Z^4/n^3 (n being the principal quantum number) and are thus very important for inner shells and high Z. The second category, is composed of corrections to the electron-electron interaction that cannot be accounted for by RMBPT or MCDF. These corrections start at the two-photon interaction and include three-body effects. The two-photon, non-radiative QED contribution has been calculated recently only for the ground state of two-electron ions and cannot be evaluated in practice for atoms with more than two or three electrons.

The radiative corrections split up into two contributions. The first contribution is composed of one-electron radiative corrections (self-energy and vacuum polarization). For the self-energy and Z>10 one must use all-order calculations (37-42). Vacuum polarization can be evaluated at the Uehling(43) and Wichmann and Kroll(44) level. Higher order effects are much smaller than for the self-energy and have been neglected. The second contribution is composed of radiative corrections to the electron-electron interaction, and scales as Z^3/n^3. *Ab initio* calculations have been performed only for few-electron ions.(45, 46) Here we use the Welton approximation that has been shown to reproduce very closely *ab initio* results in all examples that have been calculated (31, 34, 47-49).

COMPARISONS WITH EXPERIMENTS

In this section we give a sample of the new x-ray transition database restricted, in the interest of brevity, to those transitions for which highly reliable energies are available from direct or rescaled measurements. These selected results are shown in Tables 1 and 2. Entries in the tables show both theoretical (in italic type) and experimental values. Those experimental data which have particularly robust connections to the base units of the SI are indicated by use of a bold typeface.

The Z-dependence of relative differences between theory and experiment are shown in Figures 1 and 2. What is seen in these figures is a complex but rather interesting picture. To begin with the obvious, there is a significant domain in which there is relatively good concordance. This fact is somewhat suprising, considering the limitations of our procedures but, if accepted at face value, offers a convenient means for provisional substitution of theoretical values for absent experimental data in cases where the theoretical accuracy is sufficient. The occasional points of large disagreement invite re-consideration of the experimental data, and perhaps re-measurement.

The more extended regions of disagreement are evident, however, their interpretation requires some care. First, considering the case of the low-Z region, we note that a relative difference of 0.1%, corresponding to a single large division in Fig. 1, is only of the order of 1 eV at the lowest Z shown. Because of the rather large Auger widths in this region, lines are fairly broad, typically 0.1% of the transition energy, and the x-ray transitions have final hole states in the outer electronic shell or nearby. In the high-Z region, there are also rather clear differences. In this region, outer shell effects are small, the increasing radiative linewidths are precision limiters.

Turning to the L-series data shown in Fig. 2, there are evidently larger problems (note the coarser vertical scale used), the data are more sparsely distributed, and the lower Z boundary is fairly high. All of these limitations can be understood primarily in terms of measurement technology limitations in the lower Z region, stronger outer shell involvement, and larger fractional linewidths. The reader will have also noted that there are several prominent L-series lines that are not included in either the figures or the tables. This simply reflects the fact that calculations of N-shell energies have been started only recently. This same consideration also applies to the absence of $K\beta_2$ from the K-series representations. Once again, one sees sufficiently good agreement for the mid-Z region as to encourage further investigations, particularly in response to the very sparse experimental database.

OUTLOOK AND DISSEMINATION PLANS

The picture of the status of x-ray transition data shown in the following figures and tables is not changed by the inclusion of the entire (selected) experimental cohort as is given elsewhere(1). Such a synoptic represention of the entire transition database provides a good overview of the quality of the database as a function of Z and leads to the following observation. Although it would seem, at least intuitively, that the good agreement between theory and experiment 'validates' the experimental database, it is difficult to quantify such an intuition. Our original expectation was that improving theory beyond Moseley's approximation should facilitate better interpolation among robust experimental data, i.e., making a credible deviation scaling function smoother in Z. Perhaps that is what has been done, since clearly the shell boundaries are not obvious in the deviation data. On the other hand, it is both suprising and gratifying that the interpolation function that emerges is unexpectedly close to the unit multiplier.

Regarding dissemination of these results and future extensions of this work, we offer the following outlook. First, two paper publications are in process, as noted in the introduction. Second, we intend to place this resource on the NIST Physics Laboratory website for general access. If the quality of our work is found to be adequate, and the results prove useful, they are likely to be mirrored elsewhere, and propagate into secondary paper archives as well.

ACKNOWLEDGMENTS

The authors wish to express their deep appreciation for the encouragement and support that led to our starting this rather large undertaking, and has sustained it thus far. The encouragement has come from individuals such as A. J. C. Wilson, William Parrish, and others in the crystallographic community, as well as the Editor of the International Tables for Crystallography, Edward Prince. Support has come from the NIST Physics Laboratory and, more recently from the Standard Reference Data Program at NIST under the leadership of John Rumble. We are also grateful to numerous colleagues who contributed to various stages of this work and others who have generously shared their experiences and allowed us to make use of their unpublished data.

TABLE 1. Energies of directly measured and rescaled K emission lines and absorption edges in eV. See text for explanation of type faces. Standard uncertainties are given in parentheses.

Z	Sym	$K\alpha_1$	$K\alpha_2$	$K\beta_1$	$K\beta_3$	K abs. edge
12	Mg	*1254.392*	*1254.137*			
		1253.688(11)	1253.437(13)			
13	Al	*1488.018*	*1487.610*			
		1486.708(10)	1486.295(10)			
14	Si	*1741.041*	*1739.619*			
		1739.986(19)	1739.395(34)			
16	S	*2308.801*	*2307.012*			
		2307.886(34)	2306.701(38)			
17	Cl	*2624.398*	*2622.000*			
		2622.441(39)	2620.847(39)			
18	Ar	*2957.962*	*2955.957*			
		2957.682(16)	2955.566(16)			
19	K	*3314.757*	*3311.971*			
		3313.9493(50)	3311.1973(60)			
23	V					*5464.43(26)*
						5463.760(50)
24	Cr	*5414.451*	*5404.066*	*5947.066*		*5989.12(48)*
		5414.761(71)	5405.5393(74)	5946.739(86)		5989.020(40)
25	Mn					*6537.65(14)*
						6537.670(20)
26	Fe					*7110.73(41)*
						7110.750(20)
27	Co					*7708.72(71)*
						7708.780(20)
28	Ni					*8330.9(14)*
						8331.490(20)
29	Cu	*8048.141*	*8028.426*	*8905.845*		*8980.5(10)*
		8047.8264(24)	8027.915(10)	8905.341(87)		8980.480(20)
30	Zn					*9660.7(12)*
						9660.760(30)
31	Ga	*9251.899*	*9225.105*	*10265.815*	*10261.833*	
		9251.679(66)	9224.840(27)	10264.17(20)	10260.26(44)	
33	As	*10543.486*	*10507.840*	*11727.138*	*11721.620*	
		10543.2724(81)	10507.50(15)	11725.74(26)	11719.86(57)	
34	Se	*11222.551*	*11181.829*	*12496.496*	*12490.272*	
		11222.53(12)	11181.54(31)	12496.03(46)	12489.74(68)	
36	Kr	*12648.005*	*12595.556*	*14113.245*	*14105.481*	
		12648.008(52)	12595.430(56)	14112.822(80)	14104.97(11)	
39	Y					*17036.65(55)*
						17036.620(50)
40	Zr	*15774.877*	*15690.622*	*17667.451*	*17654.120*	*17996.24(79)*
		15774.922(54)	15690.653(50)	17666.586(76)	17652.636(75)	17995.880(80)
41	Nb					*18983.62(85)*
						18982.970(40)
42	Mo	*17479.450*	*17374.530*	*19607.405*	*19590.216*	*20000.5(21)*
		17479.381(10)	17374.30(29)	19608.35(42)	19590.26(41)	20000.360(20)

TABLE 1. continued

Z	Sym	Kα₁	Kα₂	Kβ₁	Kβ₃	K abs. edge
44	Ru	*19279.517*	*19150.674*	*21657.466*	*21635.742*	
		19279.17(18)	19150.50(18)	21656.77(16)	21634.66(16)	
45	Rh	*20216.000*	*20073.735*	*22724.375*	*22700.049*	*23220.15(44)*
		20216.13(20)	20073.68(20)	22723.60(17)	22698.84(17)	23222.00(30)
46	Pd	*21177.264*	*21020.659*	*23819.711*	*23792.597*	*24350.92(51)*
		21177.09(17)	21020.16(22)	23818.70(19)	23791.14(19)	24352.60(20)
47	Ag	*22163.005*	*21990.676*	*24943.084*	*24912.814*	*25515.53(48)*
		22162.928(30)	21990.31(10)	24942.43(30)	24911.56(30)	25515.60(30)
48	Cd	*23173.745*	*22984.492*	*26095.129*	*26061.420*	*26712.96(71)*
		23173.99(20)	22984.06(20)	26095.46(39)	26061.34(39)	26713.30(20)
49	In	*24209.793*	*24002.320*	*27276.149*	*27238.647*	*27940.74(69)*
		24209.76(22)	24002.04(28)	27275.57(25)	27237.51(25)	27940.40(30)
50	Sn	*25271.354*	*25044.307*	*28486.368*	*28444.728*	*29200.94(92)*
		25271.37(23)	25044.05(23)	28486.27(33)	28444.44(33)	29200.40(20)
51	Sb	*26358.706*	*26110.739*	*29726.053*	*29680.033*	*30492.00(91)*
		26358.88(25)	26110.79(25)	29725.54(22)	29679.22(29)	30490.50(20)
54	Xe	*29778.304*	*29458.404*	*33624.706*	*33563.357*	
		29778.79(10)	29458.265(50)	33624.25(12)	33563.22(12)	
56	Ba	*32192.969*	*31816.654*	*36377.591*	*36304.006*	
		32193.279(70)	31816.631(60)	36377.464(80)	36303.36(12)	
60	Nd	*37361.424*	*36848.233*	*42273.635*	*42168.527*	
		37360.758(70)	36847.521(80)	42270.92(57)	42166.26(57)	
62	Sm	*40119.371*	*39524.262*	*45418.149*	*45293.862*	
		40118.501(60)	39523.40(10)	45413.1(49)	45288.7(49)	
67	Ho	*47547.473*	*46701.022*	*53877.521*	*53696.902*	
		47547.12(77)	46700.00(15)	53877.1(70)	53711.4(69)	
68	Er	*49128.771*	*48223.028*	*55676.340*	*55482.335*	*57486.3(13)*
		49127.27(12)	48221.63(20)	55673.55(18)	55479.75(35)	57485.3(20)
69	Tm	*50742.198*	*49773.833*	*57511.051*	*57302.872*	
		50741.501(92)	49772.69(12)	57508.79(15)	57303.0(53)	
74	W	*59318.796*	*57981.895*	*67245.641*	*66952.151*	
		59318.877(51)	57982.30(54)	67245.0(11)	66952.0(11)	
79	Au	*68806.960*	*66993.072*	*77983.056*	*77577.356*	
		68804.53(18)	66990.76(22)	77979.84(38)	77575.05(61)	
82	Pb	*74970.183*	*72806.685*	*84940.499*	*84451.663*	*88004.77(69)*
		74970.14(17)	72805.46(24)	84939.12(34)	84450.49(60)	88005.7(46)
83	Bi	*77108.908*	*74816.850*	*87351.465*	*86831.932*	
		77109.3(22)	74816.25(91)	87344.1(33)	86835.8(66)	
90	Th	*93347.994*	*89956.676*	*105602.000*	*104817.266*	
		93347.42(25)	89957.08(20)	105601.56(53)	104816.59(69)	
91	Pa	*95864.796*	*92283.577*	*108422.358*	*107591.206*	
		95866.4(20)	92283.5(20)	108417.4(20)	107585.4(20)	
92	U	*98433.641*	*94653.207*	*111298.370*	*110418.482*	
		98431.62(28)	94650.89(56)	111295.14(65)	110415.72(65)	
93	Np	*101057.303*	*97067.618*	*114231.443*	*113300.184*	
		101056.4(30)	97068.4(30)	114243.3(30)	113307.3(40)	
94	Pu	*103735.445*	*99526.150*	*117224.805*	*116239.651*	
		103734.11(60)	99523.2(12)	117232.3(20)	116241.3(20)	

TABLE 1. continued

Z	Sym	$K\alpha_1$	$K\alpha_2$	$K\beta_1$	$K\beta_3$	K abs. edge
95	Am	*106469.704*	*102030.061*	*120278.917*	*119237.226*	
		106473.4(30)	102031.4(20)	120279.3(20)	119239.3(20)	
96	Cm	*109262.043*	*104580.763*	*123393.237*	*122292.030*	
		109272.4(20)	104590.4(20)	123403.3(20)	122302.3(20)	
97	Bk	*112116.052*	*107181.063*	*126573.865*	*125410.227*	
		112127.3(50)	107194.4(50)	126577.2(70)	125414.3(70)	
98	Cf	*115031.512*	*109830.353*			
		115035.3(80)	109837.3(80)			

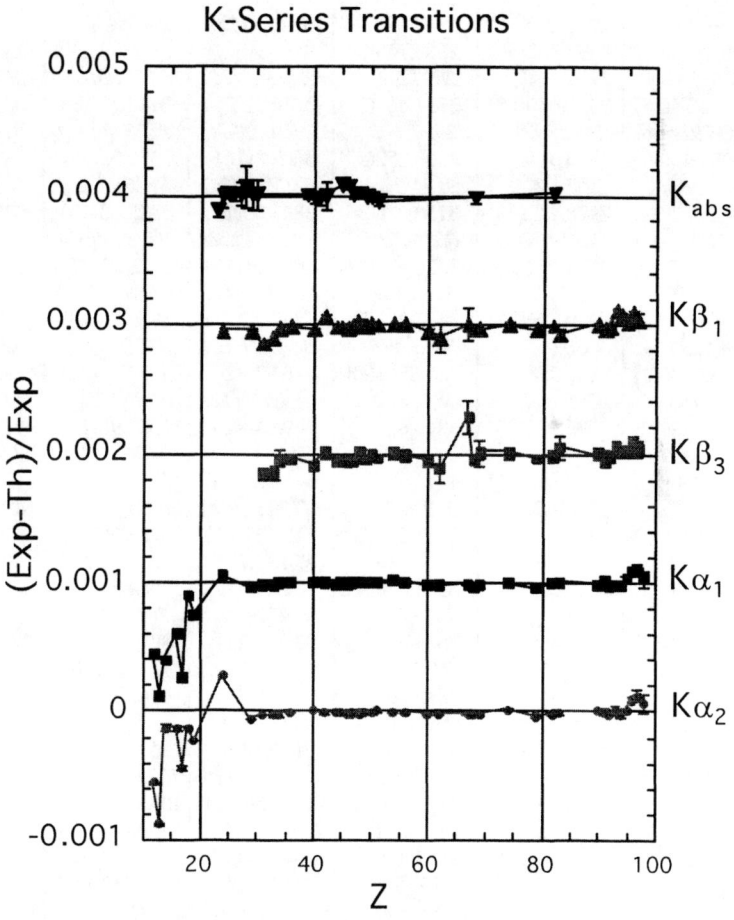

Figure 1. Relative deviations of theory and experiment for K-series transitions.

TABLE 2. Energies of directly measured and rescaled L emission lines and absorption edges in eV. See text for explanation of type faces. Standard uncertainties are given in parentheses.

Z	Sy	$L\alpha_1$	$L\alpha_2$	$L\beta_1$	L_I abs. edge	L_{II} abs. edge	L_{III} abs. edge
29	Cu	*929.653**		*949.096*	*1097.22(60)*	*952.5(10)*	*932.68(46)*
		929.68(31)*		949.84(32)		952.68(11)	933.04(10)
32	Ge	*1188.569**		*1218.948*	*1414.60(70)*	*1248.09(30)*	*1216.57(43)*
		1188.01(13)*		1218.50(18)	1413.23(24)	1249.32(19)	1217.06(18)
36	Kr	*1585.747**		*1637.114*	*1920.4(12)*	*1730.90(50)*	*1679.07(39)*
		1585.411(26)*		1636.876(21)	1916.3(44)	1729.67(36)	1677.25(34)
40	Zr	*2043.107*	*2040.340*	*2124.595*	*2530.90(21)*	*2305.67(64)*	*2221.29(57)*
		2042.490(27)	2040.19(16)	2124.395(28)	2541.1(39)	2305.36(63)	2222.31(59)
42	Mo	*2293.713*	*2289.601*	*2394.522*	*2867.20(26)*	*2625.95(33)*	*2521.09(52)*
		2293.189(50)	2289.877(50)	2394.833(55)	2880.6(50)	2627.30(82)	2523.56(76)
44	Ru	*2559.060*	*2554.235*	*2683.077*	*3224.91(26)*	*2967.47(34)*	*2838.62(19)*
		2558.580(39)	2554.332(55)	2683.265(26)	3232.9(62)	2966.1(11)	2837.78(96)
45	Rh	*2697.290*	*2692.060*	*2834.325*	*3412.22(26)*	*3146.39(44)*	*3004.00(34)*
		2696.776(78)	2692.080(78)	2834.441(38)	3416.4(70)	3144.77(59)	3002.07(54)
46	Pd	*2838.924*	*2833.707*	*2990.312*	*3604.73(64)*	*3330.66(46)*	*3173.75(50)*
		2838.640(48)	2833.314(67)	2990.252(53)	3607.3(16)	3330.35(13)	3173.02(12)
47	Ag	*2984.604*	*2978.724*	*3151.054*	*3807.41(34)*	*3525.25(26)*	*3352.58(50)*
		2984.342(32)	2978.242(53)	3150.976(36)	3807.35(17)	3525.83(15)	3350.97(13)
48	Cd	*3134.062*	*3127.457*	*3316.710*	*4019.68(22)*	*3728.54(33)*	*3538.88(47)*
		3133.757(47)	3126.952(70)	3316.607(53)	4019.01(19)	3728.01(17)	3537.60(15)
49	In	*3287.278*	*3279.875*	*3487.349*	*4238.30(27)*	*3938.71(59)*	*3730.83(45)*
		3286.984(52)	3279.325(77)	3487.246(58)	4237.27(21)	3939.33(19)	3730.25(17)
50	Sn	*3444.287*	*3436.007*	*3663.054*	*4465.01(12)*	*4156.24(78)*	*3928.94(77)*
		3444.013(42)	3437.358(56)	3662.841(48)	4464.78(24)	4157.27(21)	3928.84(18)
51	Sb	*3605.106*	*3595.894*	*3843.862*	*4699.15(12)*	*4381.22(79)*	*4132.98(88)*
		3604.758(62)	3595.360(93)	3843.617(71)	4698.44(26)	4381.9(11)	4132.33(20)
54	Xe	*4110.333*	*4097.847*	*4417.746*	*5452.57(17)*	*5106.67(20)*	*4786.48(16)*
		4110.090(20)	4097.380(30)	4417.670(30)	5452.89(35)	5103.84(31)	4782.16(27)
60	Nd	*5231.407*	*5207.636*	*5720.826*	*7129.5(10)*	*6724.77(75)*	*6211.14(60)*
		5230.242(35)	5207.7(11)	5721.448(50)	7129.52(61)	6723.56(54)	6209.37(46)
62	Sm	*5637.533*	*5608.383*	*6203.492*	*7742.25(92)*	*7315.03(57)*	*6719.4(15)*
		5635.973(33)	5609.055(61)	6204.076(93)	7747.93(72)	7313.30(64)	6717.37(54)
67	Ho	*6719.619*	*6679.958*	*7526.409*	*9395.8(20)*	*8913.9(17)*	*8068.2(14)*
		6719.678(62)	6678.487(54)	7525.67(15)	9399.7(11)	8916.39(95)	8067.57(78)
68	Er	*6947.877*	*6905.778*	*7811.520*	*9751.4(11)*	*9263.5(28)*	*8359.4(15)*
		6947.916(77)	6904.50(17)	7810.20(42)	9757.8(11)	9262.1(10)	8357.43(83)
69	Tm	*7180.105*	*7135.440*	*8103.805*	*10111.9(23)*	*9615.5(20)*	*8648.1(16)*
		7180.117(29)	7133.719(78)	8102.269(37)	10121(12)	9617.0(11)	8649.53(89)
72	Ha	*7899.136*	*7844.663*	*9022.382*	*11271.71(63)*	*10738.29(66)*	*9559.30(98)*
		7899.08(37)	7844.70(37)	9022.81(49)	11268.59(50)	10735.880(20)	9558.290(50)
74	W	*8397.620*	*8335.191*	*9672.091*	*12098.75(57)*	*11542.40(73)*	*10204.7(11)*
		8397.67(17)	8335.34(17)	9672.46(10)	12099.74(87)	11538.6(16)	10200.1(12)
78	Pt	*9442.241*	*9361.936*	*11070.537*	*13880.42(80)*	*13273.9(23)*	*11565.8(12)*
		9442.39(32)	9361.97(21)	11070.85(29)	13880.70(30)	13271.900(30)	11562.760(20)

*These values are for the unresolved $L\alpha_1$ and $L\alpha_2$ emission lines.

TABLE 2. continued

Z	Sy	$L\alpha_1$	$L\alpha_2$	$L\beta_1$	L_I abs. edge	L_{II} abs. edge	L_{III} abs. edge
79	Au	*9713.345*	*9628.141*	*11442.029*	*14353.6(14)*	*13735.68(84)*	*11920.6(13)*
		9713.44(34)	9628.06(33)	11442.45(47)	14355.30(50)	13734.200(70)	11919.700(60)
82	Pb	*10551.246*	*10449.393*	*12612.892*	*15862.1(25)*	*15198.90(74)*	*13035.10(87)*
		10551.60(27)	10449.60(65)	12613.81(57)	15858.00(10)	15199.000(30)	13035.070(30)
83	Bi	*10838.721*	*10730.807*	*13022.865*	*16388.50(94)*	*15712.02(59)*	*13419.95(85)*
		10838.95(28)	10731.06(14)	13023.66(18)	16376.0(32)	15719.8(29)	13426.7(22)
90	Th	*12966.976*	*12808.505*	*16199.823*	*20473.3(14)*	*19697.2(25)*	*16300.2(32)*
		12968.87(60)	12809.69(39)	16202.40(28)	20462.5(50)	19682.9(46)	16298.5(32)
92	U	*13614.562*	*13438.242*	*17218.676*	*21755.95(62)*	*20945.70(54)*	*17164.48(52)*
		13614.88(20)	13438.98(19)	17220.16(28)	21770.4(57)	20946.5(52)	17164.9(35)

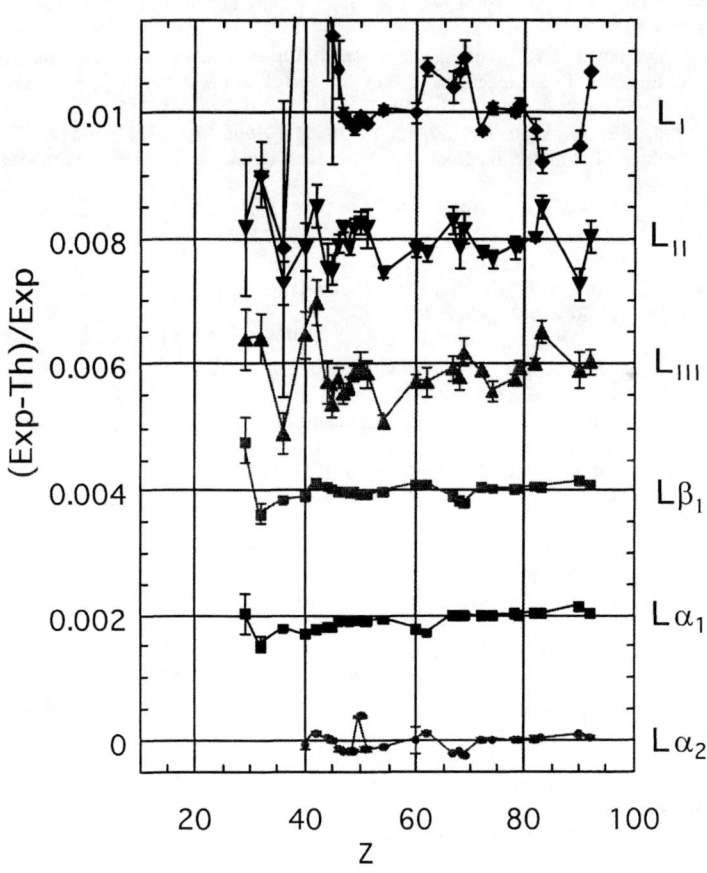

Figure 2. Relative deviations of theory and experiment for L-series transitions.

REFERENCES

1. Deslattes, R. D., Kessler, E. G., Jr., Indelicato, P. & Lindroth, E. (1998) in *International Tables for Crystallography*, eds. Wilson, A. J. C. & Prince, E. (Kluwer Academic Publishers, Dordrecht/Boston/London), Vol. C, Mathematical, Physical and Chemical Tables.
2. Deslattes, R. D., Kessler, E. G., Jr., Indelicato, P. & Lindroth, E. (1998) *J. Phys. Chem. Ref. Data* .
3. Cauchois, Y. & Hulubei, H. (1947) *Tables de constantes et donnees numeriques. I. Longueurs d'onde des emissions X et des discontinuites d'absorption X.* (Herman, Paris).
4. Cauchois, Y. & Senemaud, C. (1978) *Tables internationales de constantes selectionnees. 18. Longueurs d'onde des emissions X et des discontinuites d'absorption X.* (Pergamon Press, London).
5. Bearden, J. A. (1967) *Rev. Mod. Phys.* **39**, 78-124.
6. Deslattes, R. D. & Henins, A. (1973) *Phys. Rev. Lett.* **31**, 972-975.
7. Kessler, E. G., Jr., Deslattes, R. D., Girard, D., Schwitz, W., Jacobs, L. & Renner, O. (1982) *Phys. Rev. A* **26**, 2696-2706.
8. Beyer, H., Indelicato, P., Finlayson, H., Liesen, D. & Deslattes, R. D. (1991) *Physical Review A* **43**, 223-227.
9. Becker, P., Dorenwendt, K., Ebeling, G., Lauer, R., Lucas, W., Probst, R., Rademacher, H.-J., Reim, G., Seyfried, P. & Siegert, H. (1981) *Physical Review Letters* **46**, 1540-1543.
10. Becker, P., Seyfried, P. & Siegert, H. (1982) *Zeits. Phys. B* **48**, 17-21.
11. Härtwig, J., Hölzer, G., Wolf, J. & Förster, E. (1993) *J. Appl. Cryst.* **26**, 539-548.
12. Härtwig, Hölzer, G., Förster, E., Goetz, K., Wokulska, K. & Wolf, J. (1994) *Phys. Stat. Sol. A* **143**, 23-34.
13. Bearden, J. A., Thomsen, J. S., Burr, A. F., Yap, F. Y., Huffman, F. N., Henins, A. & Matthews, G. D. (1964) (The Johns Hopkins University, Baltimore, Maryland 21218).
14. Bearden, A. J. (1960) *Physical Review Letters* **4**, 240-241.
15. Lum, G. K., Wiegand, C. E., Kessler, E. G., Jr., Deslattes, R. D., Jacobs, L., Schwitz, W. & Seki, R. (1981) *Phys. Rev. D* **23**, 2522-2532.
16. Kraft, S., Stümpel, J., Becker, P. & Kuetgens, U. (1996) *Rev. Sci. Instrum.* **67**, 681-687.
17. Bearden, J. A. & Burr, A. F. (1967) *Rev. Mod. Phys.* **39**, 125-142.
18. Powell, C. J. (1995) *App. Surf. Sci.* **89**, 141-149.
19. Huang, K.-N., Aoyagi, M., Chen, M. H., Crasemann, B. & Mark, H. (1976) *Atomic Data and Nuclear Data Tables* **18**, 243-.
20. Deslattes, R. D. (1978) *Japan. Jour. Appl. Phys.* **17-2**, 1-6.
21. Deslattes, R. D., Kessler, E. G., Jr., Jacobs, L. & Schwitz, W. (1979) *Physics Letters* **71A**, 411-414.
22. Chen, M. H., Crasemann, B., Aoyagi, M., Huang, K.-N. & Mark, H. (1981) *Atomic Data and Nuclear Data Tables* **26**, 561-574.
23. Grant, I. P., MacKenzie, B. J., Norrington, P. H., Mayers, D. F. & Pyper, N. C. (1980) *Computer Physics Communications* **21**, 207.
24. Kessler, E. G., Jr. & Deslattes, R. D. (1986) *Physica Scripta* **34**, 408-412.
25. Indelicato, P. & Lindroth, E. (1992) *Physical Review A* **46**, 2426-2436.
26. Mooney, T., Lindroth, E., Indelicato, P., Kessler, E. & Deslattes, R. D. (1992) *Physical Review A* **45**, 1531-1543.
27. Lindroth, E. & Indelicato, P. (1993) *Physica Scripta* **T46**, 139-143.
28. Lindroth, E. & Indelicato, P. (1994) *Nuclear Instruments and Methods in Physics Research B* **87**, 222-226.
29. Indelicato, P. & Lindroth, E. (1996) *Comments in Atomic and Molecular Physics* **32**, 197-208.
30. Bosch, F. (1992) *Nuclear Instruments and Methods in Physics Research* **A314**, 269-276.
31. Indelicato, P. & Desclaux, J. P. (1990) *Physical Review A* **42**, 5139-5149.

32. Desclaux, J. P. (1975) *Computer Physics Communication* **9,** 31-45.
33. Desclaux, J. P. (1993) in *Methods and Techniques in Computational Chemistry*, ed. Clementi, E. (STEF, Cagliary), Vol. A.
34. Blundell, S. A. (1993) *Physical Review A* **47,** 1790-1803.
35. Indelicato, P. (1990) in *X-Ray and Inner-Shell Processes*, eds. Carlson, T. A., Krause, M. O. & Manson, S. T. (AIP, New-York), Vol. 215, pp. 591-601.
36. Johnson, W. R. & Soff, G. (1985) *Atomic Data and Nuclear Data Tables* **33,** 405.
37. Mohr, P. J. (1974) *Annals of Physics* **88,** 26-51.
38. Mohr, P. J. (1974) *Annals of Physics* **88,** 52-87.
39. Mohr, P. J. (1975) *Physical Review Letters* **34,** 1050-1052.
40. Mohr, P. J. (1982) *Physical Review A* **26,** 2338-2354.
41. Mohr, P. J. & Kim, Y.-K. (1992) *Physical Review A* **45,** 2727-2735.
42. Mohr, P. J. & Soff, G. (1993) *Physical Review Letters* **70,** 158-161.
43. Uehling, E. A. (1935) *Physical Review* **48,** 55-63.
44. Wichmann, E. H. & Kroll, N. M. (1956) *Physical Review* **101,** 843-859.
45. Indelicato, P. & Mohr, P. J. (1990) in *12th International Conference on Atomic Physics*, eds. Baylis, W. E., Drake, G. W. F. & McConkey, J. W., An Arbor (Michigan)).
46. Indelicato, P. & Mohr, P. J. (1991) *Theoretica Chimica Acta* **80,** 207-214.
47. Indelicato, P., Gorceix, O. & Desclaux, J. P. (1987) *Journal of Physics B: Atomic, Molecular and Optical Physics* **20,** 651.
48. Kim, Y. K., Baik, D. H., Indelicato, P. & Desclaux, J. P. (1991) *Physical Review A* **44,** 148-166.
49. Blundell, S. A., Mohr, P. J., Johnson, W. R. & Sapirstein, J. (1993) *Physical Review A* **48,** 2615-2626.

The Critical Assessment of Atomic Transition Probabilities

W. L. Wiese and D. E. Kelleher

*National Institute of Standards and Technology
Gaithersburg, Maryland 20899-0001 USA*

1. Introduction—Reference Data

Since the early 1960s, the National Institute of Standards and Technology (NIST), formerly the National Bureau of Standards, has been engaged in the critical compilation of atomic transition probabilities (1-5). The first step in the data tabulations is always the collection of all available literature sources, followed by a second, much more time consuming step of critically evaluating, comparing and tabulating the data according to an assessment procedure reviewed in this paper. Based on this assessment, we retain for our NIST data tables only the higher quality material, i.e., those data we call reference data. The assessment is complicated by the circumstance that for the data obtained from atomic structure calculations, which constitute the large majority of the existing data, no intrinsic error estimates are provided. Reference data must possess at least three characteristics:

1. They are obtained through critical assessment, and one final numerical value has been selected per spectral line, which is either from the most advanced approach or is the average of several approaches that are of comparable quality. [In principle, one may also include results of lesser quality in the average. This may be done by utilizing smaller weights for such data. However, maintaining consistency from spectrum to spectrum becomes very difficult, and the averaged results may often get worse with lesser-quality numbers mixed in.]

2. The approximate uncertainty of the selected values must be estimated in a reliable manner, despite the above noted fact that the theoretical data generally come without assigned uncertainties.

3. The selected data must be of acceptable quality. Based on the needs of numerous users, we have not included data which has a possible relative error of greater than ±50 %. This limit has been kept by us over the years, and it remains to be a useful cut-off for many applications.

2. The Principal Methods for Determining Transition Probabilities

During the period of our data assessment and compilation work from the early 1960s to the present, a significant change and consolidation has occurred in the methods to produce transition probabilities. In the 1960s 10 different methods were actively pursued (emission, absorption, anomalous dispersion ("hook" method), phase-shift lifetime, delayed-coincidence life, beam foil life, semi-empirical, and single- as well as few-configuration Hartree-Fock calculations). Nowadays theoretical methods involving many configurations deliver more than 90 % of the fair-to-excellent quality data, at least for the light elements up to about $Z=20$. Examples include multi-configuration Hartree-Fock, configuration interaction, and modified close-coupling techniques. For highly charged ions and heavier elements, fully relativistic calculations with multiple configurations become more appropriate. Of the experimental approaches only two are quite active and are often used in combination: Emission measurements (often branching-ratio measurements only) and laser-induced-fluorescence measurements, i.e., lifetime measurements by selective excitation. For the lighter elements from hydrogen to calcium, most experimental results are for the prominent lines of the neutral atoms and singly charged ions, and these data serve mainly to provide test material for the much more extensive data from the calculations. For the neutral atoms and singly charged ions of heavier elements, experimental data from emission and lifetime measurements are the data of choice and are fairly plentiful for the more common elements, such as, Ti, Cr, Fe, Ni, Cu, Kr, Mo, Ta and W.

In view of this situation, only the advanced calculational methods and the emission and lifetime experiments will be critically reviewed. We note here again that calculated data do not come with error estimates, but the experimental data do.

3. The Four General Assessment Criteria

At NIST, we have developed, and gradually refined, a system of data assessment which is based on the following four principal criteria (2,5):

1. Consideration of the *critical factors* of a method by the authors (as discussed below).

2. The authors' estimates of the uncertainties of their measured or calculated data.

3. The degree of agreement with other reliable data, based on tabular or graphical comparisons.

4. The fit of the data into systematic trends, or deviations from them.

4. The Critical Factors

Each of the approaches to determine transition probabilities contains "critical factors" which need to be adequately taken care of in order to obtain reliable results. If one or more of these critical factors are not taken into account, we consider the approach deficient and not capable of producing accurate results. We do not include such deficient experiments or calculations in the selected data if other non-deficient results are available. However, if only one or two of the critical factors is not addressed and if, furthermore, no other, better data are available, we also utilize such data with accordingly adjusted larger uncertainty estimates. But since we strive to tabulate reference data, we generally do not include data which we estimate to have a possible relative error of more than ±50 %.

We shall now discuss the critical factors for the three methods that currently produce either the very large majority of the data or the high-quality test material.

4.1 Critical Factors in the Advanced Atomic Structure Calculations

In quantum mechanical calculations, the main critical factors are the following:

4.1.1 *Configuration interaction or multi-configuration treatments.* Electron correlation (6,7) needs to be addressed in great detail in order to make accurate calculations of the energy levels and of the electric dipole moments possible. At least tens or even hundreds of the atomic basis states, which are expected to contribute to the atomic levels in question usually need to be included in the calculations in order to make the results accurate.

4.1.2 *Near-Coincidences.* Special attention must be given to cases where interacting levels nearly coincide in energy since configuration interaction then becomes especially pronounced.

4.1.3 *Cancellation.* Special attention, that is even more extensive configuration interaction treatments, must be provided for lines for which the positive and negative contributions to the transition integral severely cancel each other and which, as a result, possess rather small line strengths.

4.1.4 *Spectroscopic Coupling Schemes and Relativistic Corrections.* For calculations of the strengths of individual spectral lines in multiplets, the fine structure situation needs to be carefully addressed (7). LS-coupling is an excellent approximation when the electrostatic repulsion between the electrons is dominant, as is the case for many transitions of light-element spectra. But this needs to be checked, and generally the Breit-Pauli relativistic terms describing spin-orbit, spin-spin, spin-other-orbit, etc. interactions should be applied to calculations of individual line strengths, especially transitions involving higher angular momenta, such as p-d, d-f, etc. Fully relativistic approximations in the framework of multi-configuration Dirac-Fock theory are required for heavy atoms and highly charged ions.

4.1.5 *Convergence of Results and of the Dipole Length and Velocity Forms.* In some recent calculations, intermediate results are shown at various levels of approximation (8,9). As more interacting configurations were added, the results clearly converged. Also, the differences between the dipole length and velocity forms of the transition integral decreased gradually. Thus, this convergence as well as the close agreement of the dipole length and velocity forms are necessary, but not sufficient, conditions for accurate calculational results.

4.1.6 *Transition Energies.* The calculated transition energies should be optimized to agree with, or at least very closely match, the experimental energies. Furthermore, for the conversions of line strengths to transition probabilities and oscillator strengths, the usually available, very accurate experimental transition energies should be used.

4.2 Emission Experiments

The main critical factors here are (2,5):

4.2.1 *A Valid Plasma Model.* The plasma model applied to describing the discharge conditions must fulfill the pertinent validity criteria. In practice, the most suitable models are those of full or partial local thermodynamic equilibrium (LTE or PLTE), for which the density of free electrons is the critical quantity.

4.2.2 *Absence of Self-Absorption.* The lines must be emitted from an optically thin plasma layer, i.e., no self-absorption should occur. However, small amounts of optical depth are acceptable since accurate corrections to the case of an optically thin layer may then be carried out.

4.2.3 *Spectral Calibration*. The observed intensities must be carefully calibrated with modern radiometric standards. Accurate VUV calibrations are still difficult to achieve, as discussed in the recent literature.

4.3 Lifetime Measurements

4.3.1 *Selective Excitation*. The selective excitation of an atomic level is a critical requirement for measuring its radiative lifetime, and is usually achieved by utilizing a laser tuned to an appropriate transition which pumps electrons into this atomic level. If the excitation is non-selective, higher atomic levels will be excited, too, and cascading of electrons into the level under study will occur. In this case, a proper correction for cascading must be carried out. A successful correction procedure is the "Arbitrarily Normalized Decay Curve" (ANDC) technique (10).

4.3.2 *Collisional Effects and Radiation Trapping*. True radiative lifetimes may only be obtained when collisions and radiation trapping are eliminated. Since these two processes are strongly density-dependent, their effects may be taken into account by carrying out measurements at different densities and extrapolating to zero density.

4.3.3 *Blends*. The wavelength position of the line used for the radiative decay measurement should be closely checked for the possibility of accidental overlap with another line.

4.3.4 *Polarization Effects and Quantum Beats*. When a level is selectively excited by a tuned laser (laser induced fluorescence), quantum beats due to magnetic fields as well as polarization effects must be taken into account and minimized.

5. The NIST Assessment Procedure

5.1 General Remarks

NIST is in the process of critically re-evaluating and re-compiling transition probabilities of H to Ca (Z=1 to 20), elements published about 30 years ago in Vols. I and II of "Atomic Transition Probabilities" by W. L. Wiese et al. Many new data have been produced since then, mostly theoretical, with by far the most extensive results being provided by members of the "Opacity Project" (OP) (11). In 1996, a first installment of the new NIST compilations, "Atomic Transition Probabilities of Carbon, Nitrogen, and Oxygen" was published by W. L. Wiese, J. R. Fuhr, and T. M. Deters(5). Work on updating the other elements is continuing, and revision of Vol. II is currently focused on Na, Mg, Al, and Si (Z=11 to 14). Our new compilations contain roughly an order of magnitude more transitions than the earlier ones.

As the first step of the assessment, we obtain all pertinent literature sources from the NIST data center files. These contain every paper listed in the comprehensive NIST Bibliography on Atomic Transition Probabilities, which is constantly updated (12).

5.2 An example: O I

For the spectrum of O I we found 90 references for the allowed (electric dipole) transitions for the period from 1966, the date of our first critically evaluated tables (1), to 1996, the date of our new tables (5) containing O I. From this group of 90 papers, we excluded all data sources in which outdated approaches were used. We also excluded papers that have been superseded by more advanced approaches of the same or a similar type. Finally, we have excluded preliminary short communications or papers in conference proceedings that preceded a full, detailed publication.

This filtering greatly reduced the number of papers. A further limitation by the requirement that all pertinent "critical factors" are suitably taken care of reduces the number of publications to 5 multiconfiguration calculations and 8 experiments (4 emission, 1 absorption, 1 electron scattering and 2 lifetime experiments). Of these, the multi-configuration calculations by the Opacity Project group (11), based on the Close-Coupling (CC) approach in conjunction with the R-matrix technique, are by far the most extensive ones, delivering most of the data for the new NIST tables. The Hibbert et al. (13) configuration-interaction calculations with the CIV3 code are equally advanced but address far fewer lines. Two very similar calculations (14,15) preceded the OP and CIV3 work. Another multi-configuration approach by Biemont and Zeippen (16) has utilized still a different code, SuperStructure. All five calculations agree very well, but we have utilized only the most advanced results from each of the three codes.

Fig. 1 shows a comparison of the OP and CIV3 (11,13) results for the O I multiplet data. Clearly, the agreement is very good, with the exception of three weak multiplets. The 8 selected experiments all cover only comparatively few prominent lines. But the above cited calculations come, as usual, without intrinsic uncertainty estimates, while the experimental data do. Therefore, the experimental results serve the critical function to provide reliable comparison data to test the accuracy of the calculations. In Fig. 2 a comparison is made between the emission, absorption and electron scattering data (17-22) and theory. The great majority of the experimental data is estimated to be uncertain by no more than ±15 %. However, a comparison of the calculations with the best available (cascade-free) lifetime data (23,24) in Fig. 3, shows some discrepancies far outside the experimentally estimated uncertainties. These discrepancies remain unresolved.

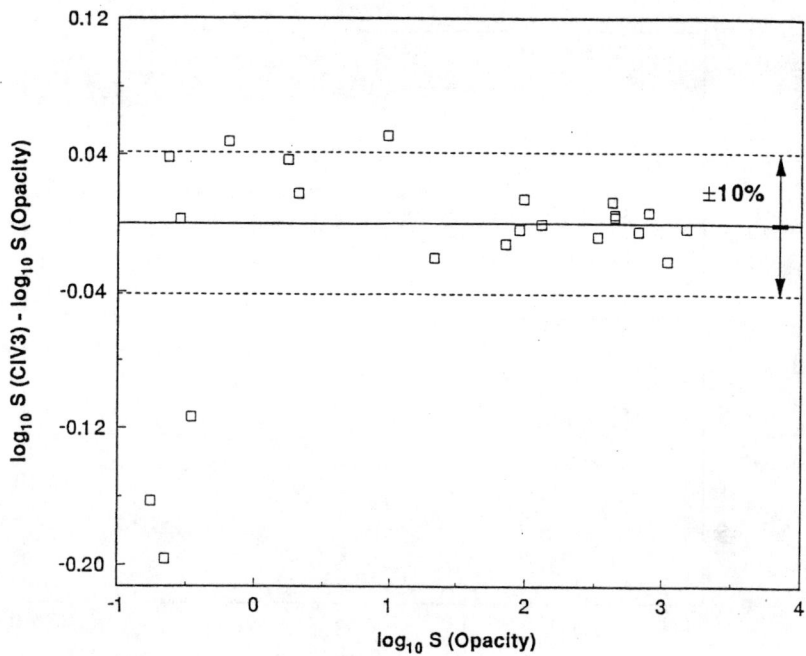

Fig. 1. Comparison of the OP multiplet data for O I by Butler and Zeippen (11) with the CIV 3 results of Hibbert et al. (13). The ratios of the multiplet line strengths, S(CIV 3)/S(OP) are plotted on a logarithmic scale versus the log S(OP). The figure indicates that for all but three weak multiplets the range of deviation for the ratios stays within ± 10%, as indicated by the broken lines.

6. A brief review of the important Opacity Project:

Opacity Project results (25) deserve a special review and assessment because they are so extensive and because they appear to be reasonably accurate in many cases. We thus rely heavily on the OP data for our new compilations. OP calculations cover an extensive range of LS-allowed transitions, and are essentially comprehensive up to n=10 and $l = 3$, with some transitions included with $l = 4$. OP results have been made electronically available for all stages of H to Ca, except P, Cl, and K, on which future work is planned.

Currently published OP calculations do not, however, include the spin-orbit or other relativistic effects, and only LS multiplet values are reported. In order to obtain data for individual spectral lines, we decompose the OP multiplets into their LS fine structure components using LS coupling rules. We include only those transitions for which experimental wavelengths or experimental energies are tabulated in the NIST

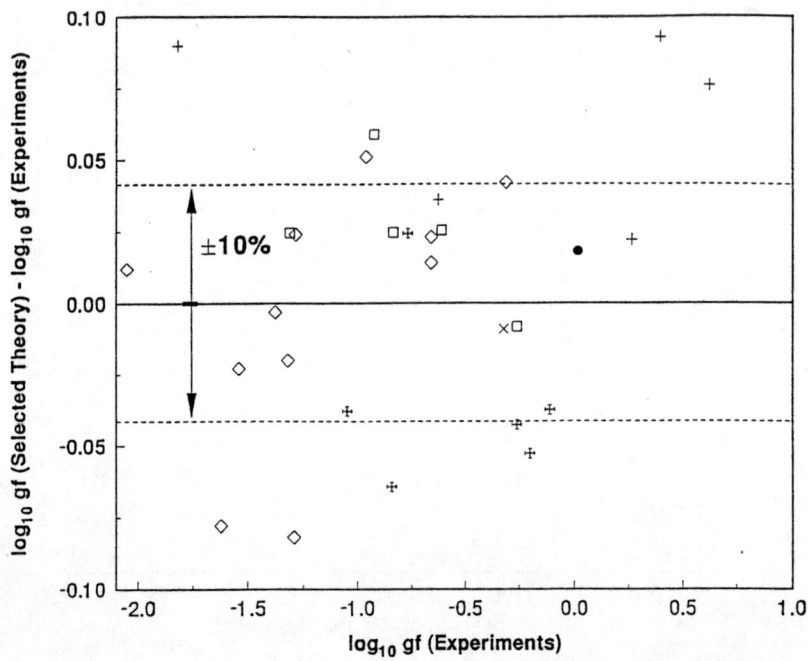

Fig. 2. Comparison of O I experimental log gf data with the selected theoretical results. The theory/experiment ratios of the gf-values are plotted on a logarithmic scale versus the log gf of the various experiments. The data from the *emission* experiments by Solarski and Wiese (18) are denoted plus signs (+), those of Ott (19) by squares; Wiese and Shumaker (20) by a filled circle; and Goldbach and Nollez (21) by diamonds. Also shown is a data point (X) from an *absorption* experiment by Jenkins (22), and the results of an *electron scattering* experiment by Doering et al. (17) (crosses). It is seen that about two-thirds of the ratios do not deviate by more than ± 10% (shown by the broken lines) from a perfect ratio of 1.00.

Atomic Energy Level (AEL) tables for both the lower and upper level, and whose terms are described using LS coupling in the AEL tables. We use the line strengths from the Opacity Project, and then apply experimental wavelengths or energy level data to convert OP line strengths to oscillator strengths and transition probabilities.

For most strong transitions as well as many weaker transitions in ionized species, reasonably good agreement is found between the Opacity Project (extended as described above) and experiments and other accurate calculations which include spin-orbit effects. Larger discrepancies are encountered for many weaker transitions in such neutral spectra as C I, N I, for lower ions such as N II and O II, and for halogen-like

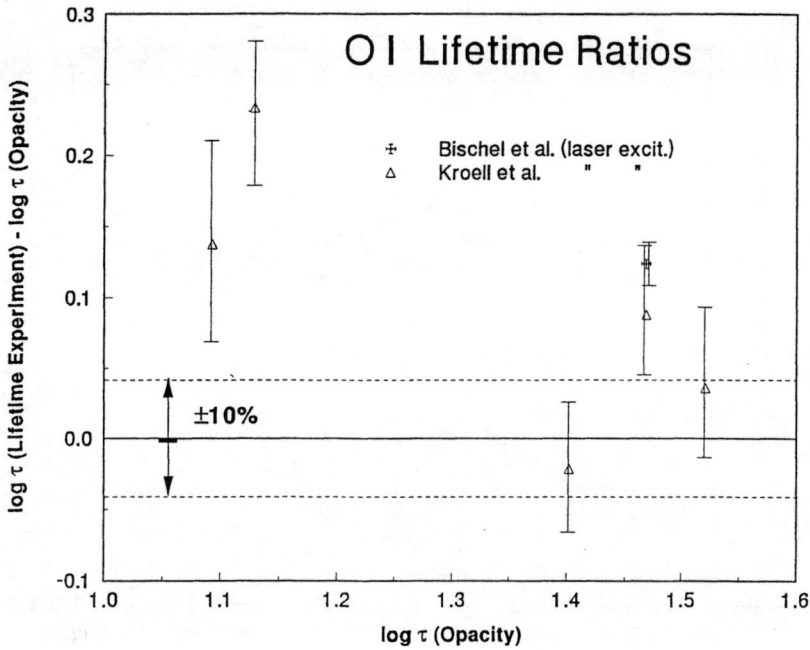

Fig 3. Comparison of O I lifetime results by Bischel et al. (23) (cross) and Kröll et al. (24) (triangles) with the OP data by Butler and Zeippen (11). The ratios of the experiment to the theoretical lifetimes are plotted on a logarithmic scale versus the log τ (OP) (τ in ns). The experimental uncertainties are indicated by error flags.

and noble-gas-like spectra. Spin-Orbit and other relativistic effects in the np^5 ion core presumably account for the bulk of the noble-gas-like discrepancies. The origin of the large discrepancies between OP(26) and CIV3(27) in fluorine-like spectra of Na III and Si VI (see Figs. 4 and 5), however, is not yet established. Comparisons in the figures only include transitions for which the outermost electron of the upper level has an n=3 principal quantum number, because only these are listed in (27).

The discrepancy in line strengths for fluorine-like spectra between OP and CIV3 results deserves special attention not only because these methods represent two of the most important methods to-date for large-scale line strength calculations, but also because the OP method relies on CIV3 (or the "SuperStructure" code, which, though quite different, gives similar results) for its foundation, as discussed below. All but the lowest three bound levels of Na III have configurations of the type $2s^2 2p^4(\ )nl$, where the parentheses indicate a parent term of either ^1S, ^3P, or ^1D. Bound perturbing levels,

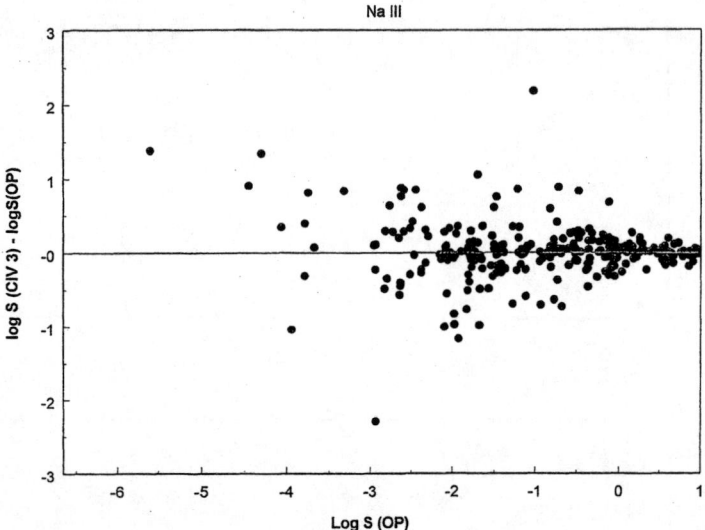

Fig. 4. Logarithms of the ratios of line strengths S(CIV 3)/S(OP) are plotted versus log S(OP) for fluorine-like Na III, from refs. (26) and (27). All transitions involve $2s^22p^43\ell$ upper levels.

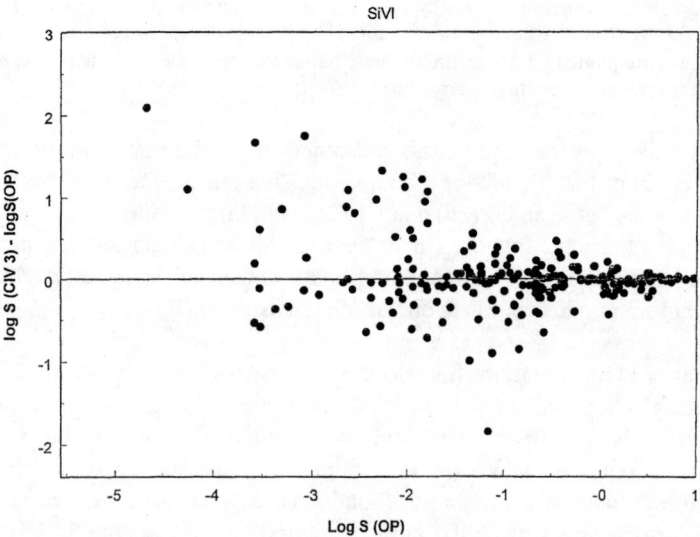

Fig. 5. Logarithms of the ratios of line strengths S(CIV 3)/S(OP) are plotted versus log S(OP) for fluorine-like Si VI, from refs. (26) and (27). All transitions involve $2s^22p^43\ell$ upper levels.

such as the $2s2p^5nl$ levels, exist only for the more highly ionized members of the isoelectronic series. The three parent configurations cited above lie sufficiently close in energy that both intermediate coupling and configuration interaction occur to noticeable and frequently comparable extents throughout the bound spectrum. Relative Landé intervals of the quartet states (relative intervals for doublets cannot be derived from energies) are typically far from those expected for LS coupling.

Because of the importance of CIV3- and CC- type calculations, and their relationship, it appears important to pursue the reasons for the surprisingly large discrepancies between the two methods for the spectra seen in the figures, as well as other light-element spectra. The agreement is good for the largest line-strengths, S, but rapidly deteriorates for $S < 1$. One question is whether this discrepancy could be due primarily to the fact that intermediate coupling is included in the CIV3 calculations, but not in the OP calculations (where in this case used the SuperStructure code was used to describe the ion core). Recently the Belfast group has included intermediate coupling via a Breit-Pauli treatment of relativistic effects into their OP codes. The differences between results from this new treatment for a subset of fluorine-like Na III transitions (28) and CIV3 results remain large.

Summary

In closing we wish to emphasize the importance of accurate measurements and state-of-the-art calculations of transition strengths, especially transitions such as spin-forbidden ones whose strengths are sensitive to cancellation effects.

References

1. W. L. Wiese, M. W. Smith and B. M. Glennon, "Atomic Transition Probabilties—Hydrogen through Neon," Vol. I, NSRDS-NIBS 4 (1966).

2. W. L. Wiese, M. W. Smith and B. M. Miles, "Atomic Transition Probabilties—Sodium through Calcium," Vol II, NSRDS-NIBS **22** (1969).

3. G. A. Martin, J. R. Fuhr, and W. L. Wiese, "Atomic Transition Probabilities—Scandium through Manganese," J. Phys. Chem. Ref. Data **17**, Supplement No. 3 (1988).

4. J. R. Fuhr, G. A. Martin, and W. L. Wiese, "Atomic Transition Probabilties—Iron through Nickel," J. Phys. Chem. Ref. Data **17**, Supplement No. 4 (1988).

5. W. L. Wiese, J. R. Fuhr and T. M. Deters, "Atomic Transition Probabilities of Carbon, Nitrogen, and Oxygen," J. Phys. Chem. Ref. Data, Monograph 7 (1996).

6. A. W. Weiss, in "Advances in Atomic and Molecular Physics" **9**, 1 Acad. Press, New York (1973).

7. R. D. Cowan, "The Theory of Atomic Structure and Spectra," Univ. of Calif. Press, Berkeley (1981).

8. A.W. Weiss, Phys. Rev. A **51**, 1067 (1995).

9. J. Olsen, M. R. Godefroid, P. Jönsson, P. A. Malmqvist, and C. Froese Fisher, Phys. Rev. E **52**, 4499 (1995).

10. L.J. Curtis, H.G. Berry, and J. Bromander, Phys. Lett. **34A**, 169 (1971).

11. K. Butler and C. J. Zeippen, J. Phys. IV Colloq. Cl, 1,C1-141 (1991).

12. J. R. Fuhr and H.R. Felrice, "Atomic Transition Probability Bibliographic Database," accessible on the WWW, address: http://physics.nist.gov/fvalbib.

13. A. Hibbert, E. Biémont, M. Godefroid, and N. Vaeck, J. Phys. **B24**, 3943 (1991).

14. K. L. Bell and A. Hibbert, J. Phys. B **23**, 2673 (1990).

15. A. K. Pradhan and H. E. Saraph, J. Phys. B **10**, 3365 (1977).

16. E. Biémont and C. J. Zeippen, Astron. Astrophys. **265**, 850 (1992).

17. J. P. Doering, E. E. Gulcicek, and S. O. Vaughan, J. Geophys. Res., Space Phys. **90**, 5279 (1985).

18. J. E. Solarski and W. L. Wiese, Phys. Rev. **135**, A1236 (1964).

19. W. R. Ott, Phys. Rev. A **4**, 245 (1971).

20. W. L. Wiese and J. B. Shumaker, Jr., J. Opt. Soc. Am. **51**, 937 (1961).

21. C. Goldbach and G. Nollez, Astron. Astrophys. **284**, 307 (1994).

22. D.B. Jenkins, J. Quant. Spectrosc. Radiat. Transfer 34, 55 (1985).

23. W. K. Bischel, B. E. Perry, and D. R. Crosley, Chem. Phys. Lett. **82**, 85 (1981).

24. S. Kröll, H. Lundberg, A. Persson, and S. Svanberg, Phys. Rev. Lett. **55**, 284 (1985).

25. The Opacity Project Team, "The Opacity Project - Vol. I," Institute of Physics, Bristol, England (1995).

26. K. Butler and C. J. Zeippen, to be published. The data is accessible via "TOPBASE", an online OP database.

27. H. M. S. Blackford and A. Hibbert, At. Data Nucl. Data Tables **58**, 101 (1994).

28. Keith Berrington, private communication.

Major Data Activities and Services

Atomic Collision Databases and Data Services – A Survey

David R. Schultz

Controlled Fusion Atomic Data Center[1], Oak Ridge National Laboratory
Oak Ridge, TN 37831-6373

Abstract. Atomic collision databases and data services constitute an important resource for scientific and engineering applications such as astrophysics, lighting, materials processing, and fusion energy, as well as an important knowledge base for current developments in atomic collision physics. Data centers and research groups provide these resources through a chain of efforts that include producing and collecting primary data, performing evaluation of the existing data, deducing scaling laws and semiempirical formulas to compactly describe and extend the data, producing recommended sets of data, and providing convenient means of maintaining, updating, and disseminating the results of this process. The latest efforts have utilized modern database, storage, and distribution technologies including the Internet and World Wide Web. Given here is an informal survey of how these resources have developed, how they are currently characterized, and what their likely evolution will lead them to become in the future.

INTRODUCTION

Around the world, data centers and other more loosely defined organizations of data producers and users are participating in a revolutionary transformation in the manner in which data is exchanged. In particular, the advent of the widespread use and great utility of the World Wide Web has given birth to efforts in which data sets and databases are being converted to, and specifically created for, on-line usage. Benefits of this trend are the great ease that contemporary electronic storage and retrieval technology affords for disseminating data and the potential it holds for improving the ability to intercompare and otherwise manipulate data from central repositories.

Indeed, at present many such sites through which data can be retrieved exist. However, the Web's ubiquity and democracy (rooted in the ability of any interested party to post or download data), drives not only the beneficial increase of readily available data, but can also suffer from the pitfalls associated with the lack of

---

[1] The CFADC is supported by the US DOE Office of Fusion Energy Sciences through a grant to ORNL, managed by Lockheed Martin Energy Research Corp. under Contract No. DE-AC05-96OR22464.

authoritative evaluation of data. Whereas pure research may thrive on data which can be treated as informed opinion, applications require data which has an assessed reliability.

Providers of atomic collision data have certainly been active participants in all aspects of this movement and surveyed here will be some of the databases and data services presently available. Also, critical issues such as the role of data centers in collecting, evaluating, and disseminating data, standardization of data formats to facilitate exchange, integration of dispersed and fragmentary data resources, and development of convenient systems for searching and retrieving data, will be emphasized (see also Ref. [1]). However, to begin with, it will be helpful to briefly outline the rationale behind the establishment of atomic collision databases and data services, to attempt to define their role in scientific and engineering endeavors, and to generally describe the evolutionary track they have followed.

Perhaps most evident is their reason for being: the collection and preservation of data which are of use both in atomic physics and in applications of atomic physics. That is, progress is facilitated if needs arising for particular atomic collision data, or, in fact, sets of atomic collision data, can be satisfied without the concomitant need to produce and evaluate the data anew. Furthermore, the collection and organization of pre-existing data facilitates the evaluation of newly created information and can be used to identify the needs for new or improved data. Beyond only collecting data, data centers and individuals or groups participating in these activities seek to provide guidance to non-expert users of data through evaluation and recommendation.

Evaluation is the assigning of a degree of expected reliability to the available data based on expert opinion and derived from consideration of a set of criteria such as the robustness of the theoretical method or soundness of experimental method through which the data were obtained and their consistency with well established physical behavior or asymptotic limits. By recommendation, one means a further level of expert evaluation, usually involving the intercomparison of all the relevant available data, resulting in data or data sets which can be utilized by the research and application communities with a high level of confidence. Often these activities result in the construction of fitted forms for tabulated reaction cross sections, scaling laws, or other useful semiempirical formulas which can be valuable to fill in gaps in the required, but not existing, data. It should also be recognized that some applications require huge amounts of atomic data which could not feasibly be individually determined by the state-of-the-art theoretical or experimental methodologies, but rather must be produced by large-scale production computer codes. In this case, it is crucial to know the expected limits of applicability and reliability of the data generated through comparison with other theory and experiment for particular benchmark tests.

Thus, the chain of activities which leads from fundamental data to those maintained by data centers or other providers begins with the collection of relevant literature references often through bibliographies, manual or electronic digitization of pertinent data, storage and organization of the data, efforts to evaluate and determine recommendations, and dissemination and periodic updating of the results. Data is distributed through a wide range of media including peer review journals, internal reports, published books, internal databases, and Internet and World Wide Web (WWW or Web) resources. As already noted, the modern tools of information technology have had a significant and growing impact on these activities. For example, bibliographies are now often on-line, electronic mail and file transfer protocol (ftp) facilitate obtaining electronic files containing data directly from authors, scanners and digitizers greatly speed up data entry, modern databases and graphical tools help in storage, manipulation, and display of data aiding evaluation efforts, and the Internet and WWW provide a practically universal method for dissemination. This trend represents not simply a change in the medium of distribution, but, moreover, the ability to implement new capabilities such as ready means for updating and interlinking data resources.

These observations may be more fully illustrated and amplified upon by considering the history, present status and organization, and expected future regarding atomic collision bibliographic services and numerical data resources.

THE EVOLUTION OF ATOMIC COLLISION BIBLIOGRAPHIC SERVICES

The starting point for many projects seeking to provide required atomic collision data, or seeking to assess or update the present status of the data available, is the primary scientific literature. Guides to this literature, annotated and categorized bibliographies, have for a long time provided the means to find the relevant publications. For example, beginning in the early 1960's the Controlled Fusion Atomic Data Center (CFADC) at Oak Ridge National Laboratory began publishing a series of reports containing bibliographic data organized into various collision reaction categories [2]. The bibliographic entries have been gathered by groups of experts who search a range of technical journals numbering up to more than one hundred and have been stored on mainframe computers as data files to facilitate the printing of the bibliographic publications. The original series was followed by several others until 1986 [3–6], merged with efforts of other data centers, and is presently extended through an on-line database.

In particular, subsequent to 1977, this data was organized into a formal relational database accessible over a local area network and was contributed to the efforts initiated at the International Atomic Energy Agency (IAEA) to publish the "International Bulletin on Atomic and Molecular Data for Fusion" [7], a series which continues to the present. The "Bulletin" also contains entries contributed through an international network of atomic data centers and has been used to compile convenient cumulative indices [8] for articles dating up to 1980, and from 1980 to 1987. In addition to these publications, the databases maintained at the CFADC and at the IAEA Atomic and Molecular Data Unit were searchable over local networks and requests for bibliographic data were answered by mailing or

FAXing results of searches to interested users. The advent during the 1990's of high speed, reliable national and international computer networks (the Internet) led to projects to allow access directly to users.

The result has been the development of a search engine, accessible through the CFADC's WWW link [9], by which the bibliography can be searched covering entries dating from 1978 to present. Archival entries from c. 1950 to 1977 are presently being converted to the appropriate contemporary data format and edited, and will be added to the on-line database in the future. IAEA provides an interface to their complete database through the telnet protocol [10] and an independent but similar on-line database covering primarily the years 1970 to present is maintained by a European group organized through the Université Paris-Sud, Orsay called GAPHYOR [11].

The obvious advantage of this type of interface is that rather than having a user contact a data center to make a search request and then waiting for the results to arrive by postal mail or facsimile, a number of searches refining the user's needs can be made remotely with no time delay. This leverages not only the user's time, but also that of the data center's staff who can therefore concentrate on improving the overall atomic database rather than being occupied by answering bibliographic requests. The user can then download the results of successful searches directly to his/her own computer for storage or printing. Furthermore, the computerized databases have advantages over the paper documents in that they are more convenient in use, cover many years cumulatively, and can be frequently updated.

Future versions of the Web-based bibliographic search engines will also have greatly expanded capabilities not possible with bibliographies printed on paper. For example, the traditional bibliographic searches by reaction category and reactants can be supplemented by flexible searches on any keyword within article titles, or by boolean searches combining these and other tokens in any combination. Perhaps even more significantly, they may be directly linked to other resources such as numerical data tabulations, applicable scaling laws or semiempirical formulas, sets of recommended data, and even resources such as these physically located elsewhere but linked over the WWW. Such linked resources might also include primary data in Web versions of research and archival journals. As a example of one such presently developing linkage, the CFADC bibliographic engine allows searching of the ALADDIN database (described below) of recommended collision cross sections and rates.

Thus, the collection and dissemination of bibliographic data has evolved from periodic publications, to computer data file storage on internal networks, to WWW and Internet resources. Along the way, advantages introduced through the coordination of efforts of contributing data centers and increased accessibility and flexibility of electronic interfaces have leveraged the available human resources devoted to these efforts and are leading to even greater utility for users.

THE EVOLUTION OF ATOMIC COLLISION DATA RESOURCES AND DATABASES

As one might readily expect, the evolution of numerical data collections and databases has followed a similar path to that of the bibliographies. That is, groups or individuals have gathered the available data from the primary literature, possibly also generating new theoretical or experimental data, and synthesized this information to form internal reports or publications containing graphical and numerical data and, often, fitting formulas and coefficients. Envisioning the capabilities that may soon be made possible by the integrated data resources mentioned above, methodologies used in this process could be significantly enhanced in the future. For example, Web resources might eventually facilitate locating, generating, gathering, displaying, and intercomparing data and scaling formulas, entirely electronically. Even if such new and remarkable systems are eventually realized, however, they will owe much to the the conversion of previous data collections and databases to contemporary storage and retrieval systems.

For example, since its inception in the late 1950's, the CFADC has produced compilations of recommended atomic collision data in addition to its bibliographic collections. Named for their red bindings, these "Redbook" series [12–16] have been produced as internal ORNL reports and distributed to on the order of a few hundred researchers. In light of the present data needs of applications and the present capabilities of electronic data systems, several shortcomings of this method of distributing recommended data are apparent. First, internal reports, even though distributed externally, have only a limited audience of users. Secondly, printed volumes of data are not readily updated. For example, when new data becomes available, revised or additional recommendations must be circulated using the same distribution as the original document. Third, while printed bibliographies may not be the most flexible and convenient format for potential users, volumes of recommended data have the additional deficiency that they must be scanned or retyped in order to be interfaced to application codes. A number of advances have been made that in one way or another address these shortcomings also inherent in the valuable works of other data centers and groups.

In order to facilitate and standardize the exchange of atomic, molecular, and particle-surface interaction data among international data centers, data users, and data producers, the IAEA Data Center Network adopted the ALADDIN (A Labeled Atomic Data INterface) format and database code in 1989. ALADDIN was originally developed by Russell Hulse [17,18] of Princeton Plasma Physics Laboratory using ANSI FORTRAN/77 and relying on ASCII-formatted data files. Presently, the ALADDIN standard is being reviewed by an *ad hoc* committee of the IAEA Data Center Network to determine how best to improve it, principally to provide greater flexibility in the data formats that it can accommodate, and to update the user interface. The present versions of the program suite, evaluation functions used to compute results from the stored fitting formulas, dictionary files, and data sets are available from the IAEA telnet site [10] or from the CFADC Web site [9]. Plans exist to set up three geographically dispersed mirror sites to maintain the official distributions of these and future files. A list of the recommended and evaluated data files is given in the accompanying table. Thus, with an accepted standard through which to exchange data, the "Redbooks" and certain other volumes of recommended data from several groups have been placed into the ALADDIN format. These files have been widely disseminated and form a backbone of recommended data used especially in significant portions of the international fusion energy research community.

TABLE 1. Available recommended/evaluated data sets in ALADDIN format

1. Atomic and Molecular Data for Fusion, Part I - Recommended Cross Sections and Rates for Electron Ionization of Light Atoms and Ions [19]
2. Recommended Data on Excitation of Carbon and Oxygen Ions by Electron Collisions [20]
3. Recommended Data on Atomic Collision Processes Involving Iron and its Ions [21]
4. Collisions of Carbon and Oxygen Ions with Electrons, H, H_2, and He, Atomic Data for Controlled Fusion Research, Vol. V [22]
5. Atomic and Molecular Data for Fusion, Part II - Recommended Cross Sections and Rates for Electron Ionization of Light Atoms and Ions: Fluorine to Nickel [23]
6. Recommended Data for Excitation Rate Coefficients of Helium Atoms and Helium-like Ions by Electron Impact [24]
7. Atomic and Molecular Data for Fusion, Part III - Recommended Cross Sections and Rates for Electron Ionization of Atoms and Ions: Copper to Uranium [25]
8. Elementary Processes in Hydrogen-Helium Plasmas [26]
9. Collisions of H, H_2, He, and Li Atoms and Ions with Atoms and Molecules, Atomic Data for Controlled Fusion Research, Vol. I [27]
10. Collisional Processes of Hydrocarbons in Hydrogen Plasmas [28]
11. Recommended Cross Sections for Collision Processes of Hydrogen Ground-State and Excited Atoms with Electrons, Protons, Multiply Charged Ions [29]
12. Particle Reflection from Surfaces: A Recommended Database [30]
13. Sputtering Data [31]
14. Recommended Data for Physical Sputtering [32]

In another measure, to widen the audience to which the original graphs, tables, and reference lists contained in the "Redbooks" are available, the two most requested volumes have been completely scanned into graphic file format and placed in a hypertext linked arrangement on the CFADC Web site [9]. This organization and storage will also allow future updates of individual reaction cross sections to be made in such as way that only the master copy residing on the Web site need be modified. Thus the latest updates will be directly accessible to all users. This system will also facilitate linking the graphic files and corresponding ALADDIN entries to the bibliographic search engine and other future tools for data mining and manipulation.

A different approach to providing primary atomic data and, moreover, derived data for modeling and interpreting radiating plasmas, is embodied in the Joint European Torus (JET) – Strathclyde University Atomic Data and Analysis Structure (ADAS), developed principally by Hugh Summers. ADAS is a suite of computer codes implementing a full collisional-radiative model developed for diagnosising fusion and astrophysical (solar) plasmas. It has an accompanying collection of atomic data files gathered from the literature, contributed by collaborators, and extended through JET contracts. ADAS was originally developed as a proprietary tool for fusion research at JET but an international consortium now exists to allow the use of ADAS by a much wider community. Detailed descriptions of ADAS can be found in the on-line manuals and periodic consortium bulletins, mirrored at the Strathclyde [33] and CFADC [34] Web sites. This system has many of the attributes of a state-of-the-art tool for applications of atomic data: a relatively complete data collection, routines to provide semiempirical or otherwise estimated data when the appropriate information does not exist, and a graphical user interface for data previewing, selection of processing codes and parameters, and display of results.

Other well known data projects and data collections exist which bear mention here especially to illustrate the range of organizations and scopes of coverage among them. For example, the Opacity Project, a huge undertaking to produce atomic data such as energy levels, f-values, and photoionization cross sections for astrophysically abundant species through extensive R-matrix calculations, was carried out by a large international group of researchers over a number of years. It has yielded information required to understand, as its name rightly implies, stellar envelope opacity but has found application in a wide range of uses. Information regarding the Opacity Project can be found on its Web site [35] and tables of data can be downloaded by anonymous ftp from the project's database, TOPbase [36]. Also, to provide detailed information for the astrophysically important element iron and its ions, the Iron Project has carried out work with similar goals [37]. Not only theoretical groups have followed the path of making their data collections available through the Internet. For example, the Atomic Physics for Fusion experimental group at ORNL has for many years produced electron impact ionization data regarding ions of particular interest in fusion energy and other applications. Until recently, the data were available only through the articles published in research journals and through internal reports which sometime contained additional, unpublished data. The extensive data from this long running experiment is now available through the WWW [9], organized in a user-friendly set of hyperlinked indices, graphs, and tables which may be easily searched, viewed, and downloaded.

Other extensive collections of primary atomic data have been stored and organized, especially regarding electron-impact excitation and ionization, and heavy-particle impact charge transfer, at the Data and Planning Center of the National Institute for Fusion Science (NIFS) in Nagoya. This and other data are contained in a modern relational database and is accessible through a Web based interface [38]. Of particular interest in astrophysics, but potentially useful in other applications as well, are the astrochemistry database at the University of Manchester [39] and the atomic data for line emission modeling and diagnosis called CHIANTI [40]. An extensive survey of atomic data available on the Web has been made by Yuri Ralchenko who maintains a current list of the URLs (Uniform Resource Locators, the formal name for the WWW addresses of various types of resources) at the Weizmann Institute of Science Plasma Laboratory Web site [41]. The latter site is also being used to develop other innovative atomic data resources, such as an on-line, restricted version of the widely used ATOM code which can now be run through the Web to compute electron-impact excitation cross sections in the Coulomb-Born-Exchange approximation.

All of these resources exist at relatively well known sites and are readily located through the links listed on most data center or research lab sites. However, the individual who has an interest in providing benchmark data, or who devotes his/her time to collecting data and performing evaluations outside of one of these larger organized data projects or services has always been a significant source of useful information. In the past, one would find their work in the published literature, but now, in addition, one can find it on the Web! For example, Y.-K. Kim and M.E. Rudd have for a long time had an interest in providing evaluated electron-impact ionization data in easily parameterized forms for use in a wide range of applications including radiation effects in matter, lighting, semiconductor processing, and fusion plasmas. Recently, they have developed a theoretical method based on well established behaviors of the ionization cross section and developed a Web site through which their results can be obtained for a range of molecules [42].

SUMMARY AND OUTLOOK

Thus, individual researchers, groups, data centers, and networks of data centers have taken to heart the message that atomic data can be best preserved and exchanged by taking advantage of modern electronic storage, database, and communications technologies. The future certainly holds a continuation of this trend, and hopefully after a time, a synthesis and synergetic amalgamation of these resources. However, to assure that this goal is reached a concerted effort must be made in a number of areas.

Efforts to coordinate the production of needed data should be enhanced through improved communication among the interested parties, namely, among data producers, data users, data centers, and funding entities. Recognition of the crucial role of atomic collision data in applications makes it imperative that the atomic physics community can responsively provide needed information, and that allowing for the necessary checks through consideration of particular data by several groups, that this be accomplished efficiently. The need for efficiency also applies to data center activities which could also be coordinated further to enhance the overall scope of existing evaluated and recommended data, and to cooperatively further the development of data collection, organization, manipulation, and dissemination technologies. These actions also call for heightened openness and fairness in sharing and citing resources and efforts made by other groups.

Data centers should also be wary to continue to provide not only raw data, but also evaluated and recommended data sets, to maintain the reliability of data for applications whenever possible. Use of production codes can also be made more reliable through comparisons and benchmarking of data produced by them in cooperation between principal users, producers, and data centers. Data centers should also seek to accelerate the trend in providing users with interfaces to integrated resources. As use of the Web becomes the paradigm of choice for users, data centers must respond by developing Web compatible tools and engines allowing access to all their data resources in a way which provides the user with the pertinent data, advises what to do to fill in gaps in the available data, and provides assessments of the expected reliability of the data found.

These efforts are particularly pertinent now since, for especially the past twenty years, reasonably accurate atomic data for a wide range of reactions has been accumulated, much of which has been evaluated or at least collected, and it would be derelict to fail to take advantage of the current technology to preserve, organize, and synthesize it. Finally, in recognition of the importance of these efforts to applications and further development of the capabilities within the field itself, the case for greater and broader support must be made clearer for data production and the chain of activities (collection-organization-preservation-synthesis-dissemination) that leads to evaluated and recommended data.

REFERENCES

1. D.R. Schultz and J.K. Nash, "On-line Atomic Data Access," *Proceedings of the Tenth Topical Conference on Atomic Processes in Plasmas*, San Francisco, CA, edited by A.L. Osterheld and W.H. Goldstein (American Institute of Physics, Woodbury, 1996), Vol. 381, p. 197.
2. C.F. Barnett, J.A. Ray, J.C. Thompson, and E.W. McDaniel, "Bibliography of Atomic and Molecular Processes for 1963," ORNL-AMPIC-1 (1965). This series continued until 1972 (cf. C.F. Barnett, J.A. Ray, I.W. Blackwell, T.C. Hale, H.B. Gilbody, J.W. Hooper, E.W. McDaniel, and E.W. Thomas, "Bibliography of Atomic and Molecular Processes for January-December 1970," ORNL-AMPIC-14 (1972)).
3. C.F. Barnett and W.L. Wiese, editors, "Atomic Data for Fusion: A Bulletin from the Controlled Fusion Atomic Data Center of Oak Ridge National Laboratory and

the National Bureau of Standards" (bimonthly, Volume 1, 1975 through Volume 8, 1982).
4. S.W. Hawthorne, E.W. McDaniel, C.F. Barnett, D.H. Crandall, H.B. Gilbody, M.I. Kirkpatrick, R.A. Phaneuf, and E.W. Thomas, "Bibliography of Heavy Particle Collisions 1950-1975," ORNL-5500 through ORNL-5503 (1979).
5. C.F. Barnett, D.H. Crandall, B.J. Farmer, H.B. Gilbody, S.W. Hawthorne, M.I. Kirkpatrick, E.W. McDaniel, R.H. McKnight, F.W. Meyer, T.J. Morgan, R.A. Phaneuf, and E.W. Thomas, "Bibliography of Atomic and Molecular Processes 1978," DOE/ER-044 (1980); C.F. Barnett, D.H. Crandall, B.J. Farmer, H.B. Gilbody, P.M. Hafford, M.I. Kirkpatrick, E.W. McDaniel, R.H. McKnight, F.W. Meyer, T.J. Morgan, R.A. Phaneuf, and E.W. Thomas, "Bibliography of Atomic and Molecular Processes 1979," DOE/ER-0074 (1980); C.F Barnett, D.H. Crandall, B.J. Farmer, H.B. Gilbody, D.C. Gregory, P.M. Hafford, M.I. Kirkpatrick, E.W. McDaniel, R.H. McKnight, F.W. Meyer, T.J. Morgan, R.A. Phaneuf, M.S. Pindzola, and E.W. Thomas, "Bibliography of Atomic and Molecular Processes 1980," DOE/ER-0118 (1982).
6. C.F Barnett, D.H. Crandall, B.J. Farmer, H.B. Gilbody, D.C. Gregory, P.M. Hafford, M.I. Kirkpatrick, E.W. McDaniel, R.H. McKnight, F.W. Meyer, T.J. Morgan, R.A. Phaneuf, M.S. Pindzola, and E.W. Thomas, "1978-1981 Bibliography of Atomic and Molecular Processes," ORNL-5921/V1 and V2 (1982); C.F Barnett, D.H. Crandall, H.B. Gilbody, D.C. Gregory, M.I. Kirkpatrick, E.W. McDaniel, R.H. McKnight, F.W. Meyer, T.J. Morgan, R.A. Phaneuf, M.S. Pindzola, and E.W. Thomas, "1982 Bibliography of Atomic and Molecular Processes," ORNL-6052 (1984); C.F Barnett, D.H. Crandall, H.B. Gilbody, D.C. Gregory, M.I. Kirkpatrick, E.W. McDaniel, R.H. McKnight, F.W. Meyer, T.J. Morgan, R.A. Phaneuf, M.S. Pindzola, and E.W. Thomas, "1983 Bibliography of Atomic and Molecular Processes," ORNL-6095 (1984); C.F Barnett, H.B. Gilbody, D.C. Gregory, D.C. Griffin, C.C. Havener, A.M. Howald, M.I. Kirkpatrick, E.W. McDaniel, F.W. Meyer, T.J. Morgan, R.A. Phaneuf, M.S. Pindzola, and E.W. Thomas, "1984 Bibliography of Atomic and Molecular Processes," ORNL-6157 (1985); C.F Barnett, H.B. Gilbody, D.C. Gregory, D.C. Griffin, C.C. Havener, A.M. Howald, M.I. Kirkpatrick, E.W. McDaniel, F.W. Meyer, T.J. Morgan, R.A. Phaneuf, M.S. Pindzola, and E.W. Thomas, "1985 Bibliography of Atomic and Molecular Processes," ORNL-6277 (1986).
7. E.C. Beaty and K. Katsonis, editors, "International Bulletin on Atomic and Molecular Data for Fusion," Vol. 1 (IAEA, Vienna, 1978). This series continues to the present (cf. J. Botero and J.A. Stephens, editors, Vol. 50-51 (IAEA, Vienna, 1996)).
8. "CIAMDA 80: An index to the Literature on Atomic and Molecular Collision Data Relevant to Fusion Research," (IAEA, Vienna, 1980); "CIAMDA 87: An index to the Literature on Atomic and Molecular Collision Data Relevant to Fusion Research," (IAEA, Vienna, 1987).
9. http://www-cfadc.phy.ornl.gov/
10. telnet://ripcrs01.iaea.or.at, username aladdin, follow directions for either guest or registered usage. A IAEA Web site is also being planned and developed.
11. http://gaphyor.lpgp.u-psud.fr/
12. C.F. Barnett, W.B. Gauster, and J.A. Ray, "Atomic and Molecular Collision Cross

Sections of Interest in Controlled Thermonuclear Research," ORNL-3113 (1961); C.F. Barnett, W.B. Gauster, and J.A. Ray, "Atomic and Molecular Collision Cross Sections of Interest in Controlled Thermonuclear Research," ORNL-3113 revised (1964).

13. C.F. Barnett, J.A. Ray, E. Ricci, M.I. Wilker, E.W. McDaniel, E.W. Thomas, and H.B. Gilbody, "Atomic Data for Controlled Fusion Research Volume I," ORNL-5206 (1977); E.W. Thomas, S.W. Hawthorne, F.W. Meyer, and B.J. Farmer, "Atomic Data for Controlled Fusion Research: Revisions and Additions to Sections D.1 Sputtering, D.5 Ion Reflections from Surfaces and D.6 Trapping of Volume I," ORNL-5207/R1 (1979); C.F. Barnett, J.A. Ray, E. Ricci, M.I. Kirkpatrick, E.W. McDaniel, E.W. Thomas, and H.B Gilbody, "Atomic Data for Controlled Fusion Research Volume II," ORNL-5207 (1980).

14. R.L. Kelly, "Atomic and Ionic Spectrum Lines Below 2000 Angstroms," ORNL-5922 (1982).

15. W.L. Wiese and A. Musgrove, "Spectroscopic Data for Titanium, Chromium and Nickel," ORNL-6551 (1989).

16. C.F. Barnett, editor, "Atomic Data for Fusion Volume 1: Collisions of H, H_2, He and Li Atoms and Ions with Atoms and Molecules," ORNL-6086 (1990); E.W. Thomas, "Atomic Data for Fusion Volume 3: Particle Interactions with Surfaces," ORNL-6088 (1985); W.L. Wiese, "Atomic Data for Fusion Volume 4: Spectroscopic Data for Iron," ORNL-6089 (1985); R.A. Phaneuf, R.K. Janev, and M.S. Pindzola, editors, "Atomic Data for Fusion Volume 5: Collisions of Carbon and Oxygen Ions with Electrons, H, H_2 and He," ORNL-6090 (1987).

17. "IAEA Consultant's Meeting on the Atomic Database and Fusion Applications Interface," INDC(NDS)-211/GA (1988).

18. R.A. Hulse, "The ALADDIN Atomic Physics Database System," *Proceedings of the Seventh Topical Conference on Atomic Processes in Plasmas*, Gaithersburg, MD (American Institute of Physics, New York, 1990), Vol. 206, p. 63.

19. K.L. Bell, H.B. Gilbody, J.G. Hughes, A.E. Kingston, and F.J. Smith, *J. Phys. Chem. Ref. Data* **12**, 891 (1983).

20. Y. Itikawa, S. Haro, T. Kato, S. Nakazaki, M.S. Pindzola, and D.H. Crandall, *At. Data Nucl. Data Tables* **33**, 149 (1985).

21. C. Bottcher, D.C. Griffin, H.T. Hunter, R.K. Janev, A.E. Kingston, M.A. Lennon, R.A. Phaneuf, M.S. Pindzola, and S.M. Younger, *Nucl. Fusion, Special Supplement* (1987).

22. R.A. Phaneuf, R.K. Janev, and M.S. Pindzola, editors, "Atomic Data for Fusion Volume 5: Collisions of Carbon and Oxygen Ions with Electrons, H, H_2 and He," ORNL-6090 (1987).

23. M.A. Lennon, K.L. Bell, H.B. Gilbody, J.G. Hughes, A.E. Kingston, M.J. Murray, and F.J. Smith, *J. Phys. Chem. Ref. Data* **17**, 1285 (1988).

24. T. Kato and S. Nakazaki, *At. Data Nucl. Data Tables* **42**, 313 (1989).

25. M.J. Higgins, M.A. Lennon, J.G. Hughes, K.L. Bell, H.B. Gilbody, A.E. Kingston, and F.J. Smith, Culham Laboratory Report CLM-R294 (1989).

26. R.K. Janev, W.D. Langer, K. Evans, and D.E. Post, *Elementary Processes in Hydrogen-Helium Plasmas* (Springer-Verlag, Berlin, 1987).

27. C.F. Barnett, editor, "Atomic Data for Fusion Volume 1: Collisions of H, H_2, He and Li Atoms and Ions with Atoms and Molecules," ORNL-6086 (1990).
28. A.B. Ehrhardt and W.D. Langer, Princeton Plasma Physics Laboratory Report PPL-2477 (1988).
29. R.K. Janev and J.J. Smith, *Atomic and Plasma-Material Data for Fusion, Supplement to Nuclear Fusion* 4 (1993).
30. E.W. Thomas, R.K. Janev, and J.J. Smith, IAEA Report INDC(NDS)-249 (1991).
31. W. Eckstein, C. Garcia-Rosales, J. Roth, and W. Ottenberger, Max-Plank-Institute für Plasmaphysik Report IPP9/82 (1993).
32. E.W. Thomas, R.K. Janev, J. Botero, J.J. Smith, and Y. Qui, IAEA Report INDC(NDS)-287 (1993).
33. http://patiala.phys.strath.ac.uk/adas/adas.html
34. http://www-cfadc.phy.ornl.gov/adas/adas.html
35. http://vizier.u-strasbg.fr/OP.html
36. http://cdsweb.u-strasbg.fr/topbase.html
37. http://www.am.qub.ac.uk/projects/iron
38. A public demonstration of the NIFS data collections can be found at http://amdata.nifs.ac.jp/amdata/index.html and the full database for registered users at http://amdata.nifs.ac.jp/test/index.html
39. http://saturn.ma.umist.ca.uk:8000/ tjm/rate/rate.html
40. http://wwwsolar.nrl.navy.mil/chianti.html
41. http://plasma-gate.weizmann.ac.il/
42. http://physics.nist.gov/PhysRefData/Ionization/Xsection.htm

Atomic and Molecular Data Activities in Japan : Production, Evaluation, Databases and Applications

H. Tawara

National Institute for Fusion Science, Toki-shi 509-5202, Japan

(e-mail : tawara@nifs.ac.jp)

Abstract. The atomic and molecular (AM) data activities including data compilation, evaluation and dissemination in Japan have been described which had started mainly with the intention of providing the reliable data for fusion plasma research. In a wide range of scientific and technological fields, the importance of the AM data has recently been recognized and their applications are growing. Some crucial issues in evaluating the AM data have been discussed.

I. Short History of AM Data Activities

The importance of atomic and molecular (AM) processes and related collision data had already been realized when the Institute of Plasma Physics (IPP) was established in 1961 at Nagoya University as a national center of high temperature plasma research for fusion. In 1970 a survey on the requirement of AM data for fusion research had started at IPP, Nagoya University, under the direction of K.Takayanagi.

Based on this survey, in 1972 a *Study group for Atomic Data for Fusion* (including 15 collision/ atomic physicists) was organized at IPP, Nagoya University, in order provide reliable AM data. The results of the compilations on basic data involving hydrogen atoms and ions were published in 1974 as "*Cross Sections for Atomic Processes*" vol.1[1] which was followed by the second volume of "*Cross Sections for Atomic Processes*" vol.2[2] (1976). This second volume contains basic data involving helium atoms and ions. These two volumes of AM data compilations and evaluation are believed to be the first devoted to high temperature fusion plasma research.

As the requirement for a wide range of AM data for fusion plasma research had begun to be internationally realized, the First International Atomic Energy Agency (IAEA) Technical Committee Meeting on Atomic and Molecular Data for Fusion[3] was held at Culham, UK, in 1976.

In order to support the international need for AM data and to provide systematic AM data including material and surface data, the Atomic and Molecular Data Center for Fusion Research was established at IPP, Nagoya, in 1978. Since the staffs were limited to only two members, many of the AM data activities were supported by scientists from other universities. At about the same time, an AM Data group for fusion research was also established at the Nuclear Data Center, Japan Atomic Energy Research Institute (JAERI). Since then, these two data centers are continuing a close collaboration and play the leading roles in the AM data activities in Japan.

In 1978, the Data Center Network (DCN) of AM Data for Fusion was established at IAEA, Vienna. Presently this DCN consist of roughly 20 institutes from 10 countries. The IAEA data center plays a key international role in collecting and providing evaluated AM data to the fusion community, a large part of which are supplied by the DCN members around the world.

In 1990, IPP was reorganized into the National Institute for Fusion Science (NIFS) and the AM data activities are being continued, as will be described later.

Recently, in a wide range of fields such as material science and technology, fluid dynamics and environmental research, the need for AM data has been recognized and several groups including not only the academic institutions but also industrial research teams are trying to establish AM databases.

II. AM Data Production in Universities/Institutes

In Japan, more than 100 small groups at institutes/universities are performing collision research and applications with the support of relatively small but continuous budgets. In order to exchange information and promote these activities, *The Society for Atomic Collision Research in Japan* was established in 1976 on the occasion of the Eleventh International Conference on Physics of Electronic and Atomic Collision (XI-ICPEAC, 1979) in Kyoto. Since then, The Society, presently with about 300 members including the graduate students, regularly publishes its annual Progress Report[4] which contains roughly 100-120 contributions every year and also a list of the published papers by the members of the Society. It should be pointed out that only some of chemistry groups are registered to this Society so far.

The scopes of the research activities of these groups range from electron collisions to photon processes, highly charged ion collisions at various collision energies ranging from thermal energies to MeV energy in both theories as well as experiments. Unfortunately it is not possible to describe all of these important activities in the present report. Some of them have been described, for example, in a series of Books of the Invited Talks of ICPEAC. Readers should see the annual Progress Report of this Society[5] in order to obtain more details of the activities in Japan. A limited number of copies are available at the Society office upon request.

III. AM Data Production in National Institutes

In addition to experimental and theoretical activities by small groups at the universities as mentioned above, the AM data are also being produced at some national institutes where relatively large and powerful facilities are available. Also some theoretical groups at national institutes such as the Institute of Space and Astronautical Sciences (electron-molecule collisions) and the Institute of Molecular Sciences (heavy ion collisions including chemical processes) are active in AM theoretical data production.

Table 1 A list of AM data production activities at national institutes in Japan

National institute	main facilities	current topics
JAERI	20 MV tandem and linear accelerator booster	zero-degree electron spectroscopy of highly charged ion collisions
PF, KEK	2.5 GeV electron photon facility	photo-dissociation, -absorption in combination with lasers
ETL	0.6 GeV electron photon facility	photo-dissociation, -absorption
RIKEN	Electron cyclotron resonance ion source, linac and cyclotron facility	collisions of highly charged ions (1 keV/amu - 100 MeV/amu)
INS	Ion storage ring with electron-cooler	dissociative recombination of molecular ions
HIMAC	Linac and synchrotrons (max. 650 MeV/amu)	relativistic collisions and medical applications
IMS	0.75 GeV electron photon facility	state-selective/time-resolved dissociation (combined with laser)
NIFS	Electron beam ion source (highly ionized ion source)	electron capture collisions at 0.1 keV/amu region
SPring-8	8 GeV electron photon facility	photon (X-ray) collision processes including photoionization of ions

The names and addresses of the national institutes cited above are as follows :
  (1) Japan Atomic Energy Research Institute (JAERI), Tokai-mura, Ibaraki
    (sataka@popsvr.tokai.jaeri.go.jp)
  (2) Photon Factory (PF), High Energy Physics Laboratory (KEK), Tsukuba, Ibaraki
    (yagisita@kekvax.kek.jp)
  (3) Electro-Technical Laboratory (ETL), Tsukuba, Ibaraki (isuzuki@etl.go.jp)
  (4) Institute of Physical Chemical Research (RIKEN), Wako-shi, Saitama
    (kambara@rikvak.riken.go.jp)
  (5) Institute for Nuclear Science (INS), University of Tokyo, Tanashi, Tokyo
  (6) HIMAC, National Institute of Radiological Sciences, Chiba (takada@nirs.go.jp)
  (7) Institute for Molecular Science (IMS), Okazaki (nakamura@ims.ac.jp)
  (8) National Institute for Fusion Science (NIFS), Toki (sakaue@nifs.ac.jp and tawara@nifs.ac.jp)
  (9) JAERI/RIKEN (SPring-8), Himeji (oura@spring8.or.jp)
 (10) Institute of Space and Astronautical Science (ISAS), Yoshinodai, Kanagawa
    (itikawa@pub.isas.ac.jp)

In Table 1 are shown some experimental AM data production activities at the national institutes, their main facilities and current topics. These experimental programs are closely supported by theoretical groups at both national institutes and universities.

III. Data Compilation, Evaluation, Recommendation, and Dissemination

The data compilation and evaluation work needs a lot of man-power and is usually a time-consuming jobs, and thus close collaborations among different experts/groups are essential. In Japan, two National Institutes closely connected with the fusion program are mainly working on the AM data compilation and evaluation in collaboration with specialists in a wide range of AM physics and collision physics :
1) Japan Atomic Energy Research Institute (JAERI), the Science and Technology Agency.
2) National Institute for Fusion Science (NIFS), the Ministry of Education.

Both groups with only small staff are intensively supported by outside collaborators. The evaluated data are stored in their data handling systems and are also simultaneously printed as data books and distributed among the users. Some of them have been published in books. Recently they have become accessible conveniently and quickly through networks on the World-Wide-Web (WWW).

1) Atomic Data Group, Nuclear Data Center, JAERI

The AM data group at JAERI is working on both collision data and spectroscopic data, together with the production of some collision cross sections involving heavy ions. All data activities at JAERI can be seen on WWW at http://wwwndc.tokai.jaeri.go.jp/jeamdl/index.html soon.
(1) Collision data and spectroscopic data :
In addition to a series of original papers, 23 data reports have been published so far.
(2) Empirical formulas for data of various collision processes :

Empirical representations (formulas) of cross sections as a function of the collision energy or temperature are most convenient in many applications such as in getting the rate coefficients in plasmas. This group is used to find nice fitting/empirical formulas[6] for various collision processes including ionization, molecular dissociation, photon emission processes etc.
(3) Atomic and molecular (collision) data storage and retrieval system (*AMSTOR*) contains the compiled original (collisional) data.
(4) The evaluated collision data are stored at *JEAMDL* (JAERI Evaluated Atomic and Molecular Data Library) as figures, together with the analytically fitted curves.
(5) Automatic production of Grotorian diagrams for spectroscopy : This is a very important development as the drawing of Grotorian diagrams was known to be time-consuming for a long time. A number of the spectroscopic reference data using this system have been published. A full report on spectroscopic databases for highly ionized ions of 10 elements over Z=22-29, Kr and Mo, completed in collaboration with the U.S. National Institute of Standards and Technology, is going to be published in the Journal of

Physical and Chemical Reference Data, Monograph No.8[7].

2) Data &Planning Center, NIFS
Most of the AM data activities including the development of the data handling systems at NIFS are supported by collaborating groups from other universities. So far, about 100 data reports (till 1997), mostly collision data, have been published and can be obtained (free of charge) upon request.

The bibliographic as well as numerical AM data obtained through these activities are stored in our data retrieval system and can be easily accessible on WWW at http://amdata.nifs.ac.jp (free access) or http://dbshino.nifs.ac.jp (ID required).

The following is some detailed description of the AM data activities at NIFS :
(1) Compilation/evaluation
a) The *total* cross sections in electron impact data including the ionization, recombination, excitation, dissociation, photon emission from atoms, ions and molecules have been compiled and evaluated. This activity is still being continued.
b) The compilations and evaluations of *differential* (in energy and angle) cross sections in electron collision data for elastic and inelastic (mostly excitation) scattering for rare gases, other atomic targets and molecular targets (including hydrocarbons, material processing molecules) are under way. It has been recently realized that the total cross sections are not sufficient but the differential cross sections are necessary, for example, to model the detailed features and behavior of the material layers deposited on the surfaces by plasma sputtering techniques.
c) The cross sections for electron capture/excitation/ionization of atoms and molecules by the impact of heavy ions are also being compiled.
d) Our systematic surveys of experimental data and simulation on the basic particle-solid collision data such as sputtering and backscattering have resulted in finding some very convenient empirical formulas which seem to be indispensable for a number of applications.
(2) Bibliographic databases (DB)
   a) *AM* (623,522 records), mostly for AM collision processes, and b) *PLASMA* (80,032) + *FUSION* (680,583), fusion-related references, both are extracted from INSPEC (since 1969) as the INSPEC itself is too large to include our data handling system and is inconvenient for our use. This bibliographic system may be one of the most extensive bibliographic DB for AM and fusion-related topics. Indeed the access to our *AM* and *FUSION* is much quicker than to all the INSPEC which includes the DB for a wide range of physics and engineering. c) The Oak Ridge National Laboratory bibliographic data have also been transferred to our system as *ORNL* (67,489).
(3) Numerical databases :
The numerical data, including the original data taken from the published papers as well as some of the data evaluated at NIFS, are accessible and can be retrieved in the form of tables as well as figures via WWW, as shown in a demonstration during the Conference. These numerical databases are categorized as follows :
a) Atomic and molecular data interactive system (*AMDIS* : 5,638) which includes mostly electron collision data including ionization, excitation, and recombination processes and

also some related empirical formulas.

b) Charge transfer data (*CHART* : 1,892) which include the cross sections of the charge transfer in collisions of ions with atomic and molecular hydrogen targets and also empirical formulas. Those for multiple ionization by heavy ion impact are being added.

c) Sputtering data (*SPUTY* : 722) which include the experimental data for monatomic solids as well as those calculated with the help of a Monte Carlo simulation (ACAT-code)[8] based on the binary collision and the empirical formulas.

d) Backscattering data (*BACKS* : 282) which include the backscattering coefficients of particles and energy of the original experimental data as well as the calculations based on the TRIM-code and also empirical formulas for > 10 eV collisions (which still is in lack of proper treatment at lower energies below 10 eV by molecular dynamics)

(4) A mirror site of ORNL databases :

This, presently under construction, is hoped to be helpful and powerful for a number of the domestic users as well as for those of Asian countries nearby.

IV. Data Evaluation

The evaluation of AM data requires a wide range of deep and accurate understanding and knowledge on atomic processes and has to proceed through some complicated procedures. In particular, careful analysis of various aspects including the critical understanding of the data obtained is required for the evaluation of the experimental and theoretical results.

1) Evaluation of experimental data

For the evaluation of *experimental* data of absolute cross sections in AM collisions, the following issues are the most important and the data as well as the experiments themselves should be carefully and closely scrutinized :
(1) primary beam preparation/specification, (2) target species, (3) collision product identification, and (4) detection system.

In each, there are a number of critical issues. Unfortunately very few papers have given a detailed discussion on such issues, and the access to them is sometimes difficult.

(1) Primary beam specification

The primary beam has to be correctly specified with respect to many aspects such as mass, charge, internal energy etc. In particular, it is difficult to specify the internal energy of the incident ions which is known to have a significant influence on the measured cross sections in many cases :

a) Mass and charge (^{4}He^{2+}/H$_2^+$; C^{3+}/O^{4+})

b) Isotope composition of ions or atoms
c) Kinetic energy/distribution/spread
d) Internal energy or excited species
e) Beam focusing and divergence
f) Beam flux through target inlet/exit holes
   The edge scattering should be minimized.
g) Effects of external fields

It is most important to know if there is no metastable content or what fraction of the metastable beam is included in the original incident ions. This is a serious issue since the cross sections involving the metastable state beam are usually larger by more than one order of magnitude compared with those for the ground state beam. Indeed, the electron transfer cross sections for the ground state $C^+(^2P) + H_2$ collisions below a few 100 eV/amu energy region are found to be more than 50 times smaller than those for the metastable state $C^+(^4P) + H_2$ collisions[9]. Also vibrationally excited molecules with relatively small internal energy, compared with that in the electronic excitation, have recently been known to strongly influence the collision cross sections.

(2) Target specification

a) Purity/contamination

It is most important to minimize any contamination of the targets. For example, a small amount of water vapor is very often contained in the gas-feeding line which can be minimized with a relatively simple liquid nitrogen trap on the line. Even 1- 0.1 % of air or N_2 in H_2 or He targets also have significant influence on the observed cross sections for most of the collision processes.

b) Effective target number density

 b-1) gas target density profile :

The "total" number of target gases determines the final absolute cross sections and, therefore, the target gas density distributions along the incident beam path have to be known accurately. Often the "effective" length corresponding to the "real" length of the target chamber is assumed. This may not be true as gases diffuse out through its inlet and exit holes. To get rid of the uncertainties due to diffusion effects, the measurements should be performed on two collision chambers with different lengths but the same sizes of holes. The signals due to the difference of lengths should give more accurate cross sections without any ambiguity due to the diffusion.

 b-2) overlapping/form factors for the colliding beam-target :

The exact knowledge of their overlapping is important for absolute cross section determinations. In particular, in gas jet or the crossed-beams experiments, very slight changes of the overlapping give serious errors/uncertainties in estimating the cross sections.

c) Internal energy :

Though most of the target gases can be in the ground state, the internal energy is critical, for example, in crossed-beam experiments, similar to those of the primary ion beam. It is important to specify the excited states (nl) of the colliding beams. Otherwise, the observed cross sections would be largely scattered.

d) Thermal motion effects (at low energy collisions) :

To overcome this problem, a cold gas target cooled down to liquid rare gas (He) temperatures has been developed and found to be a powerful tool to investigate detailed collision processes involving He targets[10].

e) pin-holes in foils :

When very thin foils are used, the observed effects should be carefully analyzed if there are no pin-holes at all.

(3) Reaction products

a) Product specification :

The mass and charge state of the collision products have to be correctly specified by the time-of-flight or electromagnetic analysis.

b) Collection and transmission efficiencies :

The dissociation products usually have relatively large kinetic energy. Therefore, a strong field is needed to collect all of them after collisions. On the other hand, this strong field deflects the primary ions which may influence the effective target interaction length and also densities. This effect is significant in low energy collisions. Also, the transmission of the products through the analyzing systems over the target region to the detector has to be known accurately. In particular, the electron transmission through the energy analyzers is known to strongly depend on the electron energy which in turn influences the energy resolution.

c) Single/double collisions (in gases) :

In gas experiments, the secondary (double) collision effects can not be avoided, in particular for small cross section measurements. Therefore, the contribution of the double (sometimes ternary) collisions should be minimized and carefully estimated by observing the product yields as a function of the target density (pressure).

d) Recoil, internal and dissociation energy :

These extra energies should be analyzed carefully. Otherwise the wrong mass or charge state may be assigned to the products. In particular, such ambiguous situations can happen in recoil or dissociation ion measurements at low energy collisions.

e) External field (ionization) effects :

If an electron is captured into a high Rydberg state of an ion, for example, in the dielectronic recombination process, it can be easily ionized under the influence of even weak external electrostatic/magnetic fields present in the collision region which clearly change the observed cross section.

f) Wall scattering effects :

The hitting by the product electrons onto the walls of an electrostatic energy analyzer produces extra electrons and often results in the wrong shapes of a spectrum. Such ambiguity can be minimized by using meshes, instead of solid plates.

(4) Detection system

a) Absolute detection efficiencies :

In order to determine absolute cross sections, the absolute detection efficiencies of the detectors should be known. This is one of the most serious difficulties in the experiments. Sometimes further complications arise due the change of the efficiencies over a long period of the measuring time.

b) Pulse height distribution/noise discrimination :

The pulse-height distributions from some multipliers often used in the detection of the products change drastically when the counting rates change. Therefore, they should be carefully checked during the measurements to make sure that the counting rates do not influence the over-all efficiencies of the detectors. Also, this type of check is important to properly discriminate the noises from the signals.

c) Secondary particle emission effects :

Secondary particles emitted from the detectors or target surfaces should be carefully suppressed. On the other hand, care should also be exercised to minimize such effects due to the "uncontrollable" neutral particles emitted from the target surfaces and arriving at the suppressor which in turn produce the ternary electrons.

d) Uniform (sensitivity/position) response :

The spatial sensitivities of the detectors and its uniformity are also important, in particular if the position information is used to analyze the characteristics of products.

e) Dark current or noise/sensitivity to stray particles/photons :

They give serious limits and errors in measuring small signals or low count rates.

f) Detector positions and solid angles :

The detector should be positioned to maximize the product detection and at the same time to minimize the stray particles to enter the detectors. The solid angles subtended by the detectors should be carefully estimated, in particular, for large-sized incident beams and products.

g) Reflection :

The reflection of the incident particles from the detector or Faraday cup surfaces may result in the loss of the detection efficiencies and thus should be minimized.

(5) Normalized data vs absolute data

Sometimes the experimental cross sections have been, for simplicity, normalized to some known (previously observed, calculated or scaled) values. For example, the experimental cross sections for atomic hydrogens, due to difficulties in preparing the targets and in determining the absolute densities, are normalized to those at high energies where the Born or other approximations are assumed to be valid. Furthermore, it should be noted that, if the collision energy is not sufficiently high, the Born approximations fail to give the correct values.

2) Evaluation of theoretical data

If no experimental data are available, we have to rely upon some theoretical calculations. There are some critically important issues in the evaluation of the calculated cross sections. Indeed it is sometimes difficult to know if they are properly treated in the references.

(1) Approximations :

There are a number of approximation techniques to treat various collision processes such as Born, distorted-wave, continuum distorted-wave approximations etc. Among them, the close-coupling methods with atomic or molecular basis, R-matrix and variational methods are believed to be the most reliable. The convergence of integration is another important issue. The interaction potentials have also to be chosen properly.

(2) Wave-function :

It is important to carefully look at the initial/final wave functions, the distorted wave functions etc. used in the calculations and also if they properly include the electron correlation.

(3) Interaction potential :

The interaction potential between the incident particle and target has to be checked whether it is a simple electrostatic potential, that with electron exchange and proper distortion or with electron correlation effect.

(4) Number of states :

Generally a sufficiently large number of the basis states are required to get high

accuracies. But it is not easy to know if the calculated results are sufficiently converged. "Large numbers" do not always guarantee to provide good results.
(5) Continuum states :
The ionization processes involving the continuum states are hard to treat. The quasi-continuum states have recently been widely used but are not well established.
(6) Valid collision energy :
It is important to know the limit of validity of the approximations such as the collision energy region or the ratios of charge-over-velocity of the incident particles which are one of the key parameters in collisions.

V. Typical Evaluation Procedures for Electron Scattering Cross Sections

It would be nice to show typical examples of how to evaluate the collision cross sections. A good example is the evaluation of the collision cross sections in electron impact on gas targets since they can be obtained by different methods and, thus, be cross-compared with each other to check the consistency of the observed data[11].

The *grand total cross section,* measured with the simple but quite reliable beam attenuation method, can be represented as the sum of the cross sections for all the collision processes involved and is believed to be more accurate than those for each collision process. There, the following relation should hold :

$$\sigma_t(E) = \sigma_{el} + \sigma_{ex} + \sigma_i + (\sigma_{rot} + \sigma_{vib})$$

where σ_t is the *grand total cross section* which can be determined with an accuracy of (2-5) % as a function of the incident electron energy E ; σ_{el}, the *total elastic scattering cross section* measurable with an accuracy of (20-30) % which represents the integrated value of the differential cross section $d\sigma_{el}/d\Omega(E,\theta)$ over the angles θ of 0°-180° ; σ_{ex}, the *total electronic excitation cross section* which is the integrated value of the differential cross section $d\sigma_{ex}/d\Omega(E,\theta)$ over the angles of 0°-180° and can be determined with an accuracy of (20-40) % ; σ_i, the *total ionization cross section* which, described in detail later, is related to the *apparent ionization cross section,* σ_{app}, measured through the condenser plate technique with an accuracy of (2-5) % ; σ_{rot}, the *total rotational excitation cross section* of the molecule which is represented by the integrated values of the differential cross sections $d\sigma_{rot}/d\Omega(E,\theta)$; σ_{vib}, the *total vibrational excitation cross section* of the molecule which is represented by the integrated values of the differential cross sections $d\sigma_{vib}/d\Omega(E,\theta)$.

Each of the "total" elastic and inelastic cross sections is usually determined through the integration of the measured differential cross section (DCS) = $d\sigma/d\Omega$ (E,θ) at various angles (θ) as a function of the electron impact energy E using the crossed-beams technique with gas jet targets. They have not always been measured at the same scattering angles and at the same incident electron energies by different authors.

Furthermore, the absolute DCS has to be normalized to the known cross section for "standard" He gas targets and then is assumed to be a smooth function of two important collision parameters, namely, 1) the *incident electron energy* (E) and 2) the *scattering angle* (θ), for a particular collision process corresponding to the particular scattered electron energy. Of course the first assumption 1) is not valid in the resonance region.

The practical evaluation procedures for the "elastic" scattering cross section can be described as follows (see Fig.1) :

(1) Smoothing of the plotted curves of the measured $d\sigma/d\Omega(E)$ as a function of the incident electron energy E at a particular scattering angle θ_s (with the proper interval of the scattering angle $\theta_s \approx 5°$).

(2) Smoothing of the plotted curves of $d\sigma/d\Omega(\theta)$ as a function of the scattering angle θ at a particular incident energy E.

Then, to get the total cross section integrated over the scattering angle, the extrapolation of DCS to 0° and 180° scattering angles is required since most of the measurements have been performed over a limited scattering angle region and very few measurements at these extreme angles have been reported due to technical difficulties (note that a convenient technique has been most recently developed[12] to measure them near 0° and 180°). But this extrapolation often results in large uncertainties in the final integrated values. Care should be exercised to avoid such unfavorable situations.

(3) Replotting $d\sigma/d\Omega(E)$ at θ_s to get a better smoothed and consistent value.

(4) Replotting $d\sigma/d\Omega(\theta)$ at E_s to get better smoothed and consistent values.

Some iterations in the procedures (1) - (4) are necessary before getting the final "smoothed" curves of $d\sigma/d\Omega(E,\theta)$. The integration of the final curves for $d\sigma/d\Omega(E,\theta)$ over the scattering angles of 0° - 180° results in the "total" cross sections which should be cross-checked with the total cross sections obtained with other techniques such as swarm experiments at low energies where the inelastic scattering process is generally negligible. Here again some iterations are not avoidable in order to get the "very final" cross section data.

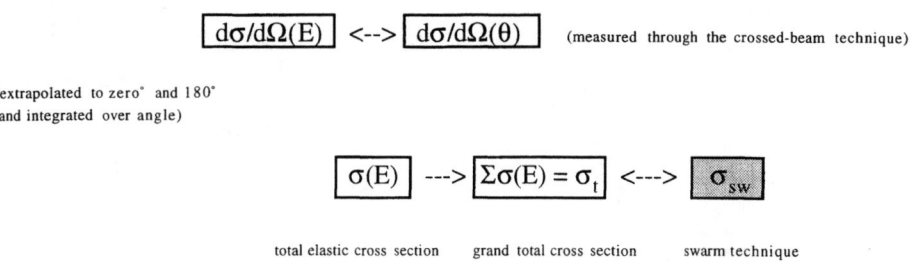

Figure 1. A scheme for the evaluation of experimental elastic scattering cross sections. Note that the cross sections can be directly compared with those observed through the swarm technique at low electron energy collisions.

A few examples of the thus evaluated cross sections are shown in Fig.2, together with the original experimental data[11].

On the other hand, the *total ionization cross section* σ_i can be determined through two different techniques. In the first method, the *apparent ionization cross section* is obtained with a condenser plate technique which collects all the secondary ions or secondary electrons and is given as $\sigma_{app} = \Sigma r \sigma_r$ where σ_r represents the "partial" ionization cross section for the production of the secondary ions with charge r. In order to get the partial ionization cross section, σ_r, it is necessary to determine relative fractions f_r of the secondary ions with different charge r using other technique such as the time-of-flight technique, which can be determined with accuracies in the range of (10-20) %. Then, the total ionization cross section is given as $\sigma_i = \Sigma \sigma_r$.

In the second method, the doubly differential cross section (DDCS) = $d^2\sigma/dEd\Omega$ is determined through the crossed-beam technique. However, as the scattered electrons have continuous energy distributions, the evaluation procedures of the DDCS for ionization are much more complicated than that for elastic scattering.

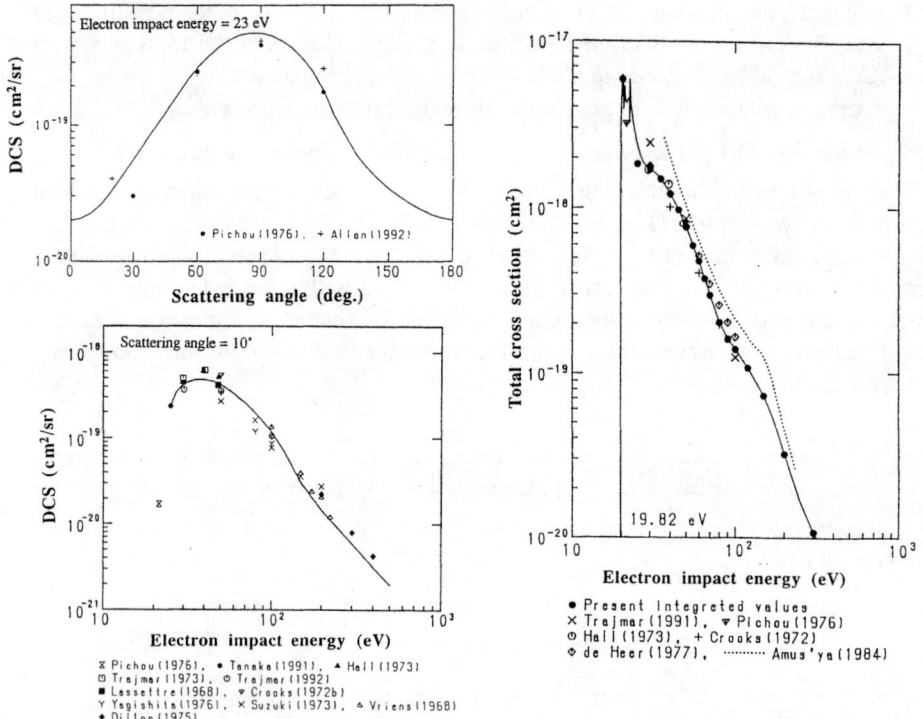

Figure 2. Examples of the original (marked points) and evaluated electron scattering cross sections (smooth curves) for the excitation to the 2^3S state (threshold energy : 19.8 eV) of He atoms[11]. Figure 2 (a) shows those at 23 eV electron impact as a function of the scattered angle. Figure 2 (b) shows those as a function of the incident electron energy at the scattering angle of 10°. Figure 2 (c) shows the total cross sections as a function of the incident electron energy.

Then, (1) the integration of the evaluated DDCS over the scattered electron energy and scattering angle results in the apparent ionization cross section σ_{app} which can be directly compared with that obtained with the condenser plate method and finally (2) the total ionization cross section σ_i can be deduced with the help of the relative fractions of ions f_r determined with other techniques. The total absolute ionization cross section thus obtained is expected to have the uncertainties of (20 - 30) %.

VI. Activities in AM Data Applications for Engineering and Industries

1) AM database development

Among a few groups which are currently working on AM data applications with the support of the academic societies as well as industrial circles, the following should be mentioned :

(1) Committee on Electron Collision Data for Discharge Plasmas[13]

This activity, supported by Japan Society of Electrical/Electronic Engineering, has been joined by about 25 participants from both academic and industrial groups. This group is planning to organize reliable collision data of electron-atoms, -molecules including those excited species which are required for understanding fundamental phenomena in gas discharge, gas insulation, gas lasers, plasma material processing, etc.

(2) Committee on Databases for Atomic/Molecular Flow Analysis[14]

This group, supported by Japan Society of Mechanical Engineering, is planning to develop and provide the AM data bases (in books as well as in electronic forms) necessary for understanding the transport phenomena of fluid and plasma (rarefied-gas) flows in material processing, vacuum technology etc. from the atomic/molecular physics basis, with 20 participants from academic and industrial groups. More information on the activities of this group can be obtained on WWW at http://www.es.mach.mie-u.ac.jp/datmolf/.

2) Industrial Applications

In Japan, a number of AM data application programs are in progress in various industrial fields. Some of them have been developed considerably and are already in practical use, though critical improvements are still necessary from the practical point of view.

(1) Environmental applications

In one of the environmental applications, where the poisonous gases in the exhausts from cars or coal/oil burning at power stations such as NO_x, SO_x are being removed through a series of chemical reactions, a device is under test at a new 200 MW power station[15]. In this system, a powerful, high energy (800 keV) electron beam is introduced into the exhausted gases through thin foil windows and, by spraying the water and ammonia vapors together, generates intense active radicals such as O, OH which in turn interact strongly with the toxic gases to form stable hydrogen sulfate or nitride. Finally, 97-98 % of these poisonous gas molecules are removed at the end of the tunnels, with the accompanying products of ammonium sulfate or ammonium nitride used as fertilizer for farming. Further improvements are necessary for optimizing various parameters in order to obtain the most economical and effective removal of such gases.

(2) Discharge phenomena

In order to prevent power failure by sporadic lightening which may result in serious blackouts, some tests of lightning control by laser beams have already been performed at some laboratories, though the present laser power has been found to be not sufficient to trigger the real lightning.

(3) Materials Industries

There are a number of the academic and industrial groups who are working on plasma surface processes for developing reliable materials processing/production techniques and surface treatments. For this work, not only the gas-phase collision data but also the particle-solid collision data are indispensable. For example, for ion implantation into some materials, very precise knowledge of the ion ranges in the materials is required, otherwise it would fail to get the uniform, high quality materials.

(4) Medical applications

It is known that in cancer therapy based on high energy heavy ion impacts basic collision data are indispensable in the detailed understanding of its effectiveness and in providing improved treatments. For this goal the energy distributions and angular distributions of the secondary electrons have been believed to be most important. On this aspect, a Japanese group has collaborated to organize collision databases for medical purposes[16]. It has been recently realized that the secondary, highly charged ions could play a more crucial role in the tumor treatments. There is still a serious lack of basic collision data applicable to such cancer therapy.

Concluding remarks

In the present paper we have tried to sketch the AM data activities in Japan. A number of groups are working on both theories and experiments and are providing good and reliable AM data in a wide range of the collision processes. The critical issues in the data evaluation processes of the AM data have been discussed in some detail.

Furthermore, it has been pointed out that the applications of AM data in various fields are expanding rapidly in Japan. In addition to the fusion community where the importance of AM data has already been realized for some time, the industrial communities are also trying to establish AM databases in order to produce reliable control of our environment and high quality for devices such as very large scale integrated circuits (VLSI) where basic AM data have been realized to be indispensable.

Acknowledgments

The author would like to thank a number of the colleagues for their support and help in the course of the AM data activities and also for providing information on their recent data activities.

References

1. "Cross sections for atomic processes", Report IPPJ-DT-48 (Institute of Plasma Physics, Nagoya University, 1974).
2. "Cross sections for atomic processes", Report IPPJ-DT-50 (Institute of Plasma Physics, Nagoya University, 1976).
3. Proc. Advisory Group Meeting on Atomic and Molecular Data for Fusion (Culham, 1976).
4. Progress Report published annually by The Society for Atomic Collision Research in Japan. So far, 22 volumes (till 1996) have been published and the 1997 issue is under printing.
5. The present office : Department of Chemistry, Tokyo Institute of Technology, O-Okayama, Tokyo 152 (present chairman : Y. Hatano : yhatano@chem.titec.jp).
6. Ito, R., and Tabata, T., "ALESQ : a code for non-linear least-square fit and TSOLVE : a code for non-linear best approximation", Tech. Report No.4 (Radiation Center, Osaka, 1983).
7. Shirai, T., Sugar, J., and Wiese, W.L., *J. Phys. Chem. Ref. Data*, Monograph no.8 (in press).
8. Yamamura, Y., and Mizuno, Y., "Low energy sputtering with the Monte Carlo program ACAT", IPPJ-AM-40 (1985, IPP, Nagoya, presently NIFS).
9. Kusakabe, T., Mizumoto, Y., Katsurayama, K., and Tawara, H., *J. Phys. Soc. Japan* **59**, 1987-1994 (1990).
10. Ullrich, J., Dörner, R., Mergel, V., Spielberger, L., and Schmidt-Böcking, H., *Comm. At. Mol. Phys.* **30**, 285-304 (1994).
11. Hayashi, M., (private communication, 1997).
12. Read, F.H., and Channing, J.M., *Rev. Sci. Instr.* **67**, 2372-2376 (1996).
13. Sakai, Y. (chairman : sakaiy@eng.hokudai.ac.jp), Department of Electrical Engineering, Hokkaido University, Sapporo (period1995-1997).
14. K.Nanbu, K. (chairman : nanbu@ifs.tohoku.ac.jp), Institute of Fluid Science, Tohoku University, Sendai (period 1995-1998).
15. Ebara Seisakusho Co., Tokyo 144.
16. "Atomic and molecular data for radiotherapy and radiation research", International Atomic Energy Agency Report IAEA-TECDOC-799 (1992).

The Russian Effort in Establishing Large Atomic And Molecular Databases

Leonid P. Presnyakov

P.N.Lebedev Physical Institute, Moscow, Russia

Abstract. The database activities in Russia have been developed in connection with UV and soft X-ray spectroscopic studies of extraterrestrial and laboratory (magnetically confined and laser-produced) plasmas. Two forms of database production are used: i) a set of computer programs to calculate radiative and collisional data for the general atom or ion, and ii) development of numeric database systems with the data stored in the computer. The first form is preferable for collisional data. At the Lebedev Physical Institute, an appropriate set of the codes has been developed. It includes all electronic processes at collision energies from the threshold up to the relativistic limit. The ion -atom (and -ion) collisional data are calculated with the methods developed recently. The program for the calculations of the level populations and line intensities is used for spectrical diagnostics of transparent plasmas. The second form of database production is widely used at the Institute of Physico-Technical Measurements (VNIIFTRI), and the Troitsk Center: the Institute of Spectroscopy and TRINITI. The main results obtained at the centers above are reviewed. Plans for future developments jointly with international collaborations are discussed.

1. INTRODUCTION

The database activities at P.N.Lebedev Physical Institute and related institutions started forty years ago. The motivation for this work was the support of spectroscopy investigations in the fusion program and in space research. The aim of the present paper is to give a brief overview of the modern status of the activities connected to fundamental research and applications for spectroscopy and diagnostics of laboratory and astrophysical plasmas. The main fields for our applications are as follows:
- Solar EUV and X-ray astronomy and astrophysics;
- Aeronomy;
- Laser-produced plasmas including plasma wall interactions
  in the expanding plasmas;
- Magnetically confined plasmas.

In these fields, we consider some results obtained by comparatively small groups at the Lebedev Physical Institute of Russian Academy of Sciences (Moscow), at the Institute of Spectroscopy of Russian Academy of Sciences (Troitsk, Moscow region), at the Institute for Physico-Tecnical Measurements, VNIIFTRI (Mendeleevo, Moscow region), and at the Troitsk Institute for Innovation and Fusion Research, TRINITI (Troitsk, Moscow region). In all these groups, the important part of the atomic data results from their own research and is found in agreement with the available data of other authors when such comparison is possible.

International collaboration is extremely useful and greatly appreciated by the groups above. The rest of the paper is organized as follows. Section 2 is dedicated to the group at Lebedev Institute. The main idea of their activities is to store the methods for calculations and the final formulas instead of storing the numbers. In Section 3, the experimental research at Institute of Spectroscopy is discussed jointly with the atomic databases created there. Sections 4 and 5 are dedicated to the activities at VNIIFTRI and TRINITI, respectively.

2. ACTIVITIES AT THE LEBEDEV PHYSICAL INSTITUTE

The spectroscopy and diagnostics of laboratory and astrophysical plasmas require reliable data for all types of radiative and collisional processes from the reaction thresholds to large (but still non-relativistic) energies. Under different plasma conditions, different processes become more important. Theoretical methods and computer codes were developed for calculating the following processes:
- atomic and ionic wavefunctions and energy levels;
- oscillator strengths and radiative probabilities;
- autoionization probabilities;
- photoionization cross sections and rate coefficients;
- electronic collisions: excitation (including resonance excitation), ionization, and three-body recombination
- radiative and dielectronic recombination;
- ionic (atomic) collisions: excitation (including transitions between Rydberg states) charge transfer, ionization, and transfer ionization;
- level populations;
- intensities of spectral lines (including satellite lines);

All the codes (both jointly and separately) were used in several experimental programs. At present, the Spectroscopy Division of Lebedev Physical Institute has three main codes for atomic and ionic processes and for line intensities in plasmas.

2.1 Main Codes For Databases

The ATOM code has been developed for calculating atomic (ionic) parameters, radiative and autoionization probabilities, and electronic cross sections. The IONCOL code is applied for ion-atom and ion-ion processes. The GKU code is employed for calculations of the ionization state in plasmas, level populations and line intensities. These three codes are interconnected.

The ATOM code, designed by L.Vainshtein as a set of general purpose programs for atomic physics, has a package of associated programs. One of them is MZ used for calculations of the eigenstates, wavelengths and other characteristics of multiply-charged ions in terms of so-called $1/Z$ expansions involving intermediate coupling. The ATOM code includes also a rather effective program for calculations of atomic wavefunctions on the basis of the modified Hartree-Fock method, SEHF (semi-empirical Hartree-Fock). The rate coefficients for radiation and autoionization processes are computed in the first-order approximation with intermediate coupling. Radiative recombination is also considered by means of perturbation theory. Dielectronic recombination being the resonant part of radiative recombination is calculated as analytic continuation of the resonance excitation. For electron-atom and electron-ion collisions, well-known BE and CBE methods are used, respectively. The important element in calculations is normalization of the S-matrix in K-matrix representation. The normalized BE and CBE cross sections are in agreement (within 25%) with the experimental data available.

The cross sections for ionization, charge transfer and excitation in ion-atom and ion-ion collisions are the main output of the IONCOL, recently designed by D.Uskov and L.Presnyakov. The non-stationary theory employed is based on the Keldysh-Coulomb continuum distorted wavefunctions and includes non-adiabatic close coupling. Analytic calculations provide the scaling laws with respect to both ionic charge and collision energy. The following rearrangement reactions have been investigated:
- charge transfer and ionization in collisions of hydrogen atoms with multiply charged ions,
- single and double ionization of helium by heavy ion impact,
- single- and double-electron removal in collisions of H^- ions with multiply charged ions,
- mutual neutralization in collisions of two H^- ions,
- quasi-resonant charge transfer between carbon and boron ions,
- charge transfer in $He^{2+} + He^+$ collisions, both total and angular differential cross sections.

The results shows agreement with recent experimental data better than 10%.

The GKU code is designed by I.Beigman for calculations of relative intensities of spectral lines from both steady and transient plasmas. As input, the GKU uses output of the ATOM and IONCOL codes. For calculations of the level populations, the kinetic equations for the electron flux have been obtained in integral form which can account for the set of discrete levels and continuum. The results have been applied to different laboratory and astrophysical plasmas. The relative line intensities were found in agreement with the observed ones within 15%.

The most important elements of the discussion above are displayed in Table 1.

2.2 Staff and Collaborations

The Spectroscopy Division of the Lebedev Institute has two theoretical groups: a group on theoretical astrophysics (L.Vainshtein, I.Beigman, A.Urnov +1 engineer and 1 student), and an atomic physics group (L.Presnyakov, D.Uskov, V.Shevelko +2 students).

The database work is developed in collaboration with IAEA (R.Janev), University of Giessen (E.Salzborn), NIFS (H.Tawara), IPP, Julich, and GSI, Darmstadt, and with several Russian centers. The research and database activities are reflected in refs.(1 - 12).

TABLE 1

Code	Query	Method	Accuracy
ATOM	wavelengths	SEHF 1/Z expansion+ QED corrections	$\Delta\lambda/\lambda<10^{-4}$ for $\lambda=1\text{-}100$ A
ATOM	electron-atom (ion) collisions	normalized BE (atom) CBE(ion)	$\Delta\sigma/\sigma<0.25$
IONCOL	ion-atom, ion-ion collisions	non-stationary Keldysh-Coulomb CDW + CC	$\Delta\sigma/\sigma<0.1$
GKU	line intensities	integral kinetic equations for the electron flux	$\Delta I/I<0.15$

3. ACTIVITIES AT THE INSTITUTE OF SPECTROSCOPY, RUSSIAN ACADEMY OF SCIENCES

3.1 Research and Data Production

A systematic study of atomic spectra is being carried out at the Institute of Spectroscopy with the aim of getting experimental and theoretical information on wavelengths and energy levels. A normal incidence spectrograph with a 6.65-m grating (1200 l/mm) is available for the region 300 - 2500 A. For shorter wavelengths, 50 - 350 A, a 3-m grazing incidence spectrograph with a 3600 l/mm grating is being used. Spectra in the X-ray region up to 1 A are taken with a variety of bent crystal spectrographs. In general, a unique array of spectrographs is in operation, which provides recording of spectra with the utmost resolution in the range 1 - 2500 A.

For the last 30 years wavelengths and energy levels were obtained for ions with several electrons, mainly for resonance series of hydrogen-and helium-like ions, using as a light source laser-produced plasmas or low-inductance vacuum sparks. The hydrogen sequence was studied up to GaXXXI, the helium sequence up to Mo XLI and the neon sequence up to Pr L.

A large class of investigated objects were the ions of elements from aluminum to arsenic with an open n = 2 shell. In combination with theoretical calculations by the 1/Z - expansion technique a complete understanding of the energy structure of n = 2 shell was achieved.

The analysis of a great number of ion spectra of extended isoelectronic sequences of copper, nickel, cobalt and iron having 3d - electrons in the outer shell for the elements from Ga to Mo (to Sn for Ni I - like ions) was done. For the Cu - like ions from Ga III to Br VII energies and widths of autionizing levels were measured.

Later the emphasis in data production was shifted to the ions of heavy elements with the 4d - and 5d -electrons in the outer shells. A project of study of moderately charged ions in platinum group elements is currently under way. The project is centered on astrophysically important Os. Studied were the Os VII - Os IV spectra as well as isoelectronic spectra of the neighboring elements Re and Ir. Preliminary data are obtained for Os III. On average about 1000 lines are identified in each spectrum. It is possible to make such analyses in a comparably short time due to a code for computer assisted identification of complex atomic spectra developed at this Laboratory. A final goal of the project is to study unknown third spectra and to improve and extend partly known second spectra of these elements.

For about the last 30 years more than 250 ion spectra were experimentally studied in this laboratory.

3.2 Bibliography on Atomic Spectra and Compilations

An atomic data bank BIBL for plasma physics, atomic physics, astrophysics etc. is being developed at the Institute for Spectroscopy.

The bibliography related to experimental and theoretical papers on identification and prediction of atomic and ionic structure should be complete from 1983, the year of the last issue of the NBS Special Publication "Bibliography on Atomic Energy Levels and Spectra", but experimental spectrum analysis can be traced back to about 1970. On other topics it is systematically maintained from 1989. For the last two years Current Contents on Diskettes is used for this purpose, with corresponding software for transforming relevant articles to the BIBLE format.

Jointly with NIST critical compilations of wavelengths, energy levels and life times for the ions of selected elements are performed. Compiled or close to final drafts are the spectra of Be I - Be II, B I - B III, F I - F VII and Ne I - Ne VIII.

3.3 Staff and Collaborations

The Division of Atomic Spectroscopy is headed by A.Ryabtsev since 1977 and at present time the permanent staff includes P.Antsiferov, A.Azarov, S.Churilov,

E.Ivanova, I.Ivanov, R.Kildiyarova, A.Kramida, K.Koshelev, L.Podobedova, and Yu.Sidelnikov. Parts of the results above have been obtained jointly with NIST, St.Francis Xavier University (Antigonish, Canada), Aime Cotton Laboratory, Meudon Observatory, Amsterdam University, and Ruhr-University Bochum. The most important results can be found in refs (13 - 19).

4. MISDC OF VNIIFTRI AND DATABASE "SPECTR"

The Multicharged Ions Spectra Data Center of VNIIFTRI is a research group, working more than 20 years in the field of X-ray spectroscopy of multicharged ions and its application for diagnostics of high-temperature plasmas. This group now consists of 7 researchers: A.Faenov, I.Skobelev, V.Palchikov plus 2 physicists and 2 engineers.

The main fields of investigations are:
- creation of Bragg and Bragg-Fresnel optic elements for high-resolution X-ray spectroscopy;
- high-precision X-ray spectroscopy of multicharged ions (emission spectra observations,
- spectral line identification, accurate wavelength measurements);
- creation of X-ray spectroscopy diagnostic methods and its applications for measurements of different plasma object parameters (laser-produced plasmas, Z- and X- pinches, plasma focus);
- creation of a database on spectra of atoms and ions (database "SPECTR").

4.1 The database "SPECTR"

The database SPECTR (DB "SPECTR"), on one hand, contains a great number of data on characteristics of atoms and ions, and, on the other hand, includes the software allowed to find fast the information needed, to browse it on the screen and to make a hard copy.

The information stored in DB SPECTR may be divided on the two parts. The first part (and now the greatest one) is the data on properties of isolated atoms and ions, that is, the data on energy levels structure, wavelengths of radiative transitions and its probabilities. The second part is the data on characteristics of collision processes. The main difference between these two parts is the following. The first part deals with a set of constants, while the second one with a set of functions. This difference causes the some difference in software used to operate with these data of different types.

The database "SPECTR" software is based on the FoxPro Database Management System. The information is stored in well-known DBF-format and can be transferred easily into another formats. The quantitative characteristics of the DB "SPECTR" are:

Data on spectral lines	397 885 records
Data on energy levels	75 100 records
Data on ionization potentials	2 292 records

Data on excitation cross sections and rates	4 932 records
Data on ionization cross sections and rates	145 records
Data on dielectronic recombination rates	276 records
Data on references	504 records
The total number of records	481 134

The qualitative characteristics:

1). "SPECTR" is a database on the characteristics of isolated atoms and ions, i.e. on spectral lines (wavelengths and radiative probabilities) and energy levels. The small number of collisional data were inputted practically only to test the software.

2). Since the MISDC research team works in the field of X-ray spectroscopy, the main part of data (about 75%) refers to the multicharged ions.

3). Data sources:

1. Published experimental data. For the X-ray region the database contains practically all published experimental data, but only some for the UV and visible regions.

2. In-house experimental data for multicharged ions. This research group has produced experimental data on X-ray spectra of multicharged ions during more than 20 years. For example during the last 5 years they produced new accurate data on satellite lines caused by radiative transitions in He-, Li-, Be-, B-, C-, N-, O-, F-, Na-, Mg- like ions and on high-n transitions in He- and Ne-like ions. Producing this kind of information is the main field of scientific interest of the MISDC group. For this purpose they create a new types of X-ray high-resolution spectrographs (with a spherically bent crystals, for example) and develop new methods to improve the accuracy of wavelength measurements.

3. Theoretical data, both published (not all, of course) and calculated especially for DB "SPECTR" in some institutes: Institute of Physics of the Lithuanian Academy of Science, Voronezh State University, Uzgorod State University. Data for ions with small number of electrons were calculated at VNIIFTRI with the help of relativistic perturbation theory with QED corrections.

Some examples are given in Table 2.

The MISDC group has collaborations with LLNL (Livermore), University of Central Florida (Orlando), Max Plank Institute (Jena), Max Born Institute (Berlin), Institute of Applied Physics and Computational Mathematics (Beijing,China) and with several Russian institutions. The results of the research activities are given in refs. 20 -30.

TABLE 2

QUERY	SEARCH TIME	THE NUMBER OF RECORDS SELECTED
Lines of copper ion in spectral region 10.1 - 10.11 A	2 s	9
Lines of silver ions in spectral region 10 - 15 A	5 s	94
Lines of H-like and He-like silicon	4 s	3192
3d-2p transitions in Ne-like and Na-like silver	2 s	313
1s2p 1P1 - 1s2 1S0 transition in He-like uranium	2 s	2
2s2p63p - 2s22p6 transitions in Ne-like ions	1 s	411
Lines of OI atom	1 s	441
Lines of O-like and N-like iron	3 s	998

5. ACTIVITIES AT TRINITI

The Troitsk Atomic Data Group is part of the Center of Theoretical Physics and Computational Mathematics, which is a division of the Troitsk Institute for Innovation and Fusion Research (TRINITI), a state Research Center subordinated to the Ministry of Atomic Energy of the Russian Federation. The staff consist of A.Godunov, P.Ivanov, V.Schipakov, Yu.Zemtsov and 1 technical assistant. Their research field is the theory of electronic and atomic collisions, that includes:
- direct and resonant ionization of atomic systems by charged particle impact, including the effects of strong electron correlations;
- two-electron transitions in collisions of atoms with charged particles;
- two-electron excitation and excitation-ionization via multiply excited states;
- dependence of the cross sections on the projectile (charge sign effects, mass effects, structured projectiles).

The group is also active in atomic data management involving development of methods and software for processing diverse atomic data; producing compact specialized banks of atomic data (excitation and ionization of atoms and ions by electron impact, autoionization, two-electron transitions in ion-atom collisions); data evaluation and approximation to obtain recommended values; supplying reliable atomic data for physical models of laboratory and space plasmas.

Their current activities will result in:
- HiBase software (version1.99) for hierarchical management of atomic data on personal computers of IBM type;
- specialized databases on excitation and ionization of atoms and ions by electron impact (databases include mainly the data from publications);
- compilation of experimental and theoretical data on the cross sections of Ne^{q+} ionization by electron impact available in the literature;
- Files with energy levels, oscillator strength and other information to be used as input for plasma calculations (H⁻ to Ne-like Ne, Al,Ar,Fe);
- specialized bibliography database for two-electron transitions in ion-atom collisions.

The group has collaborations with Paul Sabatier University, Toulouse, France; Tulane University, New Orleans, USA; Missouri Rolla University, Rolla, USA; Kansas State University, Manhattan, USA; Portland State University, Portland, OR, USA; Paris Sud University, Orsay, France. Some results of their activities are given in refs. (31 - 37).

6. CONCLUDING REMARKS

Two approaches have been discussed above. The first one is based on a set of computer programs to calculate cross sections and reaction rates for the general ion or atom and to employ them in the analysis of spectral line intensities observed from transient plasmas. The second one is the development of numerical databases with the data stored in the computer. The advantage of the first approach is obvious; however, for wavelengths and other atomic constants of complex atoms and ions only the second approach is possible.

The experience obtained can be also used in systematic studies of characteristics of atoms and molecules in strong electromagnetic fields in order to get reliable cross sections and rate coefficients for all types of electronic and atomic collisions in the presence of strong and superstrong fields.

ACKNOWLEDGMENTS

I am pleased to thank L.Vainshtein, I.Beigman, D.Uskov, A.Ryabtsev, A.Faenov, I.Skobelev, and A.Godunov for valuable discussions. The work was supported by the Russian Fund for Fundamental Research (project No. 96-02-16090).

REFERENCES

1. Sobelman, I.I., Vainshtein, L.A., and Yukov, E.A., *Excitation of Atoms and Broadening of Spectral Lines*, Berlin - Heidelberg: Springer Verlag, 1995.
2. Vainshtein, L.A., and Shevelko, V.P., *Structure and Characteristics of Ions in Hot Plasmas* (in Russian), Moscow: Nauka, 1986.
3. Shevelko, V.P., and Vainshtein, L.A., *Atomic Physics of Hot Plasmas*, Bristol-Philadelphia: Institute of Physics, 1993.
4. Presnyakov, L.P., Shevelko, V.P., and Janev, R.K., *Elementary Processes Involving Highly Charged Ions* (in Russian), Moscow: Energoatomizdat, 1986.
5. Janev, R.K., Presnyakov, L.P., and Shevelko, V.P., *Physics of Highly Charged Ions*, Berlin-Heidenberg: Springer Verlag, 1985.
6. Presnyakov, L.P. and Uskov, D.B., *Sov. Phys. JETP.* **59**, 515-523 (1984).
7. Presnyakov, L.P. and Uskov, D.B., *Proc. P.N.Lebedev Physical Inst.* **179**, 137-158 (1987).
8. Melchert, F., Benner, M., Kruedener, S., Shulze, R., Meuser, S., Huber, K., Salzborn, E., Uskov, D.B., and Presnyakov, L. P., *Phys. Rev. Lett.* **74**, 888-891 (1995).
9. Presnyakov, L.P., Tawara, H., Tolstikina I.Yu., and Uskov, D.B., *J. Phys.* **B 28**, 785-793 (1995).
10. Kruender, S., Melchert, F., Diemar, K., Pfeiffer, A., Huber, K., Salzborn, E., Uskov, D.B., and Presnyakov, L.P., *Phys. Rev. Lett.* **79**, 1002-1005 (1997)
11. Presnyakov, L.P., and Uskov, D.B., *JETP Lett.* **66**, 23-26 (1997).
12. Beigman, I.L., *Proc. P.N.Lebedev Inst.* **218**, 87-104 (1993).
13. Aglitsky E.V., Antsiferov P.S., Dricker M.N., Mandelstam S.L., Panin, A.M., *Opt.Commun.* **.55**, 97-98 (1985).
14. Aglitsky E.V., Antsiferov P.S., Mandelstam S.L., Panin A.M., Safronova U.I., UlitinS.A., Vainshtein L.A. *Phys.Scripta*, **38**, 136-142 (1988).
15. Aglitsky E.V., Antsiferov P.S., Mandelstam S.L., Panin A.M., *Can.J.Phys.* **62**, 1924-1927 (1984).
16. Ryabtsev A.N., *Nucl.Inst.Meth.Phys.Res.B* **31**, 196-205 (1988).
17. Azarov V.I., Churilov S.S., Kildiyarova R.R. et al. «Platinum group ion project», This issue.
18. A.E.Kramida, A.N.Ryabtsev, G.V.Vedeneeva, «The atomic spectroscopy bibliography database of the Institute for Spectroscopy», Troitsk. - This issue.
19. Dalton J.R., Dragoset R.A., Fuhr J.R. et al. «NIST atomic spectroscopic database.», This issue.
20. Boiko, V.A., Pal'chikov, V.G., Skobelev, I.Yu., Faenov, A.Ya., *Spectroscopic data of atoms and ions (atom spectra with one and two electrons)*, CRC Press, 1994.
21. Pikuz, S.A., Hammer, D.A., Kalantar, D.H., Faenov, A.Ya., Skobelev, I.Yu., *Phys.Rev.A* **49**, 3450-3452 (1994).
22. Bryunetkin, B.A., Faenov, A.Ya., Kalashnikov, M.P., Nickles, P.V., Schnurer, M., Skobelev, I.Yu., Abdallah, J., Clark, R.E.H., *JQSRT* **53**, 45-58 (1995).
23. Faenov, A.Ya., Hammer, D.A., Nilsen, J., Osterheld, A., Pikuz, S.A., Pikuz, T.A., Romanova, V.M., Shelkovenko, T., Skobelev, I.Yu., *Physica Scripta* **51**, 454-458 (1995).
24. Faenov, A.Ya., Skobelev, I.Yu., Pikuz, S.A., Kyrala, G.A., Gobble, J.A., Fulton, K.D., Abdallah J., Kilcrease, D.P., *Phys. Rev. A* **51**, 3529-3541 (1995).
25. Faenov, A., Bryunetkin, B., Dyakin, V., Pikuz, S., Skobelev, I., Safronova, U., Pikuz, T., Nilsen, J., Osterheld, A., *Phys.Rev.A* **52**, 3644-3661 (1995).
26. Abdallah, J., Faenov, A., Hammer, D., Pikuz, S., Csank, G., Clark, R.E.H., *Physica Scripta* **53**, 705-711 (1996).
27. Rosmej, F.B., Bryunetkin, B.A., Faenov, A.Ya., Skobelev, I.Yu., Kalashnikov, M.P., Nickles, P.V., Schnurer, M., *J.Phys.B. Letters* **29**, L299-306 (1996).
28. Magunov, A.I., Dyakin, V.M., Faenov, A.Ya., Pikuz, T.A., Skobelev, I.Yu., Osterheld, A., Goldstein, W.H., Flora, F., Di Lazzaro, P., Bollanti, S., Lizi, N., Letardi, T., Reale, A., Palladino, L., Batani, D., Mauri, A., Scafati, A., Reale L., *JETP* **83**, 267-273 (1996).

29. Magunov, A.I., Dyakin, V.M., Faenov, A.Ya., Pikuz, T.A., Skobelev, I.Yu., Pisarczyk, T., Parys, P., Wolowski, J., Makowski, J., Pikuz, S.A., Romanova, M., Shelkovenko, T.A., *Phys. Rev. A* **54**, 3971-3976 (1996).
30. Skobelev, I.Yu., Faenov, A.Ya., Dyakin, V.M., Fiedorowicz, H., Bartnik, A., Szcsurek, M., Beiersdorfer, P., Nilsen, J., Osterheld, A., *Phys.Rev.E* **55**, 3773-3781 (1997).
31. Godunov, A.L., McGuire, J.H., Schipakov, V.A., *J.Phys.B:At.Mol.Opt.Phys.* **30**, 3227-3245 (1997).
32. Godunov, A.L., Ivanov, P.B., Schipakov, V.A., *J.Phys.B:At.Mol.Opt.Phys.* **30**, 3403-3415 (1997).
33. Godunov, A.L., Schipakov, V.A., Moretto-Capelle, P., Bordenave-Montesquieu, D., Benhenni, M., Bordenave-Montesquieu A., *J.Phys.B:At.Mol.Opt.Phys.* **30**, (1997) (accepted for publication).
34. Moretto-Capelle, P., Bordenave-Montesquieu, D., Bordenave-Montesquieu, A., Godunov, A.L., Schipakov V.A., *Phys.Rev.Lett.,* (1997) (accepted for publication).
35. Godunov, A.L., Schipakov, V.A., *J.Phys.B:At.Mol.Opt.Phys.* **26**, L811-L816 (1993).
36. Godunov, A.L., Schipakov V.A., *Nucl.Instr.Meth.Phys.Res.* **B98**, 354-358 (1995).
37. Godunov, A.L., Makhrov, V.A., Sechin, A.Yu., Starostin, A.N., Stepanov A.E., *JETP* **82**, 1112-1119 (1996).

ATOMIC AND MOLECULAR DATABASES FOR PLANETARY AND TERRESTRIAL APPLICATIONS

Linda R. Brown

Jet Propulsion Laboratory, California Institute of Technology
4800 Oak Grove Drive, Pasadena, CA 91109, USA

Abstract. Characteristics of solar and planetary atmospheres ranging from chemical composition to climate can be accurately measured through spectroscopic remote sensing. Fundamental principles of physics and chemistry can then be applied to interpret the atmospheric observations (with varying degrees of success). The models of planetary atmospheres to simulate the chemistry, radiation fields, and/or dynamics require knowledge of many molecular properties ranging from photochemical and kinetic information to atomic and molecular transition parameters. Both the experimental and theoretical efforts require detailed databases of spectral parameters for use in computer simulations. The status of current databases for theoretical models and for microwave and infrared observations will be presented along with an overview of the problems associated with maintaining them.

OVERVIEW OF DATABASES FOR ATMOSPHERIC RESEARCH

Studies of a planetary atmosphere are undertaken to address important questions such as

· What is its composition?

· What is its weather and climate?

· Do our models based on "fundamental knowledge"

    a)   reproduce what we observe (to within measurement accuracies)

    b)   predict what will happen next

Answering these simply-posed questions proves very difficult because of the complexity of the dynamical processes. For example, characterizing the behavior of climate over time requires the merging of fundamental principles (physics, mathematics and chemistry). As illustrated in the lower part of Fig. 1, the development of intricate theoretical models is performed as an iterative process that is often driven by the experimental observations.

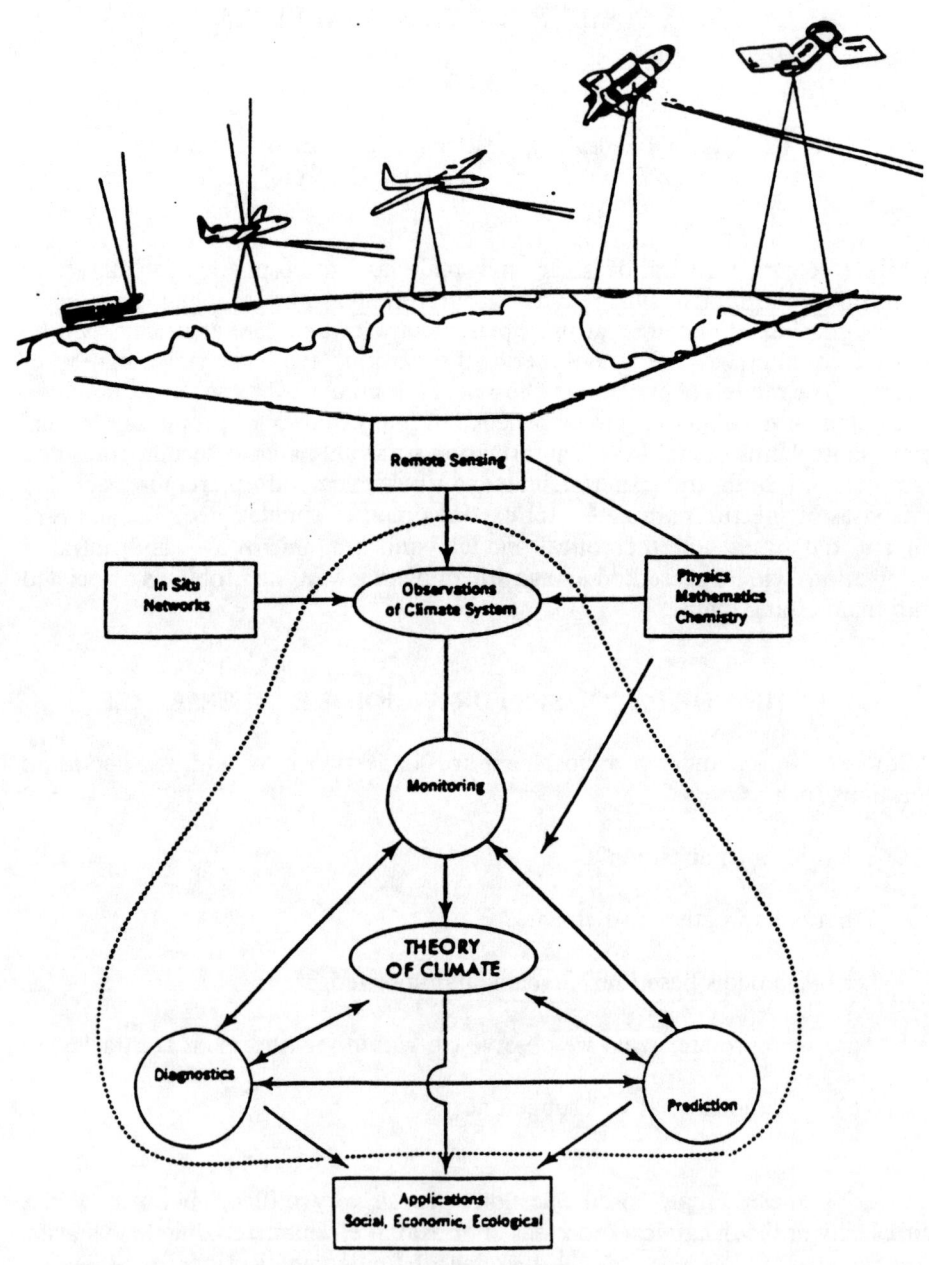

FIGURE 1. Interaction of observations and theoretical modelling of climate (1).

Extensive programs of field observations often are done through the remote sensing by spectroscopic techniques that utilize the full range of the electromagnetic spectrum from the microwave to the ultraviolet. A diverse complement of instrumentation is deployed on a variety of ground-, aircraft-, balloon-borne and space-based platforms (2,3).

Measurements from these systems then steer the development of models for both atmospheric climate and chemistry. As illustrated in Fig. 2 (4), within the Earth's atmosphere, there is a labyrinthine scheme of chemical reactions to unravel in order to understand the influence of natural and anthropogenic phenomena on the troposphere and stratosphere.

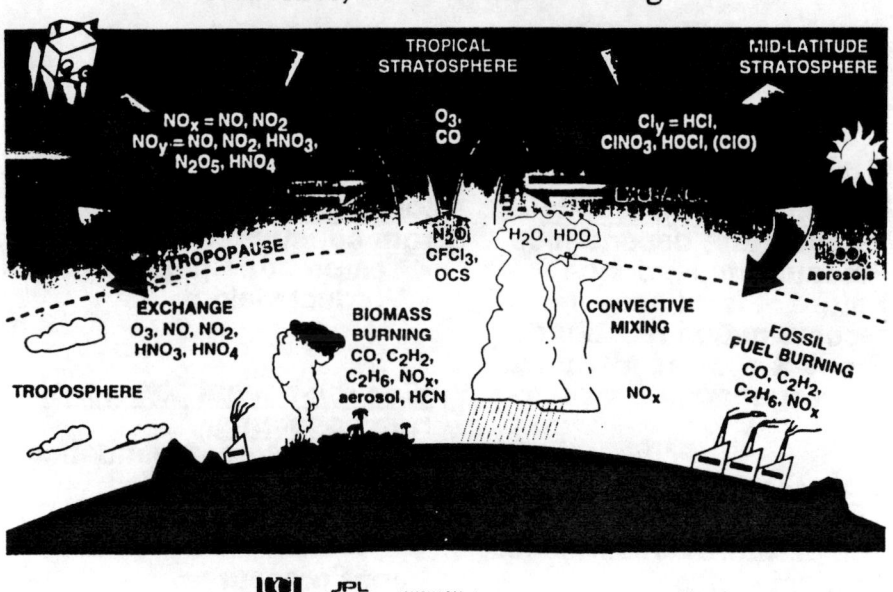

FIGURE 2. Environmental impact of industry, agriculture and biomass burning

Explaining such chemical mechanisms on a theoretical basis from fundamental principles (5) requires extensive chemical and spectroscopic information. For example, a list of the types of input needed for theoretical calculations was provided by Mark Allen at JPL (6) and is shown in Table 1; these in fact apply to any planetary or solar system body with an atmosphere. Some of the parameters are related to chemical properties such as vapor pressures, and various types of rate constants under different conditions of pressure, temperature and species combinations.

The spectroscopic parameters that we associate with molecules and atoms are indicated in the category of "photocross-sections." For some species and some spectral intervals (such as the visible and ultra-violet), these input parameters are simply absorption cross sections; they are obtained from the direct recordings of laboratory spectra and normalized using well-known and easily-applied equations. In other applications, individual features must be measured, and those retrieved numerical values are either used directly or modelled to Hamiltonian expressions to provide calculations that extrapolate beyond the content of an observed spectrum.

TABLE 1. TYPE OF INPUT CONSTANTS REQUIRED FOR MODELS

Collisional rate constants
- Temperature dependence
- Pressure dependence
- Nature of third body in recombination reactions
- Product species identified
- Heterogeneous processes

- **Photocross-sections**
  - Wavelength dependence
  - Temperature dependence
  - Product yields

Ion-electron recombination rate constants
- Temperature dependence
- Product yields

Electron-neutral collisional rate constants
- Temperature dependence
- Product yields

Vapor pressures
- Temperature dependence

Extensive efforts in laboratory studies must be made to measure and organize the required parameters. One source of information needed for chemical modeling calculations is *the Chemical Kinetics and Photochemical Data for Use in Stratospheric Modeling* (7,8). This is essentially a bibliography that gives literature sources of rate constants and UV cross sections for atmospheric models. The available information is evaluated so that recommendations can be made to the general community (see the Appendix for details about cross sections included). The primary application is the stratospheric modeling with emphasis on the ozone. With such models and input, it is possible to calculate the expected abundances of different chemicals as a function of altitude. These predictions are compared with observations such as those obtained by the ATMOS experiment (9,10). An example is shown in Fig. 3 which gives the volume mixing ratio of 19 molecular species containing Fluorine or Chlorine. The spectroscopic remote sensing provides necessary experimental measurements to validate or guide the development of reliable theoretical models.

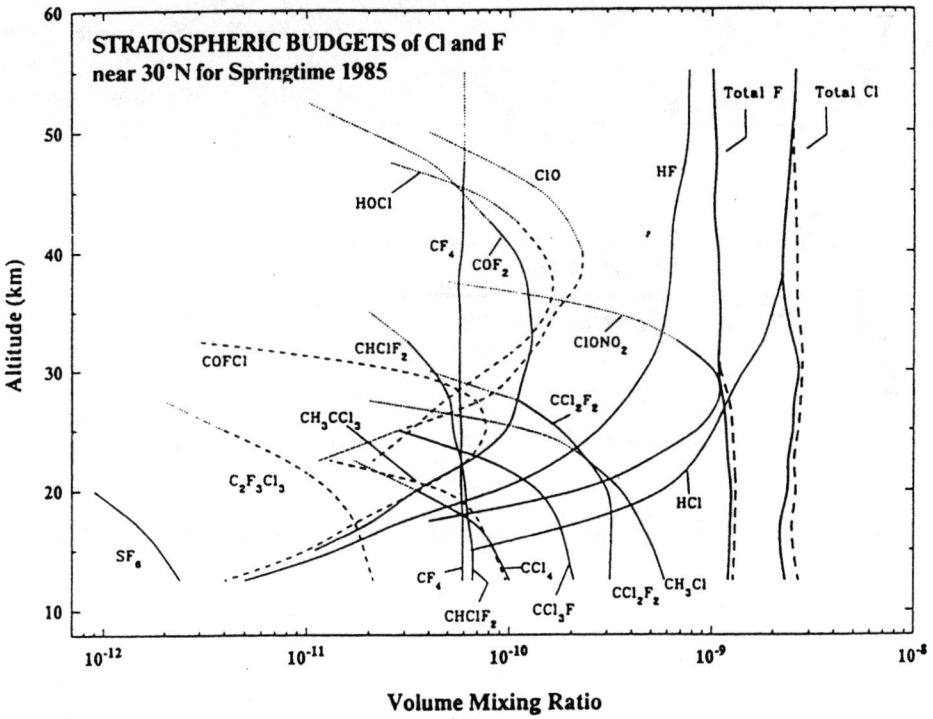

FIGURE 3. Volume mixing ratios versus pressure and geometric altitude for the Cl- and F-bearing species. The thin solid lines are measurements made with the ATMOS interferometer. The dashed and dotted lines are measurements from other sources (see Refs 9, 10 for retails).

Sometimes it seems that the input parameters for such models and observations are regarded as only a minor aspect of very extensive enterprises, but my view is just the opposite. The laboratory measurements are the true foundation of these other endeavors, as depicted in Fig. 4. The acquisition and interpretation of observations require specific atomic and molecular information and also model predictions related to radiative transfer. The models are dependent on the laboratory constants and the guidance from the field measurements. Our ultimate understanding of natural processes through model analysis and observations therefore relies on the availability of laboratory results.

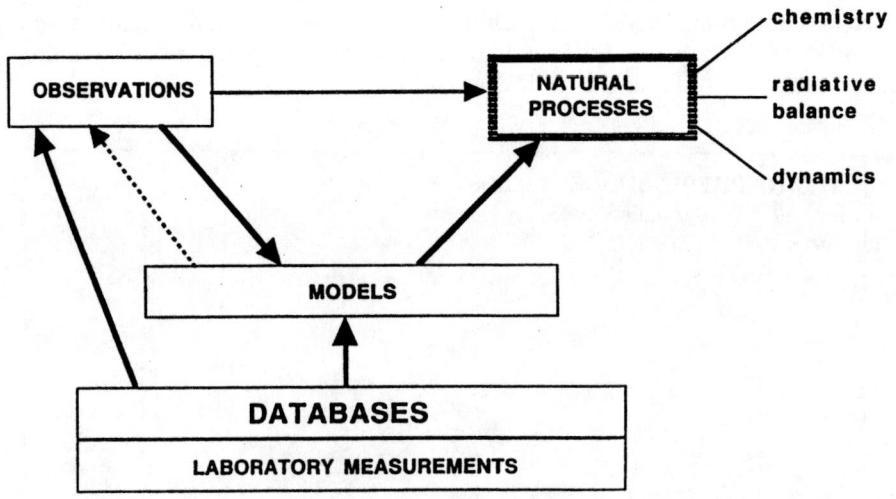

FIGURE 4. The dependence of observations and modelling on laboratory results

It is therefore essential that laboratory measurements provide information that is

- complete

- accurate

- organized in convenient form.

This means that experimental results must be available as electronic databases which have been critically reviewed and validated and then thoroughly documented.

In practice, the databases for spectroscopic remote sensing have been driven by what is needed to perform radiative transfer calculations (5) in order to compute synthetic spectra for comparison with observed spectra. The bulk of the activity has involved the infrared and microwave regions. The databases have been tailored to include specific knowledge of the molecular line parameters associated with the interaction of the electromagnetic radiation with molecules and atoms. The type of parameters required are shown in Table 2.

TABLE 2   DATABASE LINE PARAMETERS FOR REMOTE SENSING

    Rotation-vibration transition position (wavelength, frequency)

    Transition intensity
        or cross sections (for unresolvable bundles)
        including pressure-induced continua

    Line shape parameters
        Half-widths
        Pressure-induced frequency shifts
        Line-mixing parameters

    Parameters to describe how these vary with temperature and pressure
        Transition lower state
        Temperature coefficients for widths, shifts, line-mixing

---

The accuracies required depend on the application. The state-of-the-art results in experimental values are 0.0001 cm^{-1} for infrared positions, 3% for intensities, 3% of line widths, ≈ 20% pressure-induced frequency shifts and 15 - 30% for line mixing coefficients. Generally, the most important parameter for quantitative atmospheric observations is the intensity; errors in the laboratory values extrapolate directly on a one-to-one basis into the atmospheric observations.

As described by J.-M Flaud in another talk, the line positions and intensities found in a database are obtained by modelling a sampling of laboratory measurements with quantum mechanics to provide predicted values for several thousand transitions connected with a vibration-rotation band. In some cases, models fail to converge to the accuracy of the laboratory data and also to the accuracy required by the applications. In these cases, tedious and time-consuming empirical tabulations of line parameters are obtained from laboratory spectra. Some molecules give rise to unresolved bundles of transitions so that the "line-by-line" approach cannot be applied. In these cases, experimental absorption coefficients are used rather than intensities. For line shape parameters, theoretical models based on quantum mechanics are often inadequate, and so we rely on limited experimental values of a few hundred transitions extrapolated by arbitrary schemes to thousands of transitions associated with a specific molecule.

The more public databases of molecular parameters (7, 11-17) that are utilized by applications are summarized in Table 3 below. The year of the most recent release is shown before the name, and a Web site is given with the name of the database. The overall summary of region covered in cm^{-1}, number of species, number of transitions, and the references are indicated. The Appendix contains individual summaries of the contents of these databases by molecule and spectral region. Three of these (HITRAN, GEISA and the JPL catalog) have existed because they receive long-term funding to provide for specific types of applications.

TABLE 3. SPECTROSCOPIC DATABASES OF MOLECULAR PARAMETERS

DATABASE* and WEBSITES		region* cm^{-1}	number of species	millions of transitions	Ref.
1996	JPL catalog http://spec.jpl.nasa.gov	0 - 330	331 "entries"	1.5	14
1997	SAO-rotational http://firs-www.harvard.edu/dir/sao92	10 - 1600	49	0.5	16
1997	GEISA http://ara01.polytechnique.fr	0 - 22,600	50	0.7	13
1996	HITRAN http://www.hitran.com	0 - 22,600 - 58,500	46	1.0	11
1996	HITEMP	0 - 23,000?	3 or 4 ?	2.3	12
1995	ATMOS	0 - 10,000	49	0.8	15
1996	JPL VUV Reference Data http://remus.jpl.nasa.gov/jpl97	Vis-UV	≈ 65	cross sections	7
1993	Hanst Digitized IR Spectra	500 - 4000	250	lab spectra	17

acronyms:
  JPL is the Jet Propulsion Laboratory: "entries" indicate that different isotopes and bands of a species are classified as a different species

  SAO is the Smithsonian Astrophysical Observatory

  GEISA is Gestion et Etude des Informations Spectroscopiques Atmospheriques

  HITRAN is High resolution Transmission

  ATMOS is Atmospheric Trace Molecule Spectroscopy

  Vis-UV = visible and ultra-violet.

one wavenumber = 1 cm^{-1}; 5000 cm^{-1} = 2 μm and 1000 cm^{-1} = 10 μm

The particular set of molecular and atomic species that are required depends on the application. An example of the molecules for remote sensing of the Earth's atmosphere is indicated in Fig. 5. This graphic is taken from the Web site for the HITRAN (High Resolution Transmission) database (11). It shows the molecules for which there are line parameters incorporated as a function of wavelength. The most extensive information are parameters of the most abundant species like water, carbon dioxide, ozone, nitrous oxide and methane. The graph stops at 20 μm, but in fact the HITRAN list also extends through the microwave region as well. While the wavelength coverage is extensive, all parameters of all molecules in all spectral intervals are not equally well characterized. In general, the longer wavelength regions (beyond 4 μm in the infrared) are much better determined in terms of the degree of completeness and overall accuracies because there has been extensive usage of the region between 2.5 to 1000 microns by numerous field

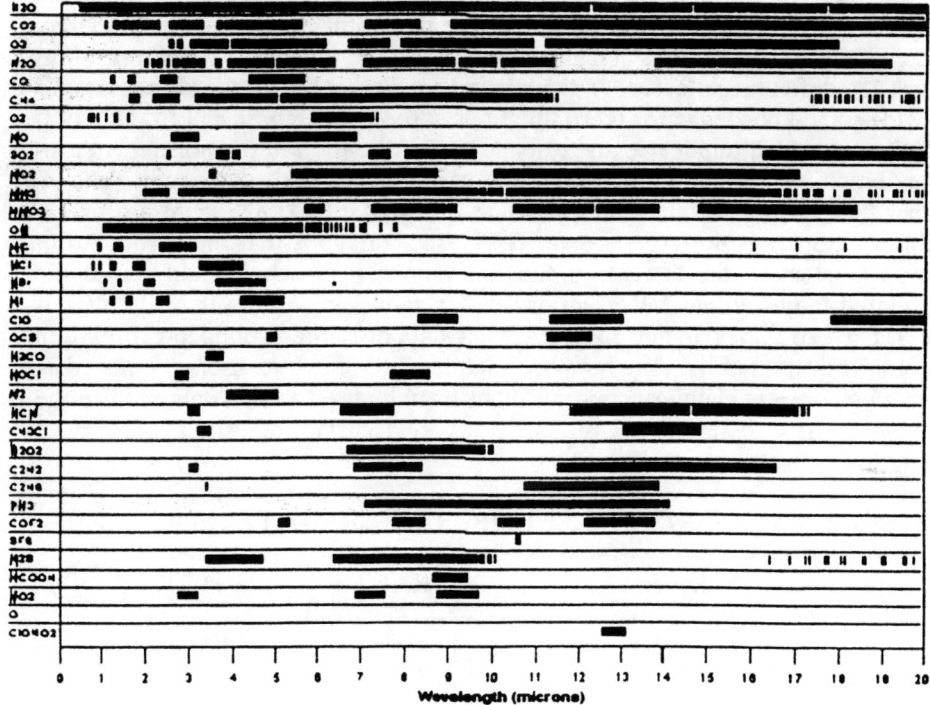

FIGURE 5. A summary of the 1996 HITRAN line parameters between 20 and 0.1 microns taken from the Web site information. The molecules from top to bottom are H_2O, CO_2, O_3, N_2O, CO CH_4, O_2, NO, SO_2, NO_2, HNO_3, OH, HF, HCl, HBr, HI, ClO, OCS, H_2CO, HOCl, N_2, HCN, CH_3Cl, H_2O_2, C_2H_2, C_2H_6, PH_3, COF_2, SF_6, H_2S, HCOOH, HO_2, ATOMIC O AND $ClONO_2$. The database also contains absorption cross sections of several species which give rise to unresolvable bands (such as CFC-11, CFC-12 and CFC-14).

experiments; also quantum mechanical models are more likely to reproduce the observed spectra in the rotational and fundamental regions. At wavelengths smaller than 2.5 microns, the compilations are generally incomplete even for the more abundant species like methane and inaccurate (for some of the bands of water).

HITRAN started in the early 1970's as the AFGL (Air Force Geophysical Laboratory) compilation (with funding from the Department of Defense) so that the transmission of the Earth's atmosphere could be understood. It has been revised and improved every two to four years by collecting new research funded by government agencies like NASA. In the past, the database organizers have also been able to support specific laboratory research in order to obtain improvements. This compilation has been the mainstay for atmospheric remote sensing.

However, for some applications, the standard HITRAN database has proved incomplete. A portion of the information about the HITRAN database given in the Appendix is repeated in Table 4 to illustrate this point. For example, planetary observations involving gases at high temperatures tend to see very weak transitions which are difficult to measure in the lab and predict by the standard models. To provide some knowledge for these situations, a second database call HITEMP (12) was formed which duplicates and extends the information in the normal HITRAN compilation for water, carbon dioxide and carbon monoxide. As seen in Table 4, rather than having 50 thousand transitions of water in HITRAN, there are over a million water transitions in HITEMP. The disadvantage is that the accuracies of the parameters in the HITEMP prediction vary widely, and in many cases the calculated intensity may be off an order of magnitude. However, if the user is observing "extra lines" in new remote sensing data or is interpreting low resolution spectra, the HITEMP list may be more useful. Alternatively, users interested in obtaining accurate chemical concentrations from high resolution data require the few (or the few thousand) transitions with very accurate intensities that can be found in HITRAN.

TABLE 4. COMPARISON OF NUMBER OF TRANSITIONS IN 1996

Mol #.	Species	HITRAN	HITEMP
1	H_2O	49,444	1,174,009
2	CO_2	60,802	1,032,259
3	O_3	275,133	0
4	N_2O	36,174	0
5	CO	4,477	113,022
6	CH_4	46,032	0

A second public database (13) is GEISA (Gestion et Etude des Informations Spectroscopiques Atmospheriques) described in the Appendix and Table 5. It was

TABLE 5. DIFFERENCES BETWEEN HITRAN 96 AND GEISA 97
(for molecules in common)
Underlined molecules have been updated for GEISA97 edition

Species	Region	v	I	γ	References
H2O	2v2-v2 lines of $H^{16}OD$ and v2 band of isotope 182 from ATMOS molecular line listin the region 8.9 - 5.7 μm (not included in HITRAN)	*	*	*	GEISA92 ATMOS line list
CO2	In the 4.3 μm region, lines of $^{18}O^{13}C^{18}O$ are included in GEISA as well as weak intensity lines of the other isotopes	*	*	*	GEISA86
O3	The 2 regions, 5.6 - 4.3 μm (bands 2v1, v1+v3, 2v3) and 4.3 - 3.6 μm [bands v1+2v2, 2v2+v3, 3v3-v2, and (v1+2v2+v3)- v2] have been updated	*	*	*	Barbe et al. Gamache et al.
N2O	The entire 19 - 16 μm region has been reworked. This includes 11 bands (band v2+v3-v2 new band)	*	*	*	Weber et al.
CO	In the whole spectral range 2929 - 1.18 μm transitions with v" ≥ 9 are present in GEISA (v" ≤ 1 in HITRAN)	*	*	*	GEISA92
CH4	New predictions in the regions 11 - 5 μm and 5.2 - 2.8 μm	*	*	*	Hilico et al. Champion et al.
NO	New spectroscopic line parameters for $^{14}N^{16}O$ in the region from microwaves to 1.07 μm	*	*	*	Goldman et al.
PH3	GEISA whole content in the region 561.6 - 4.1 μm is different from HITRAN	*	*	*	GEISA86 GEISA92
HNO3	3v9-v9 and v5+v9-v9 bands present in HITRAN96 have been replaced by new data for band v5+v9-v9 in the region 12.6 - 10.53 μm	*	*	*	Goldman et al. Perrin et al.
OH	New data for the whole content of ^{16}OH in the region from the microwaves to 2.4 μm	*	*	*	Goldman et al.
OCS	Region 20.4 - 2.4 μm entirely updated. Two new isotopes, $^{16}O^{12}C^{33}S$ and $^{16}O^{12}C^{34}S$	*	*	*	Fayt et al.
C2H6	The v9 band in the region 13 - 11 μm; The new isotope $^{12}C^{13}CH6$ band v12 in the region 13.7 - 10.9 μm	*	*	*	Weber et al.
CH3D	New list for the triad v3, v5, v6 in the region 11.8 - 5.7 μm (Independant molecule in GEISA; isotope of CH4 in HITRAN).	*	*	*	Nikitin et al.
HCN	The whole content in the 1.7 - 0.5 μm region	*	*	*	GEISA92
N2	The whole content in the 5 - 3.8 μm region	*	*	*	GEISA92
H2O2	The whole content with new predictions in the 4 μm and 2.7 μm regions	*	*	*	Perrin et al Gamache et al.
H2S	New predictions in the 4 μm and 2.7 μm regions	*	*	*	Brown et al.

started in the 1980's to support remote sensing of the planets. Its selection of included species contains species like GeH_4, AsH_3, C_2N_2 and HC_3N. These molecules are well below the detection limit for terrestrial observations, but they are abundant enough to be detected in the outer planets. Unfortunately, the GEISA organizers were unable to sustain funding for planetary studies. Later, the database became oriented toward terrestrial research as well, in a sense duplicating the efforts of HITRAN. However, the revisions of the GEISA database have generally been released in different years compared to HITRAN, and this alternation has provided users with faster access to improvements. Table 5 (18) compares differences between 1996 HITRAN and 1997 GEISA; the underlined species indicates that new laboratory studies are being included. It shows that a great deal of new information (involving 11 different species) became available in just one year after the 1996 HITRAN compilation was released. It demonstrates that we need to update and release our databases more frequently.

A third public database is the Jet Propulsion Laboratory catalog for the Submillimeter, Millimeter and Microwave Spectral region (14). This database started in the 1980's to support astrophysics observations in the frequency range between 0 and 10,000 GHz; it continues to exist because of funding from NASA. Their database contains both atomic and molecular species; a summary is given in the Appendix (19). One major difference between the JPL catalog and HITRAN or GEISA is that while the other two compilations assemble finished predictions from many different researchers, the JPL organizers gather reported laboratory measurements from many available sources and then perform their own quantum mechanical analyses in order to make the predictions for their database.

Because these three databases were started at different times and for different purposes, they have different selections of formats, parameters and units. A comparison of the parameters in the three lists is shown in Table 6. Some choices are dictated by the fact that these lists are intended as input into computer calculations rather than being reading material for human eyes. For example, the molecules and their isotopes are given identity numbers rather than names, and these become array indices in various standing software routines. Some parameter fields like the vibrational quantum numbers are rarely used in the calculations and so they are abbreviated by indices that require a lookup table to interpret. If line shape information is included, it is assumed to be air- and self-broadening values for terrestrial applications. Those applications that involve other planets must customize that portion of the lists for their own needs.

One important set of parameters on HITRAN is the scheme to indicate a range of accuracies for the positions, intensities and air-broadened widths and the sources of the data (listed as accuracy and reference indicators in Table 6). An accuracy index code appears with each transition in the database to provide users with embedded information about the data quality. This should be adopted by others.

Now, applications that implement new technology beyond the state-of-the-art inevitably find that the standard databases are lacking; new species are observed

TABLE 6. COMPARISON OF INCLUDED PARAMETERS FOR THREE DATABASES

PARAMETER	HITRAN	GEISA	JPL-cat
molecule number	x	x	x
isotope number	x	x	
frequency	x	x	x
absolute error in frequency			x
lower state energy	x	x	x
intensity	x	x	x
transition probability squared	x	≈x	
air-broadened widths	x	x	
self-broadened widths	x	x	
temperature dependence (air-width)	x	x	
pressure shifts (air-width)	x		
vibrational quanta	x	x	x
rotational quanta	x	x	x
accuracy range indicators	x	≈x	
reference indicators	x		

or accuracies of important parameters like the line intensities are too poor. For example, the ATMOS experiment (9,10,3) flew a high resolution interferometer on the space shuttle to record atmospheric and solar occultation spectra. Its investigators realized that the data to be acquired would contain features of species not already compiled in the standard databases and so they fixed the problem by creating their own specialized databases (15 and see the Appendix). Investigators at SAO (16) (Smithsonian Astrophysical Observatory) faced a similar situation, and that database is also summarized in the Appendix. Kelly Chance and Ken Jucks at SAO provided the overview of the their database (20) shown on the following page; their philosophy statements succinctly summarize what all of us are trying to do when we turn to "customizing". We start with the best available at the time (from HITRAN, GEISA, JPL CATALOG) and then do our own laboratory studies as needed. Our improvements are then fed back into the public databases. Much of what we do is to rehash existing information, but we make those changes that are important for our particular field experiment.

These public databases for remote sensing never have the objective to collect and organize all the molecular and atomic studies that have ever been published. Rather, their contents are defined by the particular applications. Those applications determine type of parameters, the selection of molecules and atoms, the wavelengths, the criteria for minimum intensities that will be included in a particular database. If an application is rather mature, then there is a good chance

---

Philosophy behind SAO database:

· To construct accurate database for analyzing FIRS-2 data

· Keep database as up-to-date as possible

· Cover the FIRS-2 spectrometer spectral range: 10 - 1600 cm^{-1}.

· Provide HITRAN with future updates in the far-infrared

Sources of data in SAO database

· HITRAN and JPL Submillimeter etc. Catalog

· "Pseudo lines" for cross section (HITRAN 96 and other sources)

· Best current laboratory data

---

that a database has been improved to accommodate the needs of that particular user community, especially if funding has been provided for focussed laboratory research.

In contrast, new or unexpected events often reveal new requirements. For example, the effect of biomass burning has received increased attention from the remote sensing community to the extent that spectra have been taken of forest fires. Figure 6 shows some aircraft data in the region near 5 microns recorded at 0.1 cm^{-1} resolution with an emission spectrometer (21). The upper panel shows the attempt to compute a synthetic spectrum using the 1992 HITRAN; it is generally matches the data with the exception of a few observed features. The modelling is improved by replacing the H$_2$O and CO parameters in HITRAN with those from HITEMP because the water vapor near the fire is at such elevated temperatures (400 to 800 K). So in this region, the available standard databases proves sufficient, but in other regions there are problems.

In general, the burning of trees and vegetation (i.e., biomass burning) generates high concentrations of some species that are not usually detected by routine atmospheric remote sensing. Interestingly, these include organic species such as acetic acid (CH$_3$OOH), methanol (CH$_3$OH), formic acid (HCOOH) and formaldehyde (H$_2$CO) for which the databases have few or no parameters in the infrared. For the remote sensing of wildfires and forest fires, new spectroscopic information is required (22, 23). For example, Table 7 shows some of the molecules and corresponding spectral intervals that are already being used to monitor these emissions. For the normal species (on the left), the compilations do not contain

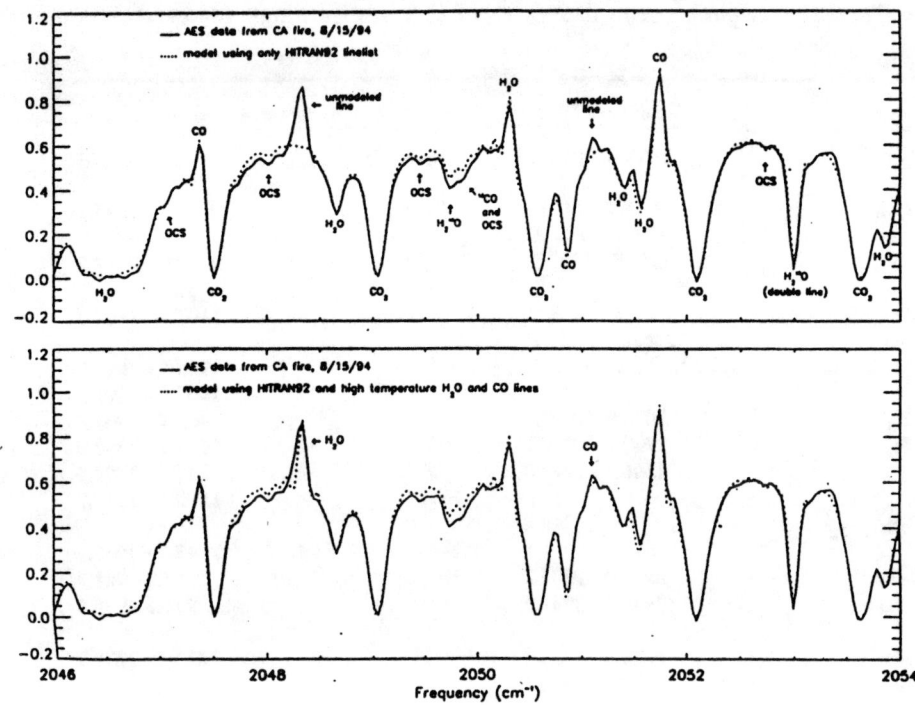

FIGURE 6. Spectrum of a California forest fire recorded at 0.1 cm^{-1} resolution by the Worden et al. (21) with AES (Aircraft Emission Spectrometer). The vertical scale is the uncalibrated flux in arbitrary units. The solid line is the observed spectrum, and the dotted line is the computed spectrum. In the upper panel, the line parameters for the calculation are taken from the 1992 HITRAN; these reproduce the observations fairly well except for two features (near 2048. and 2051 cm^{-1}). In the bottom panel, a separate calculation is made using line parameters of water and carbon monoxide from the HITEMP database; the missing lines are seen to arise from H_2O and CO respectively.

Reprinted from H. Worden et al., *Journal of Geophysical Research*, 102, pp 1287-1299, ©1997 by the American Geophysical Union.

enough hot bands. For the formic acid, a less desirable wavelength is utilized because no parameters are compiled for the other stronger fundamental bands. For the acetic acid and methanol, investigators must resort to low resolution cross sections available just at room temperature (such as the Hanst and Hanst collection). Other spectral regions for the organic molecules could be utilized to achieve better measurements and currently unknown features could be identified perhaps as species detections if complete spectroscopic information were accessible.

TABLE 7. Species seen in spectra of biomass burnings (see Ref. 23)

Molecule	Spectral Regions or Peaks	Molecule	Spectral Regions or Peaks
H_2O	3020.0 - 3007.1	CH_3COOH	1176
CO_2 and CO	2242.7 - 2210.9	CH_3OH	1033.3
CO_2	737.01 - 727.92		
CO_2	2252.8 - 2239.9[a]	CH_2O	2782.7 - 2777.3
CO	2055.8 - 2049.5[a]	HCOOH	1107.4 - 1102.7
^{14}CO	2062.2 - 2061.7[a]		
CO	2179.7, 2176.2, 2172.8[a]	NO	1913.4 - 1911.0
		NO	1907.1 - 1905.1
CH_4	3020.0 - 3007.1	NO	1903.8 - 1902.3
CH_4	2997.3 - 2976.1	NO	1901.1 - 1899.9
CH_4	1306.7 - 1299.0	NO	1906.9 - 1899.5
		NO_2	1599.5 - 1597.7
C_2H_4	949.4		
		NH_3	1047.3 - 1045.1
C_2H_4	2987.0 - 2985.7	NH_3	968.7 - 961.7
C_2H_4	2984.7 - 2982.7	NH_3	933.5 - 926.5
C_2H_2	3271.2 - 3254.7	HCN	3272.1 - 3270.7
C_2H_2	743.4 - 728.8		
		SO_2	1348.5 - 1347.7
Volatiles	2930	OCS	2071.4 - 2069.9

a) Stronger peaks used only to quantify background.
b) Regions used only to spot check the precision.

Reprinted from R. J. Yorkelson et al., *Journal of Geophysical Research*, **101**, pp 21067-21080, ©1997 by the American Geophysical Union.

The remote sensing of planets (other than the Earth) is hindered by the lack of a dedicated database. GEISA continues to accept new lists that are potentially useful, but that service does not produce the comprehensive collection required for every planet now being studied. Individual users are left to assemble their own database private collections and create customized compilations. The problem is that they must start the process at a more rudimentary level. What they produce is generally unvalidated, undocumented and undistributed.

Achieving the ideal database for the "planetary community" is difficult because the atmosphere of each planet is different. The vapor temperatures range from more than 700 K on Venus to less than 100 K in the outer planets (Jupiter and beyond). The abundances of the common species like methane and carbon dioxide vary from planet to planet by orders of magnitude, and planets have different dominant species (carbon dioxide for Venus, hydrogen and helium for the outer planets, nitrogen and oxygen on Earth). For example, Table 8 (24,25) gives a tabulation of species and mixing ratios for Jupiter. For the terrestrial molecules like CO_2, O_2, O_3 and N_2O, there have been extensive laboratory efforts to provide an adequate database for atmospheric studies, but these four species are completely missing in Jupiter. Even for a terrestrial species like methane, the standard databases are often insufficient because the abundances and optical densities in the outer planets are several orders of magnitude greater than for the Earth. Planetary investigators need a HITEMP-like collection that does not reject weak features and also provides line shape parameters for a variety of broadening species. Unfortunately, long-term funding has never been committed for the creation of a universal planetary database.

TABLE 8. THE COMPOSITION OF JUPITER: SPECIES, MIXING RATIOS

Major Species	*Mixing Ratios*	*Disequilibrium Species*	
H_2	0.865	PH_3	$(1-2) \times 10^{-7}$
He	0.156 ± 0.006		6×10^{-7} (>1bar)
		CO	2×10^{-9}
Principal Minor Constituents		GeH_4	$7 \pm 2 \times 10^{-10}$
H_2O	$\sim 2.6 \times 10^{-3}$	AsH_3	$2.2 \pm 1.1 \times 10^{-10}$
	$<10^{-6}$ (<4 bar)		
	$\leq 3.7 \times 10^{-4}$ (10 bar)	*Other Minor Constituents*	
	? (20 bar)	H	Variable
CH_4	$2.0 \pm 0.15 \times 10^{-3}$	$(H_2)_2$	Variable
	$2.5(+3/-2) \times 10^{-5}$		
C_2H_6	$1-5 \times 10^{-6}$	*Noble Gases, and Isotopic Ratios*	
C_2H_2	$2-8 \times 10^{-8}$	He	0.156 ± 0.006
	$<2.5 \times 10^{-6}$	^{20}Ne	$2.3 \pm 0.25 \times 10^{-5}$
C_2H_4	$7 \pm 3 \times 10^{-9}$	^{36}Ar	$1 \pm 0.4 \times 10^{-5}$
C_3H_4	$2.5(+2/-1) \times 10^{-9}$	^{84}Kr	$\leq 8.5 \pm 4 \times 10^{-9}$
C_6H_6	$2(+2/-1) \times 10^{-9}$	^{132}Xe	$\leq 5 \pm 2.5 \times 10^{-9}$
NH_3	$2.3-2.9 \times 10^{-4}$	$^3He/^4He$	$1.1 \pm 0.1 \times 10^{-4}$
	$\sim 0.2-1 \times 10^{-5}$	D/H	$2.2 \pm 0.5 \times 10^{-5}$
H_2S	$(1-9) \times 10^{-5}$		$3.5 \pm 1.5 \times 10^{-5}$
	$<1 \times 10^{-6}$	$^{12}C/^{13}C$	91

The lack of a complete database was dealt with temporarily last year for NIMS (Near Infrared Mapping Spectrometer) experiment (26) on the Galileo spacecraft by the creation of an interim collection. Existing calculations and predictions from work in progress were assembled together for the six species in Table 9 (the number of transitions, wavenumber range, bands, summation of the strengths and data source are shown as well). Air-broadened linewidths were overwritten with linewidths corresponding to an atmosphere composed of 90% hydrogen and 10% helium. For methane, some new predictions were made with a much lower intensity limit (10^{-27}).

Now, don't expect to see this collection advertised on a Web site! It has been useful for this one application involving low resolution spectra, but it is not yet of good enough quality for general usage. Parts of this NIMS database will give poorer results if used by high resolution experiments in place of HITRAN. I feel that we must be the "gate-keeper" and keep our poorer quality data from being widely distributed. In terms of the methane and ammonia studies indicated in Table 9, when the laboratory studies are finally concluded, new calculations will be generated, validated, documented and then submitted to the standards databases for inclusion. The point is that the individual producers have the responsibility to "do no harm".

TABLE 9.   NIMS CUSTOMIZED DATABASE

WITH 90% H_2 + 10% He LINE WIDTHS

MOL.	#LINES	FMIN - FMAX cm^{-1}	NUMBER ISO BANDS		STRENGTH x 10^{-17}	SOURCE (et al.)
CH_4	219208	0.0 - 6203.6	3	117	1.77	Hilico,Brown
NH_3	18075	0.2 - 7037.3	2	45	4.78	Kleiner,Brown,Urban
PH_3	9491	708.1 - 2478.8	1	8	2.82	GEISA,Tarrago
AsH_3	4445	1923.5 - 2289.9	1	2	4.04	GEISA, Tarrago
GeH_4	824	1937.4 - 2224.6	1	1	4.10	GEISA
H_2O	49444	0.0 -22656.5	4	137	7.28	96HITRAN + Brown,Gamache

- STRENGTH is the summation of all line intensities in $cm^{-1}/(cm^{-2} \cdot molecule)$ at 296 K.
- The database consists of separate files for each molecule covering the range shown above.
- Calculations from different sources were merged together and converted to HITRAN format.
- The air-broadened line widths in the database were replaced with widths appropriate for an atmosphere composed of 90% hydrogen and 10 % helium.
- After the release of the 1996 HITRAN, new results were obtained for methane and ammonia.
- The number of the methane transitions increased from some 47,000 on HITRAN to over 200,000 because new predictions of bands from the ground state and difference bands of both isotopes were done with a lower state intensity limit of 10^{-27}.
  Above 3400 cm^{-1}, the NIMS database contains only $^{12}CH_4$ transitions from the ground state.

In terms of new remote sensing activities, the people who study comets are having the most fun. First there was the spectacular crash of Shoemaker-Levi-9 comet fragments into Jupiter in 1994. Then there was the surprise appearance of two comets, Hyakutake and Hale-Bopp, which were bright enough to permit good spectra to be recorded at many wavelengths. Even the infrared has proved useful, as seen by the detection of CO, CH_4 and C_2H_6 in C/1996 B2 Hyakutake, shown in Fig. 1 of Mumma et al. (27). There now have been over two dozen molecular

TABLE 10. ORGANIC VOLATILES DETECTED IN COMETS

Species	Halley	Hyakutake	Hale-Bopp
H_2O	x	x	x
CH_4	x	x	x
C_2H_2		x	
C_2H_6		x	x
CH_3OH	x	x	x
H_2CO	x		x
HCOOH	x	x	x
CO_2, CO	x	x	x
HCN	x	x	x
H_2S			x
CS_2			x
OCS			x
H_2CS			x
SO_2			x

M. Mumma, private communication, Sept., 1997

and atomic species discovered in cometary vapors using wavelengths from the UV to the microwave. For example, a list of some of the detected organic species compiled by M. Mumma (27) is shown in Table 10. Surprisingly, volatiles ejected from these comets even include the "forest fire" species of methanol, formaldehyde and formic acid (listed in Table 7), as well as many sulfur-bearing species. Missing thus far, however, are unambiguous observations of nitrogen oxides such as NO, NO_2 and N_2O.

For unexpected events like these, the standard databases are lacking. Users must then survey the literature, dig out information from 30 year-old dissertations, find unpublished collections of low resolution absorption coefficients, and really scramble before they can begin to interpret some very interesting physical processes within the cometary "atmospheres."

CONCLUSION

In summary, progress made in the remote sensing applications is tied directly to quality of the database that is available. The standard databases in the infrared and microwave are very good for terrestrial applications, but not for planetary studies in general. To support just the planetary atmospheres community we should:

Organize: Define the needs. Assign tasks to avoid duplication of effort

Centralize: Collect and distribute by a dedicated staff

Generalize: Anticipate future needs

Standardize: Define selection of parameters, their units, the computer formats

Criticize: Do critical evaluations, assess data quality, detail inadequacies

Document: Provide thorough and detailed description of the database at the same time it is released

ACKNOWLEDGEMENTS

Part of the research reported in this paper was performed at the Jet Propulsion Laboratory, California Institute of Technology, under contract with the National Aeronautics and Space Administration. The author especially grateful to colleagues who readily provided information for this paper: Larry Rothman at the Phillips Laboratory; Nicole Jacquinet-Husson at Laboratoire de Meterologie Dynamique, Paris; Kelly Chance and Ken Jucks at the Smithsonian Astrophysical Observatory; Mark Allen, Ed Cohen, Bill DeMore, Mike Gunson, Glenn Orton, Herb Pickett at the Jet Propulsion Laboratory; Mike Mumma at NASA Goddard, and P. Hanst, Infrared Analysis, Inc. Anaheim, CA 92801. It should be noted that some of these provided preliminary versions of information that will be presented at a later date in their own publications.

REFERENCES

1. Peixoto, J. P., and Oort, A. H., *The Physics of Climate*, New York, AIP, 1992, ch. 1, pg. 3.
2. Grant, W. B., *J. Air Waste Mang. Ass.* **42**, 18 (1992).
3. Brown, L. R., Farmer, C. B., Rinsland, C. P., and Zander, R., pp. 97-152, in *Spectroscopy of the Earth's Atmosphere and Interstellar Medium*, Eds. K. Narahari Rao and A. Weber, Academic Press, Inc Boston, USA.
4. Gunson. M. R. (private communication).
5. Goody, R. M. and Yung, Y. L., *Atmospheric Radiation Theoretical Basis*, 2nd Edition, 1989, Oxford University Press, New York, Oxford.
6. Allen, M. (private communication).
7. DeMore, W. B., Sander, S. P., Howard, C. J., Ravishankara, A. R., Golden, D. M., Kolb, C. E., Hampson, R. F., Kurylo, M. J., Molina, M. J., *Chemical kinetics and photochemical data for use in stratospheric modeling*, JPL publication 97-4 (1997). Pasadena, Ca 91009.
8. Demore, W. B. (private communication).
9. Farmer, C. B., and Norton, R. H., A high resolution infrared spectroscopy atlas of the infrared spectrum of the Sun and the Earth from space, *NASA Reference Publication 1224*, vols 1 and 2 (1989).
10. Zander, R., Gunson, M. R., Farmer, C. B., Rinsland, C. P., *J. Atm. Chem.* **15**, 171-186 (1992).
11. Rothman, L. S., Gamache, R. R., Tipping, R. H., Rinsland, C. P., Smith, M. A. H., Benner, D. C., Devi, V. Malathy, Flaud, J. -M., Camy-Peyret, C., Perrin, A., Goldman, A., Massie, S. T., Brown, L. R., and Toth, R. A., *J. Quant. Spectrosc. Rad. Transfer* **48**, 1-780 (1992).
12. Selby, J. E. A. et al. Reims, ASA and L. S. Rothman, et al., *JQSRT* (submitted).
13. Husson, N., Bonnet, B., Scott, N. A., and Chedin, A., *J. Quant. Spectrosc. Radiat. Transfer* **48**, 509-518 (1992) and Jacquinet-Husson, N. et al., in preparation.
14. Pickett, H. M., Poynter, R. L., Cohen, E. A., Delitsky, M. L., Pearson, J. C., and Muller, H. S. P., *Submillimeter, millimeter and microwave spectral line catalog*, JPL Publication 80-23, Rev.4 (1996). Pasadena, CA.
15. Brown, L. R. et al., *Appl. Opt.* **26**, 5154-5182 (1987) and *Appl. Opt.* **35**, 2828 (1996).
16. Chance, K. V., Jucks, K. W., Johnson, D. G., and Traub, W. A., *J. Quant. Spectrosc. Radiat. Transfer* **52**, 447-457 (1994).
17. Hanst and Hanst, Digitized Reference Spectra (Infrared), copyright 1993, Infrared Analysis, Inc., Anaheim, CA 92801
18. Jacquinet-Husson, N. (private communication).
19. Cohen, E. and Pickett, H. M., (private communication).
20. Chance, K. V., and Jucks, K. W. (private communication).
21. Worden, H., Beer, R. and Rinsland, C. P., *J. Geophys. Res.* **102**, 1287-1299 (1997).
22. Griffith, D. W. T., Mankin, W. G., Coffey, M. T., Ward, D. E. and Riebau, A., J. S. Levine Ed., *Global Biomass Burning: Atmospheric, climatic and Biospheric Implications*, Cambridge, MA, MIT Press, 1991.
23. Yokelson, R. J., Griffith, D. W. T., Ward, D. E., *J. Geophys. Res.* **101**, 21067-21080 (1997).
24. Orton, G. S. (private communication).
25. Atreya, S. K., Wong, M. H., Owen, T., Niemann, H., and Mahaffey, P., *The three Galileos: The Man, The Spacecraft, The Telescope*, Rahe, J., Barbieri, C., Johnson, T., Sohus, A., Eds., Dordrecht, Kluwer Academic Publishers, 1997.
26. Carlson, R. W, Baines, K. H., Encrenaz, T., Taylor, F. W., Drossart, P., Kampl, L. W., *Science* **253**, 1541-1548 (1991).
27. Mumma, M. J., Disanti, M. A. Russo, N,. D., Fomenkova, M., Magee-Sauer, K., Kaminski C. D., and Xie, D. X., *Science* **272**, 1310-1314 (1996), and Mumma, M. J. (private communication).

a. **JPL BIBLIOGRAPHY:** DeMore, W. B., Sander, S. P., Howard, C. J., Ravishankara, A. R., Golden, D. M., Kolb, C. E., Hampson, R. F., Kurylo, M. J., Molina, M. J., *Chemical kinetics and photochemical data for use in stratospheric modeling*, JPL publication 97-4 (1997). Pasadena, Ca 91009.

JPL CHEMICAL KINETICS AND PHOTOCHEMICAL DATA 1997

Species and wavelength range (nanometers)

Species	Range	Species	Range
&O_2	205-240	CCl_4	174-275
&O_3	175-362	CCl_3F	170-260
*O_3		CCl_2F_2	170-240
HO_2	190-250	$CF_2ClCFCl_2$	184-230
H_2O	175-189	CF_2ClCF_2Cl	172-220
H_2O_2	190-350	CF_3CF_2Cl	172-204
NO		CF_4	
& NO_2	202-422	C_2F_6	
* NO_3	600-670	CCl_2O	184-229
N_2O	173-240	$CClFO$	184-229
N_2O_5	200-280	CF_2O	184-229
NH_3		# CF_3OH	
HONO	310-396	CH_3Cl	186-216
* HNO_3	190-350	CH_3CCl_3	185-240
HO_2NO_2	190-325	$CHClF_2$	174-204
CO		CH_3CF_2Cl	
CO_2		CF_3CHCl_2	
CH_4		CF_3CHFCl	
CH_2O	301-356	CH_3CFCl_2	190-220
& CH_3O_2	210-290	' $CF_3CF_2CHCl_2$	190-220
& $C_2H_5O_2$	215-295	CF_2ClCF_2CHFCl	190-220
CH_3OOH	210-280	# CH_3OCl	230-394
# $CH_3C(O)O_2NO_2$	190-350	BrO	313-374
HCN		# HOBr	300-375
CH_3CN		* $BrONO_2$ +	200-300
Cl_2	260-360	# BrCl	200-380
& OClO	272-475	CH_3Br	200-380
& ClO_3		& CF_3Br	190-272
& Cl_2O	200-320	& CF_2Br_2	190-272
& Cl_2O_2	190-386	& CF_2BrCF_2Br	190-272
* Cl_2O_3	220-320	& CF_2ClBr	190-272
Cl_2O_4	200-310	CF_3I	240-344
Cl_2O_6	200-380	SO_2	
HCl	140-220	H_2S	
HF		CS_2	
& HOCl	200-380	OCS	186-226
ClNO	190-500	SF_6	
# FNO	180-350	NaOH	
$ClNO_2$	190-370	NaCl	189-292
ClOO	220-280		
ClONO	235-400		
* $ClONO_2$	196-414		

(1) H:...son and Kieffer [132]. (2) Turco [305]. # New Entry.
* Indicates a change in the recommendation from the previous evaluation.
& Indicates a change in the note.

b. **1996 HITRAN** Rothman, L. S., Gamache, R. R., Tipping, R. H., Rinsland, C. P., Smith, M. A. H., Benner, D. C., Devi, V. Malathy, Flaud, J. -M., Camy-Peyret, C., Perrin, A., Goldman, A., Massie, S. T., Brown, L. R., and Toth, R. A., *J. Quant. Spectrosc. Rad. Transfer* **48**, 1-780 (1992).

1996 HITEMP: Selby, J. E. A. et al. Reims, ASA and L. S. Rothman, et al., *JQSRT* (submitted).

Table 2. Summary of species represented in HITRAN

Mol. No.	Specie	Number of Isotopes	Number of Bands	Number of Lines	Spectral Coverage (cm^{-1})
1	H_2O	4	137	49 444	0 - 22657
2	CO_2	8	589	60 802	442 - 9649
3	O_3	3	106	275 133	0 - 4033
4	N_2O	5	164	26 174	0 - 5132
5	CO	5	47	4 477	3 - 6418
6	CH_4	3	51	48 032	0 - 6185
7	O_2	3	19	6 292	0 - 15928
8	NO	3	50	15 331	0 - 3967
9	SO_2	2	9	38 853	0 - 4093
10	NO_2	1	12	100 680	0 - 2939
11	NH_3	2	40	11 152	0 - 5295
12	HNO_3	1	13	165 426	0 - 1770
13	OH	3	103	8 676	0 - 9997
14	HF	1	6	107	41 - 11536
15	HCl	2	17	533	20 - 13458
16	HBr	2	16	576	16 - 9759
17	HI	1	9	237	12 - 8488
18	ClO	2	12	7 230	0 - 1208
19	OCS	4	7	858	0 - 2089
20	H_2CO	3	10	2 702	0 - 2999
21	HOCl	2	6	15 565	0 - 3800
22	N_2	1	1	120	1922 - 2626
23	HCN	3	8	772	2 - 3422
24	CH_3Cl	2	8	9 355	679 - 3173
25	H_2O_2	1	2	5 444	0 - 1500
26	C_2H_2	2	11	1 668	604 - 3375
27	C_2H_6	1	2	4 749	720 - 3001
28	PH_3	1	2	2 886	708 - 1411
29	COF_2	1	7	54 866	725 - 1982
30	SF_6	1	1	11 520	940 - 953
31	H_2S		15	7 151	2 - 2892
32	HCOOH	1	1	3 388	1060 - 1162
33	HO_2	1	4	26 963	0 - 3676
34	O	1	1	2	68 - 159
35	$ClONO_2$	1	3	32 199	763 - 798
36	NO^+	1	6	1 206	1634 - 2531
37	HOBr	2	2	4 358	0 - 316

Band Stats for Supplemental files

3:High-vib O_3	184724
36:NO^+	1206
37:HOBr	4358

Band Stats for HITEMP

1:H_2O	1174009
2:CO_2	1032269
5:CO	113022

Band Stats for UV

7 O_2	110201

Cross-section Sizes

N_2O -UV	143308 Bytes
SO_2 -UV	2377758 Bytes
IR	17963586 Bytes

Aerosols

	229014 Bytes

c. 1997 GEISA: Husson, N., Bonnet, B., Scott, N. A., and Chedin, A., *J. Quant. Spectrosc. Radiat. Transfer* **48**, 509-518 (1992) and Jacquinet-Husson, N. et al., in preparation.

Detailed Content of GEISA97 Edition

Mol.	ID	# Lines	Intensity average	Alpha average	Iso. ID	# Lines	F-Min (cm^{-1})	F-Max (cm^{-1})	Int-Min $(cm\ molec^{-1})$	Int-Max $(cm.mol.^{-1})$
H2O	1	50217	1.450E-21	0.069	161	30117	0.401	22656.465	1.010E-32	2.670E-18
					162	9799	0.007	5507.548	1.240E-32	2.700E-22
					171	3744	6.471	11150.790	1.490E-27	9.830E-22
					181	6357	6.785	13900.421	1.000E-27	5.390E-21
					182	200	1231.680	1607.611	1.000E-26	7.940E-26
CO2	2	62816	1.793E-21	0.071	626	27896	442.006	9648.007	0.000E+00	3.520E-18
					636	9154	497.201	8104.666	0.000E+00	3.570E-20
					628	13554	507.860	8132.007	1.390E-36	6.850E-21
					627	6625	554.909	6961.226	1.000E-27	1.280E-21
					638	2312	567.596	4946.384	3.700E-27	7.230E-23
					637	1584	584.754	3641.072	3.710E-27	1.360E-23
					828	1107	615.974	3669.609	1.760E-40	1.310E-23
					728	288	626.438	2358.226	3.870E-27	2.500E-24
					838	296	2115.685	2276.481	4.870E-42	1.760E-25
O3	3	281607	6.533E-23	0.069	666	167755	0.026	4060.783	4.060E-29	4.200E-20
					668	19147	0.921	1177.493	4.880E-28	7.760E-23
					686	7513	1.177	1145.690	7.500E-28	7.560E-23
					667	58254	0.289	820.380	5.340E-31	5.570E-25
					676	28938	0.213	822.795	1.490E-31	6.060E-25
N2O	4	26771	2.688E-21	0.075	446	19423	0.838	5131.249	4.000E-27	1.000E-18
					447	1004	542.242	3482.917	3.430E-26	4.150E-22
					448	2034	545.179	3463.967	1.230E-25	2.050E-21
					456	2128	5.028	3462.689	5.220E-26	3.670E-21
					546	2182	4.858	3473.528	4.720E-26	3.600E-21
CO	5	13515	7.545E-22	0.047	26	5908	3.530	8464.883	7.880E-78	4.460E-19
					27	748	3.714	6338.061	8.190E-40	1.600E-22
					28	770	3.629	6266.578	7.610E-39	8.320E-22
					36	4768	3.414	8180.219	3.610E-73	4.680E-21
					38	741	3.462	6123.294	2.580E-40	8.700E-24
					37	580	1807.871	6196.551	1.030E-36	1.680E-24
CH4	6	66883	2.622E-22	0.052	211	56989	0.010	6184.492	4.060E-34	2.062E-19
					311	9894	0.032	6069.086	4.100E-34	2.329E-21
O2	7	6292	4.117E-26	0.044	66	1435	0.000	15927.806	0.000E+00	8.833E-24
					67	4186	0.000	14536.515	1.147E-47	5.337E-26
					68	671	1.572	15851.213	1.186E-35	1.710E-26
NO	8	94738	4.989E-23	0.053	46	93360	0.000	9273.214	1.401E-85	6.211E-20
					48	679	1601.909	2038.846	4.190E-28	1.390E-22
					56	699	1609.585	2060.462	4.430E-28	2.550E-22
SO2	9	38853	1.065E-21	0.114	626	38566	0.017	4092.948	1.020E-28	6.090E-20
					646	287	2463.470	2496.088	9.740E-24	3.430E-23
NO2	10	100680	6.194E-22	0.067	646	100680	0.498	2938.381	4.240E-28	1.300E-19
NH3	11	11152	4.275E-21	0.077	411	10062	0.215	5294.501	2.970E-29	5.450E-19
					511	1090	0.375	5179.786	5.460E-29	1.990E-21
PH3	12	4635	6.457E-21	0.075	131	4635	17.805	2445.553	3.690E-28	2.930E-19
HNO3	13	171504	6.879E-22	0.105	146	171504	0.035	1769.982	3.490E-27	3.020E-20
OH	14	41786	1.048E-21	0.044	61	41631	0.005	19267.869	1.401E-85	3.458E-18
					62	90	0.010	1.824	2.090E-31	5.780E-29
					81	65	0.053	6.325	1.200E-30	1.200E-26
HF	15	107	6.772E-19	0.041	19	107	41.111	11535.570	1.100E-26	1.440E-17
HCl	16	533	3.189E-20	0.040	15	284	20.270	13457.841	1.090E-26	5.030E-19
					17	249	20.240	10994.721	1.010E-26	1.610E-19
HBr	17	576	1.072E-20	0.051	19	289	16.237	9758.564	1.000E-26	1.210E-19
					11	287	16.232	9757.189	1.010E-26	1.180E-19
HI	18	237	4.623E-21	0.050	17	237	12.509	8487.305	1.020E-26	1.540E-19
ClO	19	7230	1.605E-22	0.087	56	3599	0.028	1207.639	1.520E-29	3.240E-21
					76	3631	0.015	1199.840	5.090E-30	1.030E-21

c. **1997 GEISA:** Husson, N., Bonnet, B., Scott, N. A., and Chedin, A., *J. Quant. Spectrosc. Radiat. Transfer* **48**, 509-518 (1992) and Jacquinet-Husson, N. et al., in preparation.

Detailed Content of GEISA97 Edition
(continued)

Mol.	ID	# Lines	Intensity average	Alpha average	Iso. ID	# Lines	F-Min (cm^{-1})	F-Max (cm^{-1})	Int-Min $(cm\ molec^{-1})$	Int-Max $(cm.mol.^{-1})$
OCS	20	24922	4.251E-21	0.090	622	14500	0.406	4118.004	1.560E-25	8.550E-20
					624	4764	0.396	4115.931	6.400E-27	4.720E-20
					632	2403	0.404	4012.468	1.720E-27	1.200E-20
					623	1802	509.007	4115.588	1.010E-23	8.430E-21
					822	1096	0.381	4041.565	2.620E-28	2.090E-21
					634	357	1972.188	2910.543	1.010E-23	5.240E-22
H2CO	21	2702	8.610E-21	0.120	126	1772	0.000	2998.527	1.020E-38	7.500E-20
					128	367	0.035	47.486	1.160E-30	1.110E-22
					136	563	0.037	75.745	2.020E-30	6.290E-22
C2H6	22	14981	2.686E-22	0.101	226	8944	765.027	3000.486	5.800E-27	3.210E-20
					236	6037	725.603	918.717	1.320E-28	1.770E-23
CH3D	23	11524	9.333E-25	0.060	212	11524	7.760	3146.460	5.570E-30	4.030E-22
C2H2	24	1668	2.340E-20	0.061	211	1432	604.774	3358.285	1.370E-27	1.080E-18
					231	236	613.536	3374.223	3.820E-26	1.580E-20
C2H4	25	12978	1.411E-21	0.087	211	12967	701.203	3242.172	6.940E-26	8.410E-20
					311	281	2947.832	3180.238	5.060E-24	1.620E-21
GeH4	26	824	4.978E-20	0.100	824	804	1937.371	2224.570	1.960E-22	3.680E-19
HCN	27	2575	1.205E-20	0.132	124	2275	2.956	18407.973	1.780E-28	7.100E-19
					125	115	2.870	9671.953	5.110E-26	2.730E-21
					134	185	2.880	9627.961	4.150E-26	8.290E-21
C3H8	28	9019	4.168E-23	0.080	221	9019	700.015	799.930	3.770E-24	4.310E-22
C2N2	29	2577	2.668E-21	0.080	224	2577	203.955	2181.690	6.590E-24	2.580E-20
C4H2	30	1405	3.445E-21	0.100	211	1405	190.588	654.425	2.650E-24	6.930E-20
HC3N	31	2027	2.693E-21	0.100	124	2027	474.293	690.860	6.360E-24	4.420E-20
HOCl	32	15565	2.295E-21	0.060	165	8057	0.024	3799.249	1.650E-27	3.590E-20
					167	7508	0.349	3799.682	7.220E-28	1.140E-20
N2	33	117	5.729E-29	0.047	44	117	2001.711	2619.230	2.330E-34	3.410E-28
CH3Cl	34	9355	5.584E-22	0.085	215	5311	679.050	3172.927	1.250E-25	1.130E-20
					217	4044	674.143	3161.830	4.190E-26	3.540E-21
H2O2	35	100781	5.090E-22	0.107	166	100781	0.043	1499.487	5.090E-29	5.610E-20
H2S	36	20788	2.992E-22	0.136	121	12330	2.985	4256.547	1.450E-26	1.360E-19
					131	3564	5.601	4098.234	2.020E-26	5.990E-21
					141	4894	5.615	4171.176	2.020E-26	1.080E-21
HCOOH	37	3388	5.186E-21	0.400	261	3388	1060.962	1161.251	2.140E-22	2.840E-20
COF2	38	54866	2.178E-21	0.084	269	54866	725.005	1981.273	4.740E-24	3.830E-20
SF6	39	11520	4.551E-21	0.050	236	11520	940.425	952.238	2.160E-22	1.500E-20
C3H4	40	3390	4.277E-22	(*)	341	3390	290.274	359.995	2.020E-23	1.810E-21
HO2	41	26963	9.907E-22	0.088	166	26963	0.055	3675.819	3.550E-27	2.900E-20
ClONO2	42	32199	1.093E-22	0.140	564	21988	763.641	797.741	1.250E-24	3.850E-22
					764	10211	765.212	790.805	6.340E-25	1.260E-22

TOTAL LINES IN GEISA 1997: 1,346,266

(*) missing value; set to 0.000

d. **JPL SMM CATALOG:** Pickett, H. M., Poynter, R. L., Cohen, E. A., Delitsky, M. L., Pearson, J. C., and Muller, H. S. P., *Submillimeter, millimeter and microwave spectral line catalog*, JPL Publication 80-23, Rev.4 (1996). Pasadena, CA.

ID TAG	MAXC GHz	Species	#lines	v	ID TAG	MAXC GHz	Species	#lines	v
1001	2	H-atom	1	1	29007	3550	HOC+	40	1
2001	1	D-atom	1	1	30001	6444	CO-18	60	3*
3001	2675	HD	1	2	30002	2938	HC-13-O+	34	2
4001	5539	H2D+	19	1	30003	2939	DCO+	41	2
7001	7210	Li-6-H	33	1	30004	2980	H2CO	611	1
8001	8122	LiH	40	1	30005	2975	C-13-H2NH	439	1
8002	4939	Li-6-D	40	1	30006	2994	CH2N-15-H	440	1
9001	4772	LiD	40	1	30007	2989	CH2ND	1834	1
12001	810	C-atom	2	2	30008	2994	NO	1909	1
13001	810	C-13-atom	7	2	30009	2993	NND+	41	1
13002	9926	CH	324	1	30010	3400	HOC-13+	40	1
13003	7400	CH+	9	1	30011	3675	NO+	154	1
14001	1	N-atom	2	1	30012	3032	DOC+	40	1
14002	262	N-atom-D-st	6	3	31001	2886	HCO-18+	34	2
15001	8800	NH	926	1*	31002	2985	H2C-13-O	601	1
16001	3000	O-atom	2	2	31003	9999	HDCO	4204	1
17001	20937	OH	788	3	31004	3435	HO-18-C+	40	1
17002	9497	NH3	446	3	31005	9952	HNO	10293	1
17003	2999	CH3D	80	1	32001	9928	O2	237	3
17004	11589	NH3-v2	442	1	32002	9817	O2-v1	237	4
18001	9926	OD	912	1	32003	1470	CH3OH	709	2
18002	2992	N-15-H3	235	2	32004	1424	H2CO-18	449	1
18003	29988	H2O	783	4	32005	5665	O2-snglt-dlta	67	1
18004	3000	NH2D	5036	1	32006	9999	D2CO	3682	1
18005	9993	H2O-v2	292	2	32007	9500	DNO	14739	1
19001	9839	HO-18	295	2	33001	9999	HO2	21954	3
19002	9995	HDO	1401	3	33002	9992	O-17-O	10787	1
19003	9975	H2O-17	404	1	33003	9999	SH v=0,1	646	1*
19004	9999	H3O+	420	1	34001	5961	O-18-O	400	3
20001	9998	D2O	1137	2	34002	9999	H2S	1525	1
20002	9733	HF	8	1	34003	5635	PH3	728	3
20003	9996	H2O-18	726	1	34004	9999	H2O2	38357	4
21001	9981	HDO-18	952	1	34005	9999	SD	1125	1*
21002	12474	DF	20	1	35001	2998	HDS	1138	1
25001	2351	CCH	114	1	36001	9760	HCl	137	2
25002	7073	NaH	52	1	37001	8405	DCl	228	1
26001	4000	CN, v = 0, 1	874	2	37002	3438.1	C3H	4990	1
27001	2913	HCN	40	2	37003	1618	c-C3H	2973	1
27002	4395.	HNC	49	2	38001	9746	HCl-37	137	2
27003	4322	HCN-v2	132	1	38002	9999	c-C3H2	30436	2
27004	2500	C-13-N	1218	1	38003	2650	C3D	1560	1
28001	9943	CO	91	4*	38004	2426	c-CCC-13-H	6541	1
28002	2923	HC-13-N	45	1	38005	2306	c-C-13-CCH	9753	1
28003	2998	HCN-15	35	1	38006	2274	c-C3D	6001	1
28004	2954	DCN	54	1	39001	9999	c-HC-13-CCH	17768	2
28005	2946	HNC-13	34	1	39002	9999	c-HCC-13-CH	6892	2
28006	2919	HN-15-C	33	1	39003	9999	c-C3HD	21610	2
28007	4002	DNC	53	2	39004	8382	DCl-37	228	1
28008	5406	HCNH+	74	1	40001	1345	CH3CCH	813	1
28009	3520	CO+	88	1	40002	2201	NaOH	91	1
29001	6468	C-13-O	60	3*	40003	9999	SiC	982	1
29002	3561	HCO+ v=0,1,2	246	3*	40004	9999	SiC-v1	703	1
29003	2998	CH2NH	2957	2	40005	3894	KH	40	1
29004	2992	HCO	2454	1	40006	1632	C2O	354	1*
29005	2971	NNH+	34	1	40007	1500	MgO v=0,1	88	1*
29006	5532	CO-17	50	2*	41001	1832	CH3CN v8=0,1	29304	3*

THE JPL SMM CATALOG   (continued)

ID TAG	MAXC GHz	Species	#lines	v	ID TAG	MAXC GHz	Species	#lines	v
41002	1323	CH3CC-13-H	822	1	46011	2313	DOCO+	677	1
41003	1344	CH3C-13-CH	813	1	46012	9983	HOC-13-O+	1735	1
41004	1325	C-13-H3CCH	821	1	46013	2000	O-18-CO	91	1
41005	623	CH3CCD	822	2	47001	336	H2C-13-S	110	1
41006	244	CH2DCCH	223	2	47002	1709	HC-13-OOH	1194	1
41007	9999	SiC-13	2417	1	47003	1196	DCOOH	628	1
41008	9726	CaH	387	1	47004	1319	HCOOD	612	1
41009	1972	CH3NC	1798	1	47005	2337	PO+ v=0-4	239	1*
42001	1753	CH3CN-15	2755	2*	47006	2631	PO	743	1*
42002	592	CH2CO	835	1	48001	2992	SO	330	1
42003	9299	NH2CN	13898	1	48002	2969	SO-v1	261	1
42004	4650	CaD	501	1	48003	343	H2CS-34	111	1
42005	3899	K-41-H	40	1	48004	8908	O3	7089	3
42006	1773	C-13-H3CN	9015	1*	48005	5945	O3-v2	4947	3
42007	1824	CH3C-13-N	8951	1*	48006	5163	O3-v1,3	9685	3
42008	6443	CH2DCN	19256	1*	48007	4861	O3-2v2	3031	2
42009	1700	H2CSi	670	1*	48008	7404	O3-v1,3+v2	10912	1
42010	1780	SiN	614	1*	48009	9999	NS-34	2364	1
43001	552	CHDCO	886	1	48010	2291	SO+	194	1*
43002	2997	HNCO	6078	1	49001	4603	O3-sym-O-17	26092	2
43003	2121	AlO	1023	1	49002	4681	O3-asym-O-17	52613	2
43004	2413	CP	397	1*	49003	920.	C4H	742	1
44001	2476	CS	51	2	49004	999	MgCCH	274	1
44002	1342	SiO	31	2	50001	2934	S-34-O	280	1
44003	1898	CH3CHO-a	1347	1	50002	2537	SO-18	179	1
44004	1528	N2O	61	1	50003	6529	O3-sym-O-18	3184	2
44005	2995	CH3CHO-e	691	1	50004	6209	O3-asym-O-18	7304	2
44006	2997	DNCO	5504	1	50005	4065	O3-s-O18-v2	2387	1
44007	2989	HN-15-CO	888	1	50006	3944	O3-a-O18-v2	4213	1
44008	2999	HNC-13-O	4332	1	50007	2115	CH3C1-35	6372	1
44009	1855	N2O-v2	146	1	50008	986.5	C3N	1351	1
44010	1355	HCP	34	1	50009	999	MgCN	273	1
44011	2230	AlOH	766	1	50010	999	MgNC	269	1
44012	1658	N2O-2v2	194	1*	51001	1050	HCCCN	139	3*
45001	1337	C-13-S	29	2	51002	2983	ClO	2585	3
45002	2797	Si-29-O	66	1	51003	2881	ClO-v1	2112	1
45003	1931	NH2CHO	3476	2	51004	991.05	HCCNC	563	1
45005	2709	HCS+	64	1	51005	984.98	HCCNC-v7	291	1
45006	2991	HNCO-18	4929	1	51006	983.83	HCCNC-v6	291	1
45007	2123	NN-15-O	85	1	51007	985.34	HCCNC-v5	278	1
45008	2099	N-15-NO	87	1	51008	931.415	HNCCC	574	1
45009	1154	DCP	34	1	52001	896	HCCC-13-N	152	1
45010	9994	HOCO+	1745	1	52002	895	HCC-13-CN	146	1
45011	3675	AlOD	70	1	52003	872	HC-13-CCN	144	1
45012	600	O-17-CO	285	1	52004	873	HCCCN-15	99	1
45013	3290	PN v=0-4	1637	1*	52005	835	DCCCN	156	1
46001	1633	CS-34	34	2	52006	9983	HOCl	6929	3
46002	2805	Si-30-O	67	1	52007	9999	SiCC	304	2
46003	938	H2CS	517	1	52008	950	CCCO	99	2*
46004	600	C2H5OH	12557	3	52009	2083	CH3Cl-37	6403	1
46005	2986	HCOOH	1888	1	52010	3624	CH2F2	11942	1
46006	6448	NO2	16444	2	52011	3189	CH2F2-v4	7808	1
46007	2075	N2O-18	38	1	52012	861.502	DNCCC	3098	1
46008	9999	CH3OCH3	21735	1	52013	1022	CNCN	206	1*
46009	3113.1	AlF	1188	1	53001	2000	C2H3CN gs,v's	75697	3*
46010	9999	NS	2402	1	53002	3000	Cl-37-O	2624	3

185

THE JPL SMM CATALOG   (continued)

ID TAG	MAXC GHz	Species	#lines	v	ID TAG	MAXC GHz	Species	#lines	v
53003	915	C-13-CCO	99	1	65001	3451	S-33-O2	19048	1*
53004	945	CC-13-CO	99	1	66001	2982	COF2	23090	1
53005	947	CCC-13-O	99	1	66002	6830	S-34-O2	11894	2*
53006	2834	Cl-37-O-v1	2132	1	66003	710	CaNC	259	1
53007	1000	C2H3NC	9362	1	66004	3550	SOO-18	9758	1*
53008	880	HNCCN+	99	1*	67001	3430	OCl-35-O	57232	2*
54001	349	CH2CHC-13-N	118	1	68001	572	CCCS	99	1
54002	178	CH2C-13-HCN	117	1	69001	3273	OCl-37-O	49388	2*
54003	180	C-13-H2CHCN	116	1	69002	2000	C3H7CN	131349	1
54004	800	CH2CDCN	6381	2*	70001	558	CCCS-34	99	1
54005	9999	HOCl-37	6925	3	73001	825	C6H	3031	1
54006	903	CCCO-18	99	1	74001	748.53	KCl	99	1
54007	2000	HCCCHO	8407	1	74002	2000	C2H5OOCH	60671	1
55001	3373	C2H5CN	52883	4*	75001	264	HCCCCCN	99	1
56001	530	CH3CH2C-13-N	1183	2	75002	1000	H2NCH2COOH I	27217	1
56002	519	CH3C-13-H2CN	1252	2	75003	1000	H2NCH2COOH II	26544	1
56003	524	C-13-H3CH2CN	1183	2	76001	261	HCCCCC-13-N	99	1
56004	1077	C2H5CN-15	1621	1	76002	264	HCCCC-13-CN	99	1
56005	530	CH2DCH2CN-s	1166	2	76003	264	HCCC-13-CCN	99	1
56006	480	CH2DCH2CN-a	1286	2	76004	261	HCC-13-CCCN	99	1
56007	2032	CCS	563	1	76005	257	HC-13-CCCCN	99	1
56008	2000	C2H3CHO	24051	1	76006	258	HCCCCCN-15	99	1
56009	1554	MgS	99	1*	76007	252	DCCCCCN	99	1
57001	1232	C-13CS	1013	1	76008	727.608	KCl-37	99	1
57002	1283	CC-13S	1015	1	76009	700	C4Si	229	1
58001	1988	CCS-34	565	1	76010	271	C5O	99	1*
58002	1229.9	NaCl	99	1	79001	2078	HOONO2	50775	1
60001	1200	OCS	99	2	80001	9999	HBr-79	143	2
60002	1740	SiS	97	1	81001	600	Cl-35-NO2	3520	1*
60003	9999	HCO(O)CH3-A	16153	1	82001	9999	HBr-81	143	2
60004	9999	HCO(O)CH3-E	17260	1	88001	208	C6O	567	1*
60005	1227.19	NaCl-37	99	1	89001	6115	Sr-88-H	391	1
61001	1196	OC-13-S	99	2	90001	4352	Sr-88-D	922	1
61002	1725	Si-29-S	98	1	92001	183	C5S	99	1
61003	1343	C5H	2594	1	94001	178	C5-34-S	99	1
62001	1170	OC-34-S	99	2	95001	2172	Br-79-O	1892	2*
62002	1126	O-18-CS	99	2	96001	9471	HOBr-79	9898	1
62003	1713	Si-30-S	99	1	97001	2163	Br-81-O	1892	2*
62004	1726	SiS-34	99	1	97002	1000	Cl-35-ONO2	78323	2*
62005	1290.1	AlCl	11525	1	98001	524	H2SO4	5690	2
62006	1303	C5D	4436	1	98002	9469	HOBr-81	9920	1
63001	2114	HNO3	36551	3	99001	889	Cl-37-ONO2	49505	2*
63002	1000	HNO3-v7	16478	1	99002	112.7862	HC7N	518	1
63003	1000	HNO3-v9	32404	1	100001	114	1 C7O	99	1*
63004	1000	HNO3-v6	15704	1	102001	1000	1 ClOOCl	17266	1
63005	1000	HNO3-v8	14537	1	104001	1000	1 Cl-37-OOCl	17482	1
63006	999	HNO3-v5	13269	1	111001	2000	1 OBr-79-O	52631	1*
63007	1733	PS	2340	1*	112001	2510	1 Se-80-O2	7484	1
63008	1999	PO2	7323	1*	112002	207	1 C8O	644	1*
64001	2823	S2	174	2	113001	2000	1 OBr-81-O	52840	1*
64002	7682	SO2	13573	4*	123001	57.52	1 HC9N	99	1
64003	1260.2	AlCl-37	11326	1	124001	59	1 C9O	100	1*
64004	478	C4O	535	1*	147001	33.5	1 HC11N	99	2*
64005	5900	SO2-v2	9225	1*					

e. **1995 ATMOS**: Brown, L. R. et al., *Appl. Opt.* **26**, 5154-5182 (1987) and *Appl. Opt.* **35**, 2828 (1996).

mol#	MOLECULE	#LINES	FMIN cm^{-1}	FMAX cm^{-1}	#ISO	#VIB	#BANDS	Total Intensity Sum cm^{-1}/molec·cm^{-2} 296K
1	H_2O	31016	0.40	9999.49	3	45	73	7.279 x10^{-17}
2	CO_2	60790	442.00	9648.00	8	319	589	1.126 x10^{-16}
3	O_3	196415	0.02	4060.78	5	92	104	1.836 x10^{-17}
4	N_2O	26249	0.83	5131.24	5	113	162	7.196 x10^{-17}
5	CO	3600	3.46	8464.88	5	9	41	1.059 x10^{-17}
6	CH_4	45456	0.01	6184.49	3	40	51	1.772 x10^{-17}
7	O_2	1619	0.00	9468.43	3	6	9	9.107 x10^{-24}
s 7	O_2	162	1432.50	1676.96	1	1	1	1.470 x10^{-27}
8	NO	7385	0.00	3966.00	3	42	50	4.730 x10^{-18}
9	SO_2	26225	0.01	2526.03	2	6	7	4.122 x10^{-17}
10	NO_2	44975	0.49	2938.37	1	9	9	6.241 x10^{-17}
11	NH_3	5817	0.21	2153.76	2	14	16	4.457 x10^{-17}
12	HNO_3	129741	0.03	1769.98	1	11	11	1.204 x10^{-16}
s 12	HNO_3	8751	845.11	908.83	1	2	2	3.246 x10^{-16}
13	OH	3168	0.00	9997.35	3	39	43	4.554 x10^{-17}
14	HF	84	41.11	7994.58	1	5	5	7.326 x10^{-17}
15	HCl	324	20.24	8454.45	2	7	14	1.709 x10^{-17}
16	HBr	398	16.23	9758.56	2	8	16	6.187 x10^{-18}
17	HI	237	12.50	8487.30	1	9	9	1.096 x10^{-18}
18	ClO	6038	0.01	886.21	2	4	8	1.160 x10^{-18}
19	OCS	4096	493.26	4117.99	3	19	21	1.068 x10^{-16}
20	H_2CO	2702	0.00	2998.52	3	8	10	2.326 x10^{-17}
s 20	H_2CO	587	1707.09	1781.38	1	1	1	7.195 x10^{-18}
21	HOCl	15371	0.02	3799.68	2	4	8	3.235 x10^{-17}
s 22	HO_2	6820	1032.06	3675.81	1	4	4	1.795 x10^{-17}
23	H_2O_2	5444	0.15	1499.48	1	2	2	1.872 x10^{-17}
s 24	HONO	2348	779.12	1711.70	1	1	1	2.059 x10^{-17}
c 25	NO_2NO_2		802.	804. (at 220K)				1.66 x10^{-17}
c 26	N_2O_5		1210.	1275.				4.21 x10^{-17}
s 27	$ClONO_2$	23264	763.64	1309.99	2	3	4	3.113 x10^{-17}
28	HCN	772	2.87	3421.96	3	6	8	3.086 x10^{-17}
s 29	CH_3F	739	987.91	1089.13	1	1	1	1.806 x10^{-17}
30	CH_3Cl	10432	661.79	3172.92	2	4	8	5.783 x10^{-18}
s 30	CH_3Cl	621	1324.69	1460.14	2	2	3	1.163 x10^{-18}
c 31	CF_4		1255.0	1289.				1.61 x10^{-18}
c 32	CCl_2F_2		810.	965.				5.95 x10^{-17}
c 33	CCl_3F		810.	880.				6.95 x10^{-17}
s 34	CH_3CCl_3	251	1382.50	1385.00	1	1	1	1.25 x10^{-19}
c 35	CCl_4		770.00	809.99				5.14 x10^{-17}
36	COF_2	33932	725.00	1981.27	1	7	7	1.195 x10^{-18}
s 37	CFClO	2449	764.11	1907.92	1	2	2	3.957 x10^{-17}
38	C_2H_6	10000	745.22	951.65	1	1	1	6.967 x10^{-18}
s 38	C_2H_6	115	2976.63	2996.98	1	1	1	2.875 x10^{-18}
40	C_2H_2	1267	638.25	3374.22	2	8	9	4.103 x10^{-17}
41	N_2	120	1992.62	2625.40	1	1	1	6.719 x10^{-27}
c 42	CHF_2Cl (FTS)		780.	839.				2.05 x10^{-17}
c 42	CHF_2Cl (TDL)		828.95	829.16				3.11 x10^{-19}
s 43	$COCl_2$	32871	828.54	1862.45	1	2	2	3.35 x10^{-17}
s 44	CH_3Br	3421	871.32	1038.06	1	1	1	6.427 x10^{-19}
s 45	CH_3I	2528	793.46	971.65	1	1	1	1.207 x10^{-18}
s 46	HCOOH	3388	1060.96	1161.25	1	1	1	1.757 x10^{-17}
47	H_2S	661	994.12	1573.81	1	1	1	7.959 x10^{-20}
s 48	$CHFCl_2$	9583	785.00	1099.99	1	1	1	5.284 x10^{-17}
49	HDO	8505	0.01	5507.54	2	9	10	7.694 x10^{-21}
c 50	SF_6		940.42	952.23				5.243 x10^{-17}

The ATMOS Main	list	:	694359 lines		from	0	to	9999 cm^{-1}
The ATMOS Supplemental	list	s :	116016 lines		from	725	to	3675 cm^{-1}
The ATMOS Cross Section	list	c :	8 species		from	770	to	1289 cm^{-1}

f. **1997 SAO:** Chance, K. V., Jucks, K. W., Johnson, D. G., and Traub, W. A., *J. Quant. Spectrosc. Radiat. Transfer* **52**, 447-457 (1994).

Table 1. SAO database statistics

Mol.	Num. lines	Low Freq.	High Freq.	Strongest	Comment
H_2O	3896	10	1600	2.67×10^{-18}	161 Iso.
CO_2	16792	449	1415	2.98×10^{-19}	
O_3	27351	10	1495	4.20×10^{-20}	666 Iso., $v''=0$
N_2O	8554	10	1600	1.66×10^{-19}	
CO	480	10	190	1.45×10^{-21}	
CH_4	12056	73	1600	9.68×10^{-20}	
O_2	910	10	1600	5.21×10^{-25}	
NO	422	11	99	2.95×10^{-22}	
SO_2	5571	11	1397	6.09×10^{-20}	
NO_2	33339	10	1600	1.13×10^{-19}	
NH_3	445	19	1600	5.50×10^{-19}	
HNO_3	78371	10	1371	3.02×10^{-20}	
OH	1395	31	990	1.90×10^{-18}	
HF	22	41	615	1.44×10^{-17}	
HCl	230	20	438	4.60×10^{-19}	
HBr	256	16	383	1.21×10^{-19}	
HI	44	12	320	1.54×10^{-19}	
ClO	3672	10	887	3.24×10^{-21}	
OCS	6173	10	1090	1.55×10^{-20}	
H_2CO	492	10	100	6.12×10^{-20}	
HOCl	6525	10	1307	3.59×10^{-20}	
HCN	671	11	1537	7.10×10^{-19}	
CH_3Cl	2668	674	768	1.23×10^{-20}	
H_2O_2	9083	12	1330	5.61×10^{-20}	
C_2H_2	752	646	1441	1.08×10^{-18}	
C_2H_6	2019	748	903	3.42×10^{-21}	
COF_2	20991	745	1283	3.64×10^{-20}	
SF_6	3001	925	955	6.30×10^{-19}	Pseudo lines
H_2S	1264	10	1444	1.36×10^{-19}	
HCOOH	2205	1063	1139	2.84×10^{-20}	
HO_2	14131	10	1458	2.74×10^{-20}	
O^3P	2	68	158	1.13×10^{-21}	
$ClNO_3$	33889	403	831	1.25×10^{-20}	Pseudo, real
HOBr	4358	10	316	1.73×10^{-20}	
H_2OIS	2864	10	1600	5.39×10^{-21}	171 and 181
HDO	2668	10	1600	2.76×10^{-22}	
O_3ISO	30393	10	1178	7.76×10^{-22}	668, 686, 667, 676
$HotO_3$	53687	10	1434	1.28×10^{-21}	666 Iso., $v'' > 0$
$O_2{}^1D$	66	11	189	2.80×10^{-23}	
N_2O_5	13347	257	1286	9.02×10^{-19}	Pseudo lines
HNO_4	341	802	804	5.98×10^{-21}	Pseudo lines
CFC12	25001	850	1200	1.40×10^{-19}	Pseudo lines
CFC11	14001	810	1120	5.00×10^{-20}	Pseudo lines
CCl4	4001	770	810	4.44×10^{-20}	Pseudo lines
CF4	9001	1250	1295	5.88×10^{-19}	Pseudo lines
CFC113	408	786	990	6.17×10^{-19}	Pseudo lines
HCFC22	21479	776	1335	9.80×10^{-20}	Pseudo lines
HCFC23	9583	785	1100	2.16×10^{-20}	Pseudo lines
HCFC140	251	1382	1385	2.07×10^{-21}	Pseudo lines

g. LAB SPECTRA CATALOG: Hanst and Hanst, Digitized Reference Spectra (Infrared), copyright 1993, Infrared Analysis, Inc., Anaheim, CA 92801

ALPHABETICAL INDEX   Compounds, Computer File Names*, and Chapters.

* Computer file names are given here without prefix. Each digitized file has a prefix Z, H, O, or T indicating the resolution of the file.

Acetaldehyde, 314AA, H
Acetic acid, monomer, 55AAC, G
Acetic acid, monomer and dimer, 110AAC, G
Acetone, 236ACT, I
Aceto nitrile (Methyl cyanide), 456AN, N
Acetyl chloride, ACCL350, T
Acetylene, 16AC, A
Acrolein, ACRO340, H
Acrylic acid, monomer and dimer, 240ACYA, G
Acrylic acid, mostly monomer, 160ACYA, G
Acrylo nitrile, ACN1652, N
Allyl alcohol, 200AA, J
Alpha-Pinene, 100APIN, C
Ammonia, 85NH3, E
Arsine, AS3-700, E
Benzaldehyde, 396BZA, H
Benzene, 10BZ, D
Beta-Pinene, 100BPIN, C
Boron trichloride, BCL150, T
Bromobenzene, 200BRBZ, S
Bromochloromethane, 300BCM, P
Bromoform, 304BRF, P
Bromo methane, 950BRM, P
1,3 Butadiene, 103BD, A
Butane, 151BU, A
2-Butanone, MEK762, I
n-Butyl acetate, 60BAC, K
n-Butyl alcohol, 300BALC, J
n-Butyraldehyde, 190BAL, H

Carbon dioxide, 100CO2, F
Carbon disulfide, 50CS2, O
Carbon monoxide, 225CO, F
Carbon tetrachloride, 27CT, P
Carbonyl fluoride, 39COF2, T
Carbonyl sulfide, 39COS, O
Chlorobenzene, 114CBZ, S
1-Chloro-1,1 difluoro ethane, 60GT, Q
Chlorodifluoromethane, 64F22, P
Chloroethane, 551CE, Q
Chloroform, 45CLF, P
Chloromethane, 652CLM, P
2-Chlorotoluene, 100CLTO, S
4-Chlorotoluene, 100CLTL, S
Chlorotrifluoroethylene, 80CTFE, R
Chlorotrifluoromethane, F13-43, P
Crotonaldehyde, 114CRA, H
Cyclohexane, 48CHX, C
Cyclohexene, 114CHE, C
Cyclopentene, CYPE282, B
Cyclopropane, CYPR615, A
1,2 Dibromo ethane, 228EDB, Q
Dibromomethane, 300DBM, P
m-Dichloro benzene, 200MDCB, S
o-Dichloro benzene, 100ODCB, S
p-Dichloro benzene, 200PDCB, S
Dichlorodifluoromethane, 20F12, P
1,1 Dichloroethane, 11DC474, Q
1,2 Dichloroethane, 250DCE, Q
cis-1,2 Dichloroethylene, 100DCEL, R

Dichloro-1-fluoro ethane (Genesolv), 125GS, Q
Dichloromethane, 160DCM, P
1,2-Dichloropropane, 300DCP, T
1,3-Dichloropropane, 300DCPR, T
1,2 Dichlorotetrafluoroethane, 114-318, Q
Diethyl amine, 248DEAM, N
Diethyl ether, 144DEE, L
Diethyl ketone, 320DEK, I
Di-isopropyl ether, 114DIE, L
Dimethyl ether, 240DME, L
1,1 Dimethyl hydrazine, DMHZ846, N
Dimethyl sulfide, 365DMS, O
Dinitrogen pentoxide, 175N2O5, M
Ethane, 260ET, A
Ethanol, 320ETA, J
Ethyl acetate, 93EAC, K
Ethyl acrylate, 93EAY, K
Ethyl benzene, 200EBZ, D
Ethyl butyrate, 91EBU, K
Ethyl formate, 103ETF, K
Ethylene, 82ETY, A
Ethylene oxide, 125EO, F
Ethyl vinyl ether, 160EVE, L
Exxon 87 gasoline, 250GAS, T
Fluoro benzene, 64FBZ, S
Formaldehyde, 176FA, H
Formic acid, momer, 140FAC, G
Furan, FURA6133, F
n-Heptane, 213HEP, C
n-Hexane, 121HX, C
Hydrazine, 400HZ, E
Hydrogen bromide, 635HBR, E
Hydrogen chloride, 285HCL, E
Hydrogen cyanide, 36HCN, M
Hydrogen fluoride, 30HF, E

Hydrogen Iodide, 10KHI, E
Hydrogen peroxide, 1KH202, F
Hydrogen sulfide, 32KH2S, O
Isobutane, 148IBU, A
Iodo methane, 798IM, P
Isobutanol, 196IBA, J
Isobutylene, 190IBY, A
Isoprene, 147IP, B
Isopropanol, 228IPA, J
Isopropylbenzene, 200IPB, D
K-1 Kerosene, 200KER, T
Mesitylene, 285MTY, D
Methane, 128ME, A
Methanol, 141MA, J
Methyl acetate, 100MAC, K
Methyl acrylate, 135MACR, K
Methyl amine, 400MAM, N
2-Methyl butane, 110BUT, B
2-Methyl-2-butane, 2M2BU474, B
3-Methyl-1-butene, 3M1BU474, B
Methyl formate, 57MF, K
Methyl isobutyl ketone, H228MBK, I
Methyl mercaptan, 1500MMC, O
Methyl methacrylate, 114MM, K
Methyl nitrite, MNI1300, N
2-Methyl pentane, 2MP114, C
3-Methyl pentane, 3MP218, C
2-Methyl-1-pentane, 185MPE, C
2-Methyl-2-pentane, 2M2P282, C
4-Methyl-2-pentane, 4M2P790, C
Methyl vinyl ether, MVET474, L
Methyl vinyl ketone, 356MVK, I
Nitric acid, 140HNO3, M
Nitric oxide, 625NO, M
Nitro benzene, NBZ23, N

Nitro ethane, 114NE, N
Nitro methane, 1400NM, N
Nitrogen dioxide, 125NO2, M
Nitrogen dioxide (and N2O4), 100NO2, M
Nitroso benzene, 30N0BZ, N
Nitrous acid, HNO2-10, M
Nitrous oxide, 80N2O, M
n-Octane, 219OCT, C
Ozone, 32IOZ, F
n-Pentane, 152P, B
1-Pentene, 1PENT564, B
2-Pentene, 2PENT564, B
Perfluoro butane, 20KPFBU, T
Peroxy acetyl nitrate, 80PAN, N
Phosgene (Carbonyl chloride), 60PHG, T
Phosphine, PH3-945, E
Phosphorus trichloride, PTC700, T
Propane, 118PP, A
Propionaldehyde, 274PA, H
Propionic acid, monomer and dimer, 257PAC, G
Propionic acid, mostly monomer, 170PAC, G
n-Propyl acetate, 60PAC, K
n-Propylbenzene, 300PBZ, D
Propylene, 128PPY, A
Propylene oxide, 388PYO, F
Silicon tetrafluoride, 165IF4, T
Styrene, 145STY, D
Sulfur dioxide, 125SO2, O
Sulfur hexafluoride, 4SF6, O
Tertiary butyl benzene, 100TBBZ, D
1,1,1,2 Tetrachloroethane, TETC336, Q
1,1,2,2-Tetrachloroethane, 406STCE, Q
Tetrachloroethylene, 125TCE, R
Tetrafluoromethane, 5TFM, P
Tetrahydrofuran, 200THF, F

Tetrahydrothiophene, 200THT, O
Thiophene, 200TP, O
Toluene, 137TO, D
1,1,1 Trichloroethane, 95TC1, Q
1,1,2 Trichloroethane, 125TC2, Q
Trichloroethylene, 160TCE, R
Trichlorofluoromethane, 38F11, P
1,2,3-Trichloropropane, 300TCP, T
Trichlorotrifluoroethane, 113-141, Q
1,2,4-Trimethyl benzene, 2.0TMBZ, D
Vinyl acetate, VIAC474, K
Vinyl chloride, 200VC, R
Vinylidene chloride, VC282, R
Water, 1KH2O, E
m-Xylene, 200MXL, D
o-Xylene, 1000XL, D
p-Xylene, 200PXL, D

Astrophysical and Atmospheric
Data Needs

Atomic and Molecular Data for Cosmology

A. Dalgarno

Harvard-Smithsonian Center for Astrophysics
Cambridge, MA 02138

Abstract. An account is given of some of the atomic and molecular data needed in the interpretation of cosmological observations and in the construction of models of the evolution of the early Universe through the recombination era.

INTRODUCTION

Atomic and Molecular (AM) Data are critical inputs to a wide range of scientific and technological endeavors. The diversity of the areas in which AM data are employed should be a strength and intellectually that is so. Yet in seeking funds to support data activities, it seems that the very breadth of the subject is a weakness, because it has led to a fragmentation of support in which occasional individual projects that meet specific needs are funded, but the long term secure investment that effective data-related activities require is generally lacking. With few exceptions, agencies are unwilling or unable to take a broader view beyond their own domain and agencies usually lack the internal administrative structure that would encourage the pursuit of a shared responsibility. Applied research in industrial settings would clearly benefit from ready access to a reliable data base but the cooperation needed to provide adequate support is difficult to bring about because of the competitive nature of industrial enterprises. Action is further inhibited because data needs are not confined to one country. International collaboration programs exist to address problems recognized as global in their impact, but are usually restricted to issues that immediately effect the health of society.

Our intent in initiating this Conference Series is to make more clearly apparent the enormous value of reliable, accessible data bases to potential sources of funding and to achieve a unification of our separate activities by exploiting developments in computer networking. The Conference will address questions relating to the acquisition, assessment and dissemination of AM data. Data can be in the form of tables or graphs or of computer programs that generate tables and graphs.

Data have limited value without an assessment of quality. Ludwig Boltzmann is reported to have said "There is nothing more practical than a good theory," to which one might add "There are no good theories without good data". Quality assessment is a demanding, long-term project requiring broad expertise and extensive experience, and should be developed as a cooperative effort.

This Conference Series came about through a suggestion initially put forward by Ratko Janev. We decided the suggestion should be explored further in a gathering of interested individuals. This was done at a meeting sponsored by the Harvard-Smithsonian Institute for Theoretical Atomic and Molecular Physics held at the Center for Astrophysics in Cambridge, Massachusetts in June 1996. At that meeting, it was agreed to initiate an International Conference Series and Wolfgang Weise bravely stepped forward and volunteered to organize the first of the series. Our debt to Wolfgang, to Peter Mohr and to many others at the National Institute of Standards and Technology who have helped in the organization is profound.

Because of the extraordinary range of physical conditions of temperature, density and radiation environments encountered in astrophysical objects and events, there is perhaps no field in which the need for access to reliable AM data is greater. Almost all the information about the Universe is brought to us by photons, and photons and the processes by which they are created, modified and detected, belong largely to the domain of Atomic, Molecular and Optical physics. The topic of data needs in astronomy and astrophysics is vast. I will restrict myself to AM data that relate to cosmology and the evolution of the early Universe.

AM DATA FOR COSMOLOGY

The helium abundance

According to the standard cosmology, after a brief period of inflation, the space of the Universe expanded from a singularity. Matter and radiation shared a common temperature. The Universe was initially very hot, too hot for any composite particles to survive except as transitory events. As the Universe expanded, it cooled and after about 100s, the formation of protons from the combination of quarks was not reversed by collisions or by radiation. Neutrons were also created by electron capture and a brief period of nucleosynthesis occurred in which nuclear reactions produced a significant abundance of ^{4}He nuclei and trace amounts of ^{3}He, ^{2}D and ^{7}Li.

The fractional abundance of ^{4}He depended on the neutron lifetime and the rate of expansion of the Universe which in turn is controlled by the number of types of light neutrinos. The calculated value of the mass ratio Y_p of ^{4}He/H is 0.238 if there are three light neutrino species and 0.250 if there are four (1). The derived observational value of Y_p is 0.232 ± 0.007 (2) but there are arguments that it is too low (3,4).

The observational determination of Y_p is made from optical spectra of less evolved metal-poor extragalactic nebulae (5-7). The relationship between the relative intensities of the recombination lines of helium and hydrogen and the helium-hydrogen abundance ratio involves a consideration of the ionization distribution of H^+, He^+, and He^{2+} in the nebulae and of the many processes that populate the levels from which the observed radiation is emitted. Complexities arise from effects of radiation transfer (8), of collisions (9, 10) and of singlet-triplet mixing. The influence of the long-lived metastable $He(2^3S)$ state which can affect the ionization balance and excitation which can enhance the emission lines, presents additional complications. The rate coefficients for electron impact excitation of the 2^3S state to other states, singlet and triplet, and for electron impact ionization are crucial parameters in the theoretical interpretation of the emission line intensities. Different conclusions about the importance of electron collisions have been drawn using the cross section data of Berrington et al. (11), (9) and the data of Berrington and Kingston (12) (10). No calculations have yet been reported based on the more refined studies of electron impact of Sawey et al. (13), Sawey and Berrington (14) and Fon et al. (15). The calculations for astrophysical plasmas appear to omit collisions of protons with excited helium atoms though proton impacts are potentially efficient in producing changes in angular momentum. Proton collisions are important in the positive column of glow discharges (Fujimoto 1979). Other processes involving heavy particles, such as charge transfer

$$H^+ + He(2^3S) \rightarrow H + He^+$$

(17), radiative charge transfer and radiative association

$$He^+ + H \rightarrow He + H^+ + \nu$$
$$\rightarrow HeH + \nu$$

(18-20) and Penning and associative ionization

$$He(2^3S) + H \rightarrow He + H^+ + e$$
$$\rightarrow HeH^+ + e$$

(21) may not be negligible at the level of accuracy needed to draw definite conclusions about the number of light neutrino species.

The deuterium abundance

The era of early nucleosynthesis produced deuterium and because deuterium was destroyed in collisions with other deuterium nuclei, the fractional abundance of D is a measure of the primordial baryon density (1). The [D]/[H] ratio has been measured in the interstellar medium in the solar neighborhood by the absorptions produced by H and D atoms of radiation emitted by stars. The absorptions appear as narrow features in He stellar spectra. For the region to the nearby stars it may be safely assumed that all the hydrogen and deuterium is in the form of atoms or ions and that the cosmic abundance ratio [D]/[H] is given by the atom ratio D/H. The measured ratio for the interstellar medium is $1.6 \pm 0.4 \times 10^{-5}$ (22-24).

The [D]/[H] ratio can also be inferred from measurements of the ratio of abundances of DCO^+ and HCO^+. The molecule DCO^+ is chemically fractionated and its abundance in the cold interstellar gas relative to HCO^+ is greatly enhanced (25). The major sources of the ions are the reactions

$$H_3^+ + CO \rightarrow HCO^+ + H_2$$
$$H_2D^+ + CO \rightarrow DCO^+ + H_2.$$

Now H_2D^+ is formed by

$$H_3^+ + HD \rightarrow H_2D^+ + H_2$$

and

$$H_3^+ + D \rightarrow H_2D^+ + H$$

and removed by the reverse reaction

$$H_2D^+ + H_2 \rightarrow H_3^+ + HD,$$

dissociative recombination

$$H_2D^+ + e \rightarrow H + H + D$$
$$\rightarrow H + HD$$
$$\rightarrow H_2 + D$$

and reactions with O,

$$H_2D^+ + O \rightarrow OH^+ + HD$$
$$\rightarrow OD^+ + H_2$$

and CO,

$$H_2D^+ + CO \rightarrow DCO^+ + H_2$$
$$\rightarrow HCO^+ + HD.$$

Dissociative recombination of the chemically enhanced ions H_2D^+ and DCO^+ provides a significant supply of deuterium atoms and exchange reactions such as

$$D + HCO^+ \to H + DCO^+,$$

are an important source of deuteration in molecular clouds (26). Using recent data on the dissociation recombination of H_2D^+, Larsson et al. (27) have employed these reactions and their rate coefficients in an analysis of the DCO^+/HCO^+ line intensity ratio measured in the molecular cloud L1529 (28) and have derived a cosmic abundance ratio [D]/[H] of 1.7×10^{-5}, consistent with the value obtained for the local interstellar medium.

Deuterium is consumed in stars so that the value for our galaxy is less than the primordial value and theoretical corrections are necessary However, deuterium atoms have now been detected in absorption towards distant quasars (29-34), though its interpretation is confused by the possibility of interloper hydrogen atoms. Nevertheless the conclusion seems firm that baryonic matter cannot close the Universe.

After the primordial nucleosynthesis, the Universe coasted for about 200, 000 years, becoming less dense and becoming colder. As it cooled, it ran out of energetic photons and neither photoionization nor electron impact ionization could reverse radiative recombination. The positive ions were neutralized in the order of their ionization potentials. Li^{3+} was converted into Li^{2+} and then to Li^+. Soon after He^{2+} became He^+ and then neutral helium was formed to be followed by the recombination of H^+ to neutral hydrogen. The Universe was transformed from a completely ionized plasma into an almost neutral gas. Because of the scarcity of electrons after hydrogen recombination, lithium remained as Li^+ and some additional relict ionization survived as the recombination time scale lengthened.

The formation of molecules

The recombination era saw the dawn of chemistry as the first molecular entities were created through the co-existence of H^+ with neutral He leading by radiative association

$$H^+ + He \to HeH^+ + \nu$$

to the molecular ion HeH^+.

A diverse array of atomic and molecular processes then occurred, including spontaneous and stimulated radiative attachment and association, charge transfer, proton transfer, associative detachment, dissociative attachment, associative ionization, excitation transfer, dissociative recombination and mutual neutralization. Reactions involving excited states also played a role.

There are no cosmological data on the period of early molecule formation. It had been suggested that molecules with large dipole moments might through

Thomson scattering, which is enhanced at resonance wavelengths, leave an imprint on the cosmic background blackbody radiation (35). The molecule LiH was a potential candidate (36-41). However a comprehensive chemical model (42-46) indicates that the conversion of lithium into LiH is very inefficient.

The cosmic background temperature

At the time of recombination, matter and radiation decoupled and the Universe became transparent to the cosmic blackbody background radiation. We see it today at a temperature of 2.726 K, observed directly with the COBE satellite, but it was first measured in a study of the visible absorption of stellar radiation by the molecule CN present in an interstellar cloud along the line of sight (47-49).

Three lines were detected, one arising from the lowest rotational level N=0 and the other two from the N=1 level in transitions from the ground $X^2\sum^+$(v= 0) electronic state to the $B^2\sum^+$(v= 0) excited state. The relative populations $N(1)/N(0)$ of the rotational levels could be determined from the observed absorption strengths. The excitation temperature T_{exc} may be derived from the Boltzman equation $N(1)/N(0) = 3\ exp\ (-\Delta E/kT_{exc})$ where ΔE is the energy difference. The value obtained was 2.3 K and though unrecognized at the time, this is the approximate temperature of the cosmic blackbody background radiation.

The levels are populated not only by the absorption and stimulated emission of the background radiation and by spontaneous emission but also by ultraviolet pumping and by collisions with electrons and protons. These give rise to a small correction of about 0.1 K and the value obtained from modern measurements of CN is 2.729 ± 0.03 K (50).

The possibility of measuring the cosmic background temperature spectroscopically is of more than historical interest because the method is applicable to external galaxies which we observe as they were at earlier times when the background temperature was higher. Indeed some measurements of absorption at millimeter wavelengths by the molecules HCO^+, HCN and HNC in interstellar clouds of a galaxy in the direction of the radio source PKS 1830-211 have been reported (51). The derived excitation temperature is approximately 5 K, the value expected at the redshift of 0.886 of the source.

The background radiation temperature in external galaxies can also be derived from atomic absorption line measurements. The ground electronic states of C and C^+ have fine-structure levels and individual absorption lines arising from them have been detected. From them the excitation temperature can be derived. The excitation temperature derived from observations of the absorption lines originating in the $^3P_0$ and $^3P_1$ levels of neutral carbon at rest wavelengths of 165.693 nm and 165.627 nm in the damped Lyman alpha system of the QSO 0013-04 at a redshift $z = 1.9731$ is 11.6±1.0 K (52).

There are substantial additional contributions to the excitation of the fine-structure levels and the interpretation of the inferred excitation temperature requires reliable data on the ultraviolet pumping and on fine-structure excitation by electrons, protons, H, H_2 and He. Ge et al. (1977) concluded that the cosmic background temperature is 7.9±1.0 K which is consistent with the value of 8.105±0.0030 K given at $z=1.9731$ by the standard cosmology.

Constancy of the fundamental constants

The interpretation of cosmological phenomena makes use of atomic and molecular data. Cosmology also contributes to atomic and molecular physics. By comparing the frequencies of atomic and molecular transitions in objects at different redshifts, a measure can be obtained of the rate of change of the fundamental constants. Recently Cowie and Sougalia (53) have compared ultraviolet transition wavelengths of molecular hydrogen obtained from redshifted absorption lines detected in the quasar PKS 0528-250 at a redshift of 2.811. They conclude that the variation of the ratio μ of the electron mass to the proton mass satisfies

$$\frac{\dot{\mu}}{\mu} = \left(0.0 \, ^{+9.7}_{-7.6}\right) \times 10^{-14} \, \text{yr}^{-1}.$$

It is a challenge to laboratory atomic and molecular physics to improve on this upper limit.

ACKNOWLEDGEMENT

This work has been supported by the Division of Astronomy of the National Science Foundation.

REFERENCES

1. Walker, T. P., Steigman, G., Schramm, D.N., Olive, K.A., and Kang, H-S., *ApJ* **376**, 51 (1991).
2. Olive, K.A. and Steigman, G. *ApJS* **97**, 49 (1995).
3. Hata, N., Scherrer, R.J., Steigman, G., Thomas, D., Walker, T.P., Bludman, S., and Langecker, P.. *Phys. Rev. Lett* **75**, 3977 (1995).
4. Copi, C.J., Schramm, D.N. and Turner, M.S., *Phys. Rev. Lett* **75**, 3981 (1995).
5. Campbell, A., *ApJ* **401**, 157 (1992).
6. Pagel, B.E.J., Simonson, E.A., Terlevich, R.J., and Edmunds, M., *MNRAS* **255**, 325 (1992).
7. Skillman, E. Terlevich, R.J., Kennicutt, R.C., Garnett, D.R., and Terlevich, E. *ApJ* **431**, 172 (1994).

8. Sasselov, D. and Goldwirth, D., *ApJ* **444**, L5 (1995).
9. Ferland, G.J., *ApJ* **310**, L67 (1986).
10. Clegg, R.E.S., and Harrington, J.P. *MNRAS* **239**, 869 (1989).
11. Berrington, K.A., Burke, P.G., Freitas, L.C.G., and Kingston, A.E. *J. Phys. B* **18**, 4135 (1985).
12. Berrington, K.A., and Kingston, A.E., *J. Phys. B.* **20**, 6631 (1987).
13. Sawey, P.M.J., Berrington, K.A., Burke, P.G., and Kingston, A.E., *J. Phys. B.* **23**, 3939 (1990).
14. Sawey, P.M.J., and Berrington, K.A., *Atom. Data Nucl. Data Tables* **55**, 81 (1993).
15. Fon, W.C., Lee, T.G., Lui, K.P., and Berrington, K.A., 1995, *J. Phys. B.* **28**, 4129 (1995).
16. Fujimoto, T., *J. Quant. Spectr. Rad. Tans.* **21**, 439 (1979).
17. Clegg, R.E.S., *MNRAS* **229**, 31P (1987).
18. Jura, M., and Dalgarno, A., *A&A* **14**, 243, (1971).
19. Zygelman, B., and Dalgarno, A, *ApJ* **365**, 239 (1990).
20. Stancil, P.C., and Zygelman B., *ApJ* **472**, 102 (1996).
21. Dalgarno, A., and Roberge, W.G., *ApJ* **255**, 489 (1982).
22. McCullough, P. R., *ApJ* **390**, 213 (1992).
23. Linsky, J. L., Brown, A. Gayley, K., Diplas, A., Savage, B.D., Ayres, T.R., Landsman, W., Shore, S.N., and Heap, S.R., *ApJ* **402**, 694 (1993).
24. Lemoine, M., Vidal-Madjar, A., Bertin, P., Ferlet, R., Gry, C., and Lallement, R., *A&A* **308**, 601 (1996).
25. Watson, W.D., *Rev. Mod. Phys.* **48**, 513 (1976).
26. Dalgarno, A., and Lepp, S., *ApJ* **287**, L47 (1984).
27. Larsson, M. et al., *A&A* **309**, L1 (1996).
28. Wootten, A., Loren, R.B., and Snell, R.L., *ApJ* **255**, 160 (1982),
29. Songaila, A., Cowie, L.L., Hogan, C., and Rogers, M., *Nature* **368**, 590 (1994).
30. Carswell, R.F., Weymann, R.J., Cooke, A.J., and J. R. Webb, *MNRAS* **268**, P1 (1994).
31. Wampler, E.J., Williger, G.M., Baldwin, J.A., Carswell, R.F., Hazard, C., and McMahon, R.G., *A&A* **316**, 33 (1996).
32. Rugers, M., and Hogan, C.J., *ApJ* **459**, L1 (1996).
33. Songaila, A., Wampler, E.J., and Cowie, L.L., *Nature* **385**, 127 (1997).
34. Tytler, D., Burles, S. and Kirkman, D., 1998 in press.
35. Dubrovich, V.K., *Astron. Lett.* **19**, 83 (1993).
36. DeBernardis, P. et al. *A&A*, 269, 1 (1993).
37. Melchiorri, B., and Melchiorri, F., *Riv. Nuovo Cimento* **17**, 1 (1994).
38. Maoli, R., Melchiorri, F., and Tosti, D., *ApJ* **425**, 425, 372 (1994).
39. Maoli, R. Ferrucci, V., Melchiorri, F., Signore, M., and Tosti, D., *ApJ* **457**, 1 (1996).
40. Signore, M. et al. *ApJS* **92**, 535 (1994).
41. Signore, M., *Astro. Lett. & Comm.* **35**, 349 (1997).
42. Stancil, P.C., Lepp, S., and Dalgarno, A., *ApJ* **458**, 401 (1996).
43. Bougleux, E., and Galli D., *MNRAS* **288**, 638 (1997).

44. Galli, D., and Palla, F, *A&A* in press (1998).
45. Stancil, P.C., and Dalgarno, A. *ApJ* **490**, 76 (1997).
46. Puy, D., and Signore, M., *New Astronomy* **2**, 299 (1997).
47. McKellar, A., *PASP* **52**, 307 (1940).
48. McKellar, A., *Publ. Dominion Astrophys. Obs.* **7**, 251 (1941).
49. Adams, W.S., *ApJ* **93**, 11 (1951).
50. Roth, K.C., and Meyer, D.M., *ApJ* **441**, 129 (1995).
51. Wiklind, T., and Combes, F., *Nature* **379**, 139 (1996).
52. Ge, J., Bechtold, J., and Black, J.H., *ApJ* **474**, 67 (1997).
53. Cowie, L.L., and Songaila, A., ApJ **453**, 596. (1995).

Atomic Data and Modelling in X-ray Astronomy

T. R. Kallman

Laboratory for High Energy Astrophysics
NASA/Goddard Space Flight Center

Abstract. In the next 3 years the study of astronomical X-ray spectra will make a significant advance owing to the launch of several new instruments. These will represent an order of magnitude improvement in both spectral resolving power and sensitivity over current instruments, and will allow use of plasma diagnostics which have not been previously accesible. In this talk I will illustrate some of the uses of such diagnostics. I will discuss the status of the atomic data available, and point out areas where work is most urgently needed.

INTRODUCTION

X-ray astronomy dates from the beginning of the space program. In a short time experimental techniques have progressed by an amount comparable to what required 150 years in optical astronomy. The next few years are likely to see this trend continue, and diagnostic techniques are needed to make the best use of the new data. In order to do this work it is necessary to bring astrophysical modelling techniqes closer to those used in modelling laboratory plasmas, particularly in the area of the atomic physics assumptions and the various rate and cross section data incorporated in the codes. The development of new comprehensive computational programs and databases for dissemination of these quantities is particularly welcome in this regard.

An illustration of the rapid evolution of X-ray astronomy can be gained by comparing the figures of merit characterizing astronomical X-ray detectors: A=collecting area, and R=resolving power. Early X-ray instruments such as the proportional counters on UHURU and HEAO-1 had A=10-100, R=1-10, (low resolution), which might be regarded as the analog of filter photometry on a small optical telescope. The current generation of istruments, such as the BBXRT, $ASCA$ and Einstein (HEAO-2) experiments, utilize CCD or solid state detectors and focussing X-ray optics. $ASCA$ has A$\sim$100 and R$\sim$10-50, which we will refer to in what follows as medium resolution. In the next 3 years a significant improvement will occur in the resolving power and collecting area with the launch of the AXAF,

XMM and Astro-E experiements. These utilize gratings spectrographs in the case of AXAF and XMM, and a calorimeter in the case of Astro-E, and have A=100-1000, R=50-1000. In addition, AXAF will have CCD spectrometers which will obtain medium resolution spectra of many obects. Figure 1 illustrates the effective areas and resolving powers anticipated for the AXAF instruments (e.g. Zombeck, 1983). These will allow spectra of unprecedented quality to be obtained from a wide variety of stars and galaxies. In fact, it is difficult to predict with certainty what we are likely to find when these instruments are first turned on. However, we can guess based on past observations using the CCDs on *ASCA* and on the few observations made with high resolution on the *Einstein* observatory satellite. In the remainder of this paper I will discuss several examples of what we expect to learn in the near future from the new X-ray instruments, and what this implies for the atomic data needs associated with the analysis of the new data.

ABSORPTION SPECTRA OF ACTIVE GALAXIES

All astrophysical X-ray spectra are affected by photoelectric absorption by material along the line of sight to the photon source. Such material may be quite far from the source and thus unaffected by the radiation from the source. Photoelectric absorption by neutral material is affected by both the shape of the photoionization cross section, and by the relative abundances of the various elements in the absorbing gas. For abundances close to those inferred for the sun (e.g. Withbroe, 1972) elements heavier than helium never contribute more than approximately 50% of the total opacity at any energy between 1 Ry and 1 KeV (e.g. Morrison and McCammon, 1978). If so, the opacity distribution in this range of photon energies can be characterized as an approximately ε^{-3} (where ε is photon energy) decrease punctuated by modest discontinuities at the K shell thresholds of the various abundant trace elements such as C, O, Ne, Si, and S. This universal shape has served as an adequate description of the data from many objects.

If, on the other hand, absorbing gas is close enough to the source to be appreciably heated and ionized, then the opacity can be considerably affected. Measurements of the opacity then can serve to constrain the ionization state of the absorbing gas and hence, via modelling, the conditions in the absorber. The opacity distributions from partially ionized absorbers resemble the neutral absorber at the highest energies where absorption is detected, typically near 1 KeV, but have reduced opacities at lower energies. This is interpreted as being due to ionization of the lighter elements responsible for opacities at low energies, H, He, and C, while heavier elements which dominate the opacities at high energies, O, Ne, and Mg, are more nearly neutral. This is consistent with the predictions of ionization balance models in which lighter elements are more easily ionized under a given set of conditions than are heavier elements (Krolik and Kallman, 1984). Such ionization may come from either electron impact collisions (coronal case) or from photoionization from a strong source of continuum radiation; the qualitative difference between the two cases is negligible for most astrophysical purposes. However, photoionization is a more plausible mechanism for producing an ionized absorber since the ionization can be produced by the same continuum which is passing through the absorber.

An example of the use of photoelectric absorption in diagnosing photoionized plasmas comes from the *ASCA* observations of the Seyfert 1 galaxy NGC 3516. This object is thought to contain an active nucleus which is not hidden from our direct line of sight (owing to the presence of strong unpolarized lines in the UV

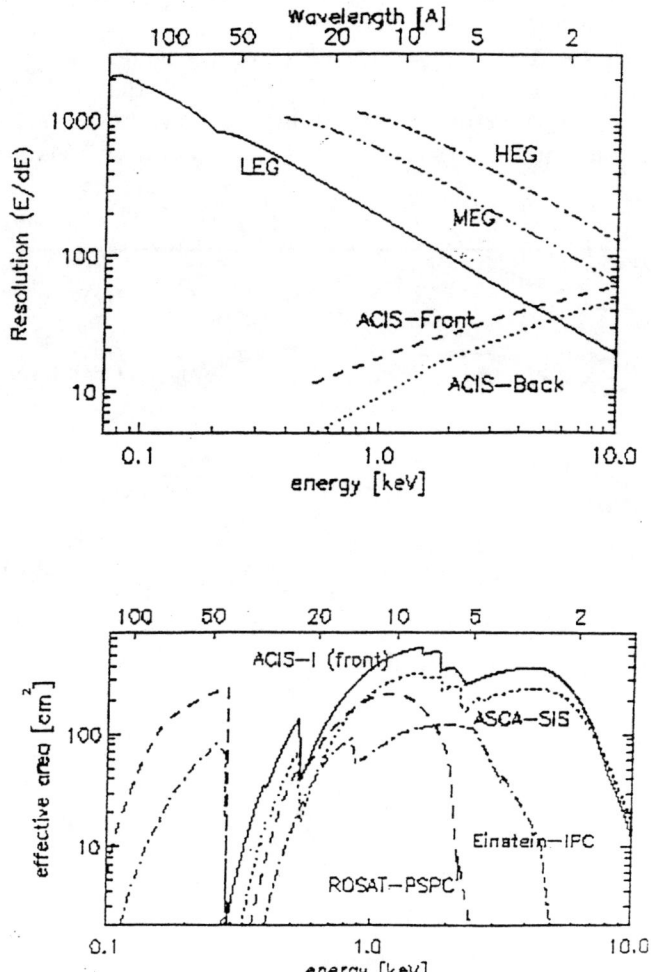

FIGURE 1. Resolving powers (upper panel) and effective collecting areas (lower panel) of various AXAF instruments. HEG=AXAF High energy grating, MEG=AXAF medium energy grating, LEG=AXAF low energy grating, ACIS=AXAF CCD, ASCA-SIS=*ASCA* CCD.

spectrum). Figure 2 shows the count spectrum (Kriss et al., 1996) observed with *ASCA* together with a fit to an ionized absorber. The fits constrain the ionization parameter of the gas, defined as the ratio of the ionizing radiation flux (F) to the gas density (n), to be $\log(\xi)=\log(4\pi F/n)=1.7$, and the column density of the absorber is $N = 2.5 \times 10^{21}$ cm^{-2}. Together with a luminosity estimate based on extrapolating that observed in the X-ray band, $L \simeq 10^{43}$ erg s^{-1}, these require that the absorber have a density $n \simeq 31.$ cm^{-3} and a distance from the continuum source of $\simeq 8 \times 10^{19}$ cm. This is approximately consistent with the expected position of the gas responsible for the emission of UV lines.

When viewed with AXAF this object is expected to reveal information about the velocity structure of the gas. This is because AXAF will have the spectral resolving power required to detect the resonance line absorption features that are expected to accompany the bound-free edge features. A simulation of an AXAF observation

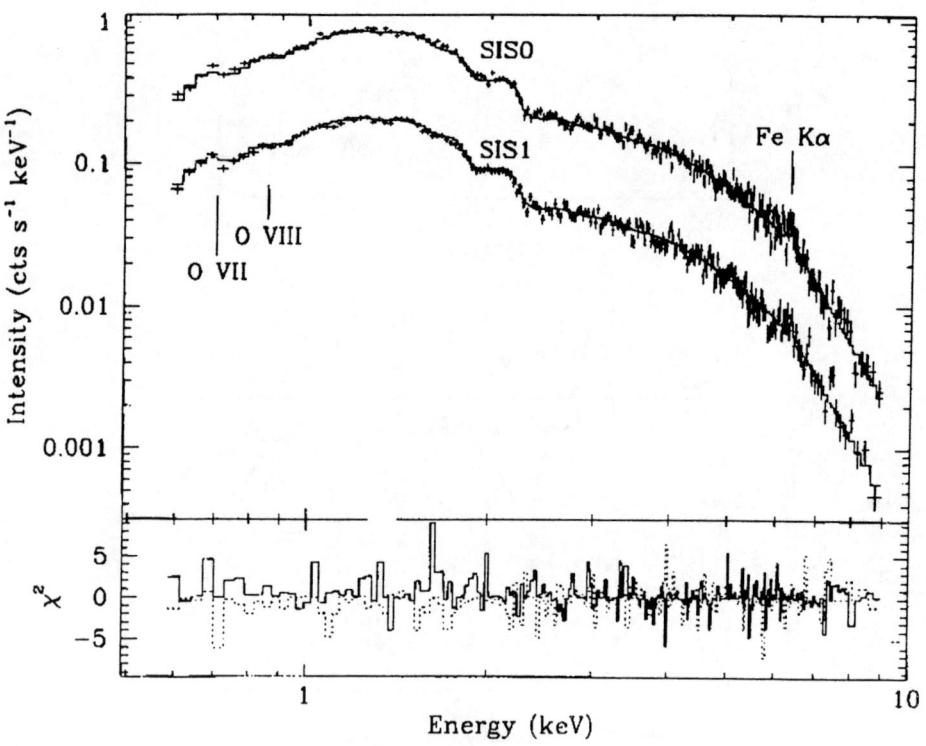

FIGURE 2. *ASCA* CCD spectrum of the Seyfert galaxy NGC 3516 (Kriss et al., 1996). Features due to O VII and O VIII are indications of absorption by partially ionized material.

of the same source is shown in Figure 3, clearly demonstrating detections of the Ca XVIII (0.83 keV), Ne IX (0.91 keV), and Fe XV (0.906 keV). This simulation assumes a width for this line of 300 km s^{-1}, significantly greater than would be produced by thermal Doppler broadening alone. AXAF will be capable of measuring widths as small as 100 km s^{-1} at this energy; velocities smaller than this value would be surprising owing to the likely proximity of a massive compact object to the absorbing gas. This will provide new constraints on the dynamics of the gas near these enigmatic objects.

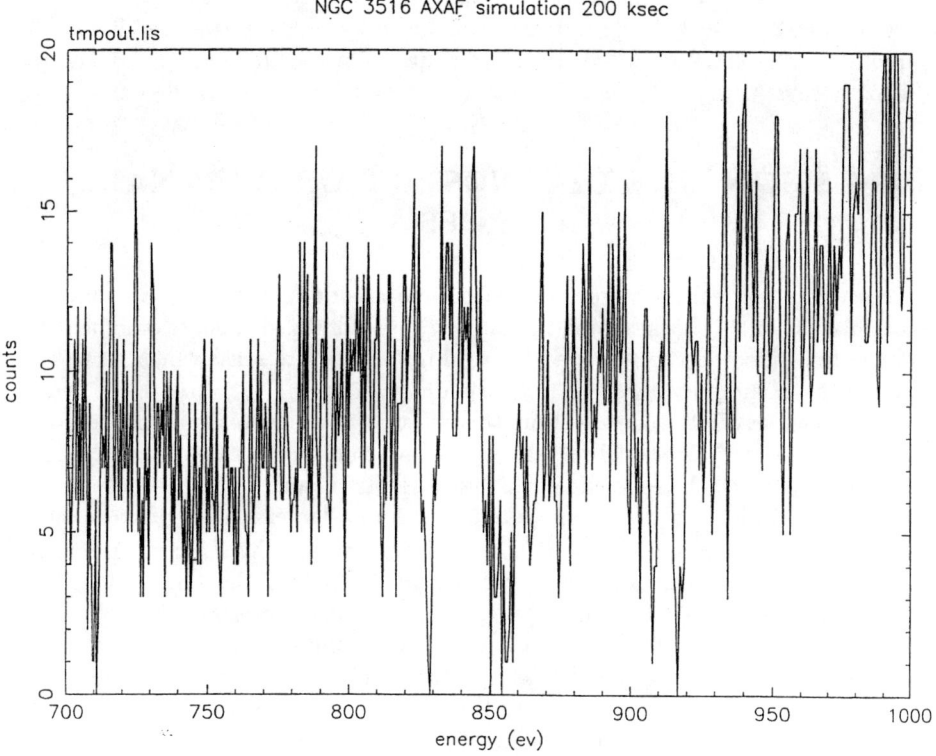

FIGURE 3. Simulated spectrum of NGC 3516 as viewed by AXAF medium energy grating in the 0.7 - 1 keV energy band.

RECOMBINATION EMISSION FROM BINARY X-RAY SOURCES

X-ray binary stars are expected to contain gas exposed to a very strong flux of ionizing radiation. Under these conditions the primary mechanism for emission of reprocessed radiation is recombination. Emission features corresponding to the free-bound continua from highly ionized O, Si, and S are seen in several of these sources, and provide a sensitive diagnostic of temperature in the recombining gas. Figure 4 shows the *ASCA* spectrum of binary star Cyg X-3 (Liedahl and Paerels, 1996), demonstrating the detection of the free-bound feature due to S XVI near 3.5 keV. Figure 5 shows a simulation of an AXAF observation of this same object, demonstrating the detection of additional free-bound features due to Si XIV (2.66 keV) and Si XIII (2.43 keV), in addition to many other features. This is of particular interest because Cyg X-3, along with many other binary X-ray sources, is highly obscured by interstellar gas and dust, and so is unobservable in energy bands other than hard X-rays.

EMISSION LINE DIAGNOSTICS OF CORONAL SOURCES

Sources where coronal processes dominate the emission include the outer layers of stars, in addition to shock heated interstellar gas and gas in clusters of galaxies. Diagnostic lines include the so-called 'wxyz' lines of He-like ions, which serve to constrain density and temperature (e.g. Pradhan 1982). These lines have been measured extensively in the sun and from the the supernova Puppis A by the crystal spectrometer on the *Einstein* Observatory satellite from the ions O VII and Ne IX (Winkler, et al., 1983). AXAF will be capable of resolving and detecting these lines and will allow detailed diagnostic measurements from a large number of sources.

In spite of the high resolution likely to be achieved by the AXAF gratings, many observations will be made at resolutions comparable to those of the ASCA satellite, using the non-dispersive CCD detectors ($R \simeq 20$). In order to interpret these data it is necessary to resolve the problems that have affected analysis of ASCA data – that of line blending introduced by the limited spectral resolution of the instrument. An example of the difficulties associated with this procedure comes from the spectrum of the star Capella as obtained by ASCA. Past attempts to model this spectrum consistently produced a deficit of photons at 1.2 keV (10 Å) relative to those at 1 keV (12 Å). This was resolved by the realization that the combined contributions of decays from levels with high principle quantum number ($n \geq 6$) into $n=2$ in the ions Fe XVII, XVIII and XIX could account for the deficit (Brickhouse, et al., 1997). This illustrates the need for comprehensive atomic models, including a larger selection of levels than might naively be anticipated, in modelling medium resolution spectra.

IRON K LINES

The K lines of iron have a unique importance in X-ray astronomy. Observationally, they fall in a region of the spectrum which is relatively free of confusion due to lines from other elements, photoelectric absorption, and features due to absorption in the instrument; iron is the most abundant heavy ($Z \geq 10$) element in the sun. In photoionized plasmas lines are formed as a direct result of K shell ionization, and the rate of line formation is only weakly dependent on the ionization state and temperature of the gas over a wide range of conditions.

The most easily observable quantities are the line strength and the line energy. In a photoionized gas the strength depends primarily on the number of photons absorbed by ionizing iron; the lines are an effective 'bolometer' for exciting photons near 10 KeV. The energies of the iron K lines are probes of the ionization state in the gas near the compact object. Lines near 6.4 KeV (1.94 Å) are emitted by

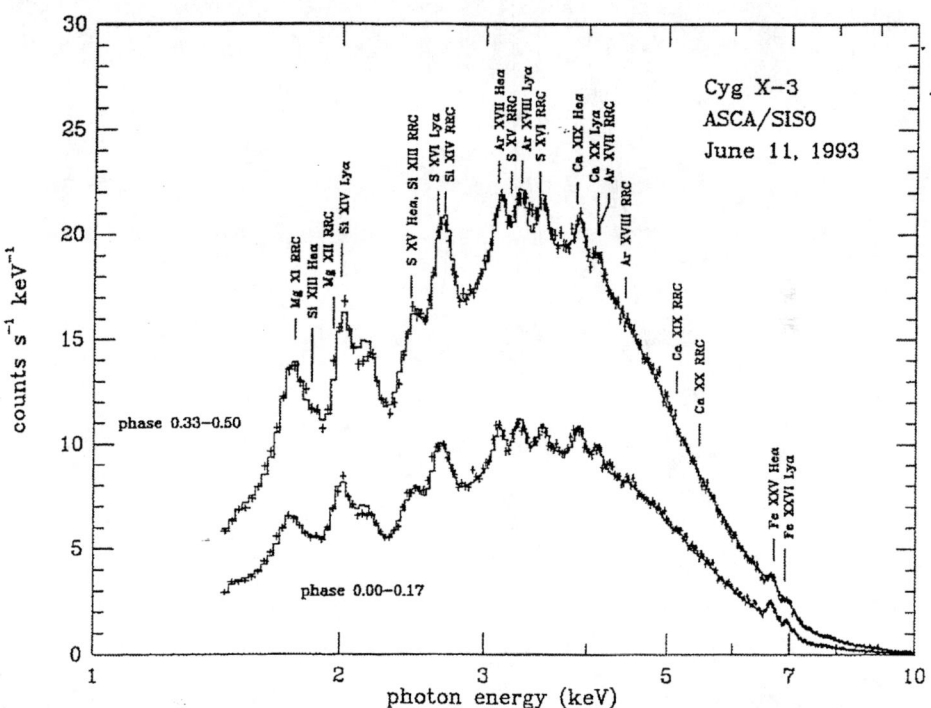

FIGURE 4. Spectrum of Cyg X-3 as viewed by ASCA (Liedahl and Paerels, 1996).

ions between neutral and neon-like (Fe XVII). The lines emitted by higher ion stages range in energy up to ∼7 KeV for Fe XXVI. The complexity of the iron line spectra is illustrated in figure 6, which shows a set of simulations of emission from photoionized gas in a binary X-ray source, in which the gas is flowing onto a neutron star in an accretion disk. The left panels show the spectrum emitted at various radii in the disk when viewed at a resolution corresponding (approximately) to that of AXAF. Noise introduced by counting statistics are not included in this simulation. The contributions from several ion stages of iron are clearly apparent in these figures. The broad pedestals under the lines which appear at the smallest radius are due to Compton scattering in an ionized corona above the disk (Kallman and White, 1989). The right panels show the same spectra when observed with an instrument with a spectral resolving power of ≃10, similar to that of the early proportional counter detectors.

Broadened and redshifted iron K lines have been detected from several active

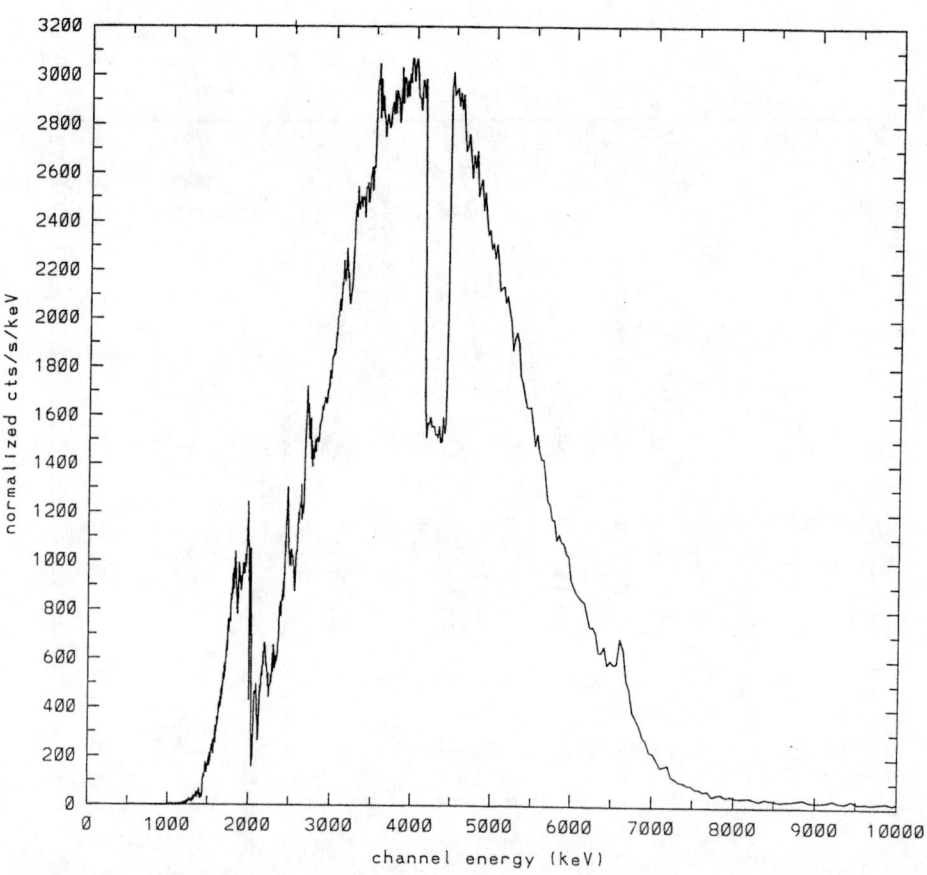

FIGURE 5. Simulated spectrum of Cyg X-3 as viewed by the medium energy grating on AXAF.

galaxies (e.g. Tanaka, et al., 1996). These provide evidence for either large Doppler shifts due to relativistic motion (e.g. orbital motion) or Compton downscattering. Accurate use of this diagnostic information requires understanding of the intrinsic spectra emitted by the gas: accurate energies and fluorescence yields under a variety of conditions.

ATOMIC DATA NEEDS

The examples so far serve to illustrate some of the astrophysical problems which can be addressed with the next generation of X-ray astronomy experiments, and

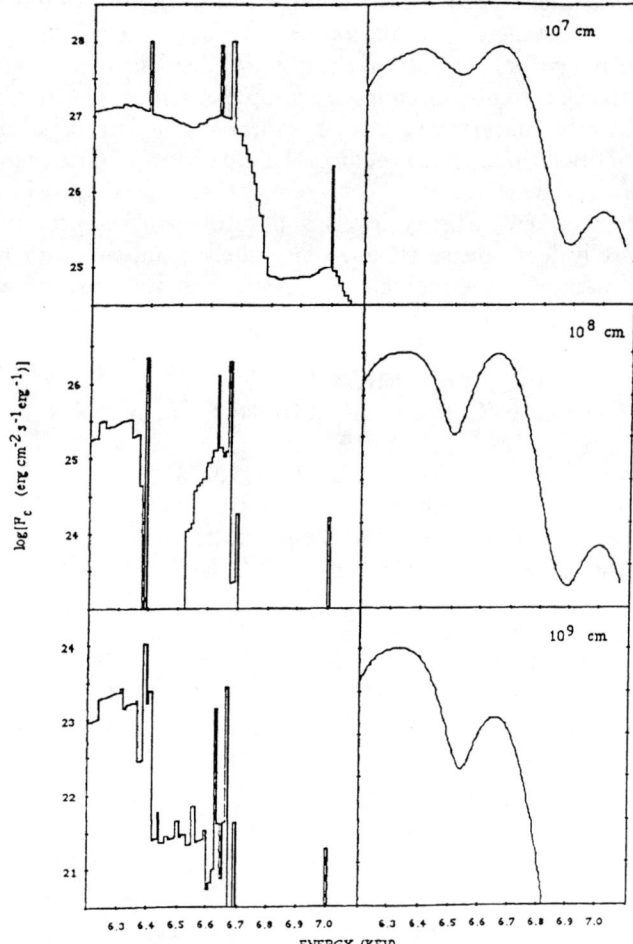

FIGURE 6. Model spectra of iron K line emission from an accretion disk in a binary X-ray source (Kallman and White 1989).

also the challenges which they pose for modelling and the atomic data needed to carry out such modelling: (i) X-ray observations at higher spectral resolution than has previously been available will require atomic energy levels, line energies and bound-free continuum thresholds to a higher precision (i.e. as high as 0.1 %) than is currently available for many lines. This is particularly important in crowded spectral regions. (ii) Recombination is an emission mechanism which has previously received less attention in the X-ray spectral region than electron impact excitation, and one which will provide useful diagnostics from the next generation instruments. Rate coefficients, energy levels, and cascade calculations are needed to intepret recombination emission features. (iii) Traditional coronal diagnostics, such as those which have been applied to the sun, will be accessible from many objects. Line blending will continue to require the use of spectrum synthesis models. Such models require the simultaneous calculation of rates affecting ionization balance and emission from many ions, and comprehensive collections of rate coefficients are crucial. Dielectronic recombination is an example of a process which continues to introduce considerable uncertainty into spectrum synthesis calculations. Line blending in medium resolution spectra requires the use of atomic datasets which are comprehensive, but not necessarily as accurate as those required for interpreting high resolution spectra. (iv) Energy levels and rate coefficients for inner shell processes are generally less precise than valence shell quantities, and these will be observed from many sources at high resolution with the next generation of instruments.

REFERENCES

Brickhouse, N. S., Dupree, A. K., Edgar, R. J., Drake, S. A., White, N. E., Liedahl, D. A., and Singh, K. P., 1997, *BAAS*, **191**,2.513

Kallman, T. R., and White, N. E., 1989, *Ap. J.*, **341**,9.55

Kriss, G., et al., 1996, *Ap. J.*, **467**,6.29

Krolik, J.H. and Kallman, T.R., 1984, *Astrophys. J.*, **286**,3.66

Morrison, J., and McCammon, D., 1983, *Ap. J.*, **270**,1.19

Pradhan, A., 1982, *Ap. J.*, **263**,4.77

Tanaka, Y., et al., 1995, *Nature*, **375**,6.59

Winkler, F., et al., 1981, *Ap. J.*, **245**,5.74

White, N.E., Peacock, A., and Taylor, B.G. 1985, *Ap. J.*, **296**,4.75

Withbroe, G., 1971, in, *The Menzel Symposium on Solar Physics, Spectra, and Gaseous Nebulae*, e.d. by K. B. Gebbie (NBS SP353)

Zombeck, M., 1983, *Adv. Sp. Res.*, **2**,2.59

A New Perspective on the Sun from SOHO–Challenges for Atomic Physics

Kenneth Dere* and Helen Mason[†]

Naval Research Laboratory, Washington DC 20375
[†]*University of Cambridge, Cambridge UK CB3 9EW*

Abstract.
The Solar and Heliospheric Observatory (SOHO) carries several ultraviolet spectrometers to diagnose the plasma conditions of the solar atmosphere in order to address several outstanding questions in solar research: why is the corona hot and how is the solar wind accelerated. Diagnostic techniques for determining the magnetohydrodynamic state (magnetic field, density, temperature, velocity) are discussed. We describe the CHIANTI database which has been designed to provide the necessary spectroscopic diagnostics for these data as well as for other spectroscopic missions. The database is built on 3 main files for each ion: atomic energy levels, radiative data, and electron collisional data. The electron collision strengths are assessed and scaled using the Burgess and Tully scaling laws. Analyses of EUV line intensities often reveal discrepancies between observed and predicted line intensities of about of 2 when considering lines of the same ion. A more recent analysis has shown that it is possible to arrive at a subset of EUV spectral lines where the line intensities are reproducible to within about 25%. For X-ray lines, as will be observed by AXAF, there is still considerable need for energy levels, radiative data and collision strengths, particularly for the $\Delta n \geq 1$ transitions.

INTRODUCTION

Two major goals of research in solar physics are to understand the heating of the solar corona and the acceleration of the solar wind. In December 1995, the ESA/NASA Solar and Heliospheric Observatory (SOHO) was launched in order to make significant progress in our understanding of these physical processes. SOHO is currently in orbit at the L1 Lagrange point where it is continuously observing the Sun with an array of remote sensing telescopes and spectrometers and *in situ* plasma detectors. Further information and images obtained with the SOHO instruments can be found on its web page:

http://sohowww.nascom.nasa.gov

CP434, *Atomic and Molecular Data and Their Applications*
edited by P.J. Mohr and W.L. Wiese
© 1998 The American Institute of Physics 1-56396-751-0/98/$15.00

The spectroscopic instruments, in particular, require a reliable set of analytical techniques for determining the properties of the solar corona. Some of these techniques and the atomic data needed to implement them are considered here.

THE SOLAR CORONA: A MAGNETOHYDRODYNAMIC MEDIUM

The solar corona is a low-density, highly-ionized medium containing a pervasive magnetic field. Since the electrical conductivity of the solar corona is very high, it is governed by the laws of magnetohydrodynamics (MHD). The state of an MHD medium is described by the magnetic field, the particle density, velocity and temperature, all as a function of position. In addition, the thermal conductivity, electrical conductivity and viscosity can be important transport coefficients.

SPECTROSCOPIC DIAGNOSTICS OF THE SOLAR ATMOSPHERE

Magnetic field measurements

The magnetic field strength in the solar photosphere is measured on a regular basis, both on the ground (NOAO Kitt Peak, for example) and with the MDI instrument on SOHO, by means of the Zeeman effect. Both the separation in wavelength and the polarization properties of the different Zeeman components are used in the measurement. It is not currently possible to measure the magnetic field in the corona. First, the field strength in the corona is considerably less than in the photosphere. This is because the field at the photospheric level is essentially bipolar so that it becomes weaker with height. Second, strong spectral lines formed at coronal temperatures (10^6K) are generally found at wavelengths below 1000Å. Since the Zeeman splitting goes as the square of the wavelength, the coronal lines will be considerably less sensitive than those at visible wavelengths used for the photospheric measurements. Finally, materials and techniques do not exist for measuring the polarization of lines at short wavelengths with high sensitivity. Consequently, our knowledge of the magnetic field in the corona is based on potential and force-free field extrapolations and MHD models.

Coronal spectroscopy

The solar corona is optically thin for most lines emitted at temperatures of 10^5K and above. Consequently, the observed emission is given by,

$$\epsilon_{i,j} = N_j(X^{+m}) A_{j,i} \frac{hc}{\lambda_{i,j}} \qquad (1)$$

The density of the upper level $N_j(X^{+m})$ is given by the product of the hydrogen density N_H, the relative elemental abundance of the element X, the relative ion abundance of the ion X^{+m} and the relative population of the upper level $N_j(X^{+m})$. Solar abundances are usually taken as a fixed set of numbers of the elemental abundance relative to hydrogen. However, it has been shown that there is a significant difference between abundances in the photosphere and in the corona [6]. This difference appears to be a function of the first ionization potential (FIP) of the element. Furthermore, it is now clear that the FIP differential can vary from feature to feature [8].

In the corona, the ion populations are determined by the balance between the ionization rate, which increases with temperature, and the recombination rate, which generally decreases with temperature. For example, the most abundant iron ion at 4×10^6K is Fe XVI. It has a peak fraction of 0.27 and the FWHM is about a factor of 2 in temperature. Several calculations of ionization equilibria are now available [2,1].

The relative populations of the upper levels is controlled by the electron collisional excitation rate from the lower levels. The electron collisional excitation rate coefficient (cm^3 s^{-1}) from level i to level j for a Maxwellian electron velocity distribution with a temperature T_e (K) is given by:

$$C_{i,j}^e = \frac{8.63 \times 10^{-6}}{T_e^{1/2}} \frac{\Upsilon_{i,j}(T_e)}{\omega_i} \, exp\left(\frac{-E_{i,j}}{kT_e}\right) \qquad (2)$$

where ω_i is the statistical weight of level i; $E_{i,j}$ is the energy difference between levels i and j; k is the Boltzmann constant and $\Upsilon_{i,j}$ is the thermally-averaged collision strength:

$$\Upsilon_{i,j}(T_e) = \int_0^\infty \Omega_{i,j} \, exp\left(-\frac{E_j}{kT_e}\right) d\left(\frac{E_j}{kT_e}\right) \qquad (3)$$

where $\Omega_{i,j}$ is the collision strength, and E_j is the energy of the scattered electron relative to the final energy state of the ion. Collision strengths are generally derived from calculations. For many of the strongest lines, the only level that is significantly populated is the ground level so that the equations become simplified.

Temperature measurements

A commonly used temperature estimate is simply to take the temperature of the maximum ionization fraction pertaining to a given ion. For example, it is common to characterize structures seen in Fe XVI lines as characteristic of 4×10^6K plasmas.

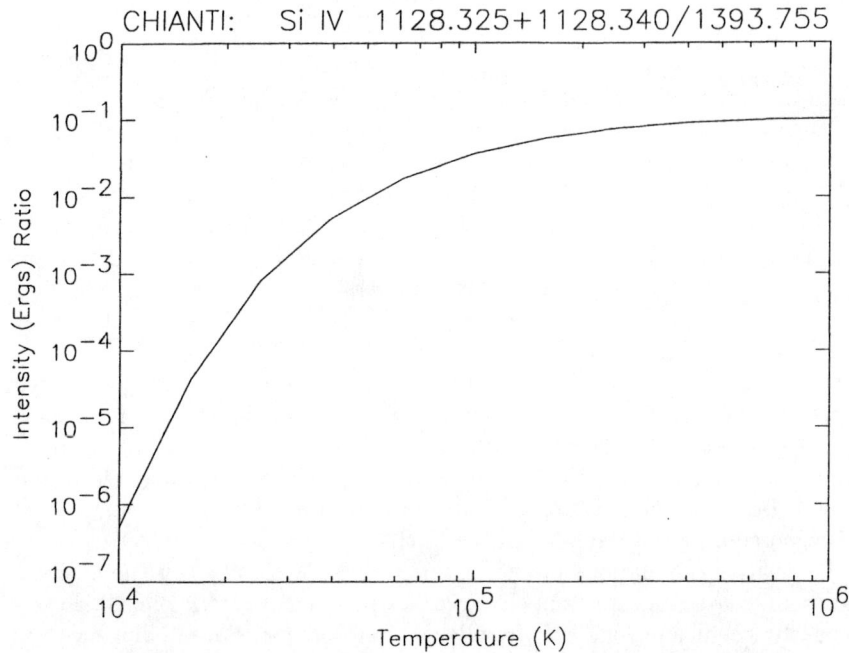

FIGURE 1. Temperature-sensitive line ratio of Si IV calculated from the CHIANTI database

More quantitative diagnostic methods are also available. From Eq. (2), it can be seen that the intensity ratio of two lines from the same ion will be a strong function of temperature if the excitation energies of the upper levels are considerably different or if there is a significant difference between the temperature (energy) dependence of the collision strengths to the upper level. An example of such a ratio involving lines of Si IV is shown in Figure 1. This ion is in the sodium isoelectronic sequence. The chosen lines correspond to 3p-3d and 3s-3p transitions so that the lines are relatively nearby in wavelength but with different upper energy levels. The use of lines close in wavelength is prompted by the practical consideration that spectrographs have limited wavelength coverage and the calibration of X-ray and ultraviolet instruments is not always well known.

Density measurements

For ions where the predominant population is in the ground level, the intensity of a spectral line is directly proportional to the emission measure of

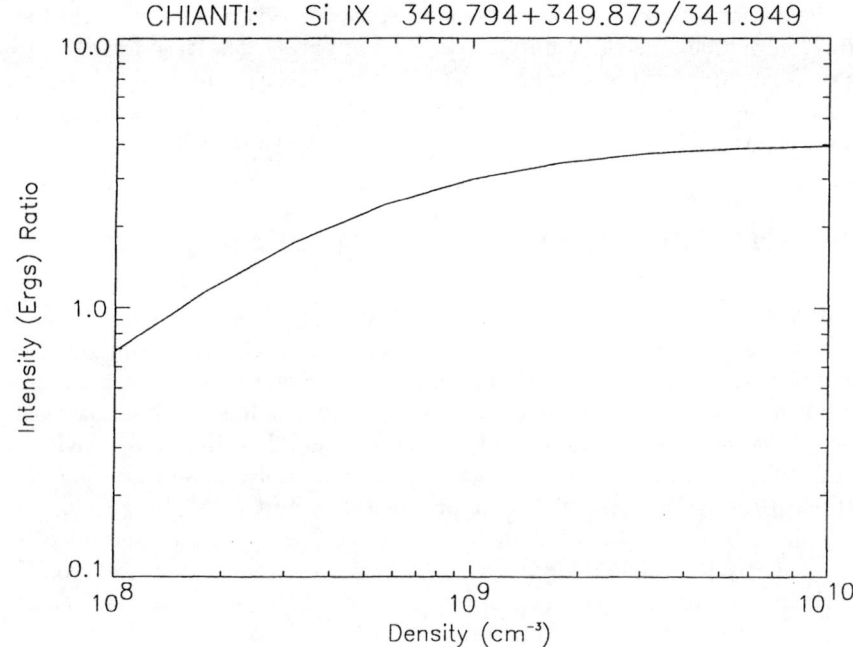

FIGURE 2. Density sensitive line ratio of Si IX calculated from the CHIANTI database

the observed plasma, either the line-of-sight emission measure ($\int N_e N_H dl$) or the volumetric emission measure ($\int N_e N_H dV$). If there is information about the spatial extent of the object then an average density can be obtained from the emission measure. For example, a coronal loop can be approximated as a cylinder.

For some ions, one or more metastable levels exist, often in the ground configuration. In this case, the relative populations of these levels are linear functions of density at low densities, but at higher densities the levels assume a quasi-Boltzmann equilibrium. For example, the boron sequence contains $^2P_{1/2}$ level and $^2P_{3/2}$ levels in the ground configuration. At low densities, only the $^2P_{1/2}$ level has any significant population and the population of the $^2P_{3/2}$ level varies linearly with density. At higher densities, the levels become populated roughly in a 2:1 ratio, following the ratio of their statistical weights. Relative changes in the populations of the lower levels are often reflected by changes in the populations of the upper levels so that the intensity ratios of some line pairs become functions of the electron density. An example of a density-sensitive line pair is shown in Fig. 2.

Many models of coronal heating suggest that coronal heating takes place in the dissipation of currents or motions on spatial scales much below the resolution of current instrumentation [7]. By comparing densities derived with the emission measure and volume method with those derived from density-sensitive lines, it is possible to determine whether there is any need for sub-resolution structure to explain the observed line intensities.

The CHIANTI database for astrophysical spectroscopy

CHIANTI [5] consists of a database of atomic parameters and a set of Interactive Data Language (IDL) routines for the calculation of emission line intensities from hot, optically-thin plasmas. The construction of this package has taken an approach different from that of previous line emission packages which assemble the atomic data for a known emission line and provide a lookup table to calculate the intensity as a function of temperature, perhaps with corrections for density. The approach taken by CHIANTI is to simply assemble all of the relevant data for any given ion so that the emissivities can be naturally calculated as a function of density and temperature. In addition, the user can specify a set of elemental abundances and their choice of ionization equilibria. In practice, we construct 3 ascii data files for each ion, one specifying the energy levels, another containing the radiative data (wavelengths, oscillator strengths and A values), and a third containing fits to the electron collision strengths.

The energy level files make use of energy levels values derived from observations, largely from the NIST database. Theoretical values are used to fill in for missing observed values. Since the theoretical values are usually of little use for line identification, they are tabulated in a column separately from the observed energy levels. The collision strengths are stored as fits to calculated collision strengths scaled according to the formulation of Burgess and Tully [4]. The CHIANTI database is freely available at:

http://wwwsolar.nrl.navy.mil/chianti.html

The current version of the database is essentially complete for wavelengths greater than 50 Å but also contains a large number of spectral lines below 50 Å. We are rapidly filling in the gaps that will allow us to calculate spectra at all wavelengths accessible to AXAF and expect to be finished by the launch of AXAF.

One of the advantages of CHIANTI is that it has predictive capabilities. It makes no assumptions about what produces any particular line but simply calculates the intensity from all lines in the database. In other words, it predicts line blends in a natural way. Blending is often a real problem when using density-sensitive line ratios or other diagnostics.

ATOMIC DATA FOR CORONAL SPECTROSCOPY

The atomic data needed for detailed diagnostics of coronal spectra continues to improve, both in terms of accuracy and completeness. Brickhouse et al. [3] and Young et al. [9] have compared calculations of line intensities with the SERTS observations of a solar active region in the 170-450 Å wavelength range. This is one of the best observed regions of the solar spectrum and the atomic data is quite complete. Brickhouse et al.considered the iron ions and found an overall average deviation of about 35% for the first order lines with most within a factor of 2. When these data are used to investigate density-sensitive line ratios, the errors are magnified and the errors in the deduced densities can be as large as an order of magnitude or more.

Young et al. [9] compared the line intensities predicted by the CHIANTI database with the same set of observed SERTS line intensities. They proceeded by first comparing relative intensities of lines originating from the same upper level of the same ion so that their ratio would be a function only dependent on the respective A values, which should be less subject to error than collision strength values. Next, they considered line intensity ratios that were insensitive to density and temperature. By considering these lines, they were able to eliminate what they considered to be redundant lines and adopted a reduced set of line ratio pairs for deriving electron densities. Using these ratios, the scatter became more acceptable. They find that it is possible to derive a subset of the observed lines that are strong and unblended and agree with their predicted relative intensities to within about 15%.

Since these two studies consider only the ratios of lines produced from the same ion, the conclusions pertain to the A values and collision strengths which control the line intensities. It is generally considered that the collision strengths are by far the more unreliable of the two. When comparing intensities from several ions, such as in the derivation of the differential emission measure, the analysis depends on both the ionization equilibrium and the elemental abundances, if lines from more than one element are involved. These analyses suggest that even larger errors may exist in the ionization equilibrium calculations which are based on the ionization and recombination rates. It is possible that some of these differences can be due to time-dependent processes.

At X-ray wavelengths which will be observed by AXAF, we expect the situation with regard to the atomic data to be considerably different. In the EUV, most lines are the result of $\Delta n=0$ transitions for which the energy levels, radiative data and electron collision data are fairly well known. In contrast, X-ray transitions are primarily $\Delta n \geq 0$ and many involve inner-shell electrons. For these transitions, the energy level values, radiative rates and collision rates are often incomplete and not especially accurate. Analysis of the AXAF spectra should provide a real challenge to both the astronomers and atomic physicists.

REFERENCES

1. Arnaud, M., and Raymond, J.C., 1992, *ApJ*, **398**, 39
2. Arnaud, M., and Rothenflug, R. 1985, *A & AS*, **60**, 425
3. Brickhouse, N. S., Raymond, J. C., and Smith, B. W., 1995, *ApJSS*, **97**, 551.
4. Burgess A., and Tully J. A. 1992, *A & AS*, **254**, 436
5. Dere, K.P., Landi, E., Mason, H.E., Monsignori Fossi, B.C., and Young, P.R., 1997, *A & AS*, **149**, 149
6. Meyer, J.-P., 1985, *A & AS*, **57**, 173.
7. Parker, E. N., 1988, *ApJ*, **330**, 474.
8. Widing, K. G., and Feldman, U., 1989, *ApJ*, **344**, 1046.
9. Young, P. R., Landi, E., and Thomas, R. J., 1998, *A & AS*, **329**, 291.

Some Examples of Spectroscopic Data Needs for Optical Remote Sensing in the Atmosphere

Jean-Marie Flaud

*Laboratoire de Photophysique Moléculaire, Bâtiment 210,
Université de Paris-Sud 91405 Orsay, France*

Abstract. Among the various methods used to measure temperature profiles as well as concentration profiles of minor constituents in the terrestrial atmosphere, optical remote sensing is the most widely used. This method which is based on various experimental techniques (Grating spectroscopy, Fourier transform spectroscopy, diode laser spectroscopy, microwave spectroscopy...) may use different platforms (ground, aircrafts, balloons, satellites...) but in all cases it requires an accurate knowledge of the spectral parameters (line positions, intensities, widths, absorption cross-sections ...) in order to derive reliable profiles. It is worth noticing that nowadays this requirement is even more critical given the recent progress in experimental techniques.

In this paper we present and discuss two selected examples for which the ultimate limitation on the accuracy is due to the spectral parameters. In particular recent results concerning the infrared spectral parameters of HNO_3 and $C\lambda ONO_2$ will be described.

INTRODUCTION

The purpose of this paper is neither to review all the spectroscopic problems linked to remote sensing measurements in the atmosphere nor to give an overview of the optical techniques used to sound the atmosphere. Instead the choice has been made to discuss on two selected examples progress recently achieved in the spectroscopic knowledge of some molecules of atmospheric interest, namely HNO_3 and $C\lambda ONO_2$ as well as deficiencies still existing for these molecules. Also, we discuss briefly the difficulties faced during laboratory measurements and the need for more sophisticated theoretical methods able to generate a reliable line list of spectral parameters. Finally it is important to point out that some spectral parameters such as linewidths or collisional narrowing effects are not discussed within the frame of this paper.

HNO₃ MOLECULE

HNO$_3$ is a reservoir molecule for the NO$_x$ species which play a catalytic role in the ozone destruction. It is measured in the atmosphere by optical remote sensing techniques working either in the absorption or in the emission mode. Various bands absorbing in different spectral regions in the far infrared or the infrared are used for the atmospheric measurements : The v_9 band at 25 μm (1), the $2v_9$ and v_5 bands at 11 μm (2,3), the v_3 and v_4 bands at 7.5 μm (3) or the v_2 band at 5.8 μm (4). Other absorptions from HNO$_3$ have also been observed in atmospheric spectra : $v_8 + v_9$ around 8.3 μm and v_1 around 2.8 μm (5) but, to our knowledge, they have not been used for atmospheric remote sensing of this molecule.

There are various spectroscopic problems which can affect the derivation of reliable atmospheric concentration profiles of HNO$_3$

- Within a given band, the relative line intensities should be consistent. For example Figure 1 (from ref. (6)) presents the modeling of a high resolution atmospheric spectrum around 878 cm^{-1} using two sets of spectroscopic data. The first one (7) was generated taking into account the Fermi interaction linking the rotational levels of the v_5 and $2v_9$ states. The second calculation (8, 9) takes into account not only this Fermi interaction but also the Coriolis one. Given this theoretical improvement in the spectroscopic model, better line positions and intensities were generated. This can be seen clearly when looking at Figure 1 where both the residuals and the standard deviations are noticeably reduced when using the new spectroscopic data.

- The relative band intensities between different spectral regions have to be consistent too. This is indeed crucial for meaningful comparisons of concentration profiles using different spectral regions.

For a molecule such as HNO$_3$ whose concentration in a laboratory sample is difficult to measure precisely, deriving experimentally consistent intensities for bands absorbing in different spectral regions is not an easy task. As an example one can compare on a relative basis intensities measured at high resolution in the 11 μm and 7.6 μm regions. One measurement (9) was performed using Fourier transform spectrometry. In this way it was possible to record simultaneously both spectral regions and this eliminates any problem linked to the composition of the sample. On the other hand, an experiment at 11 μm (10) as well as one at 7.6 μm (11) used diode laser spectrometry and only lines appearing in one of the two spectral regions were measured. When comparing the results of (9), which, given the simultaneous recording of the two spectral regions are consistent (This is corroborated by low resolution measurements, see next paragraph), with those of ref. (10) and (11) one finds a difference of $\cong + 20$ % for the 11 μm region and of $\cong - 20$ % for the 7.6 μm region. It is then clear that the two last measurements are not consistent. The spectroscopic results of ref. (9) have been used to derive HNO$_3$ concentration profiles. The measurements were made in emission using a

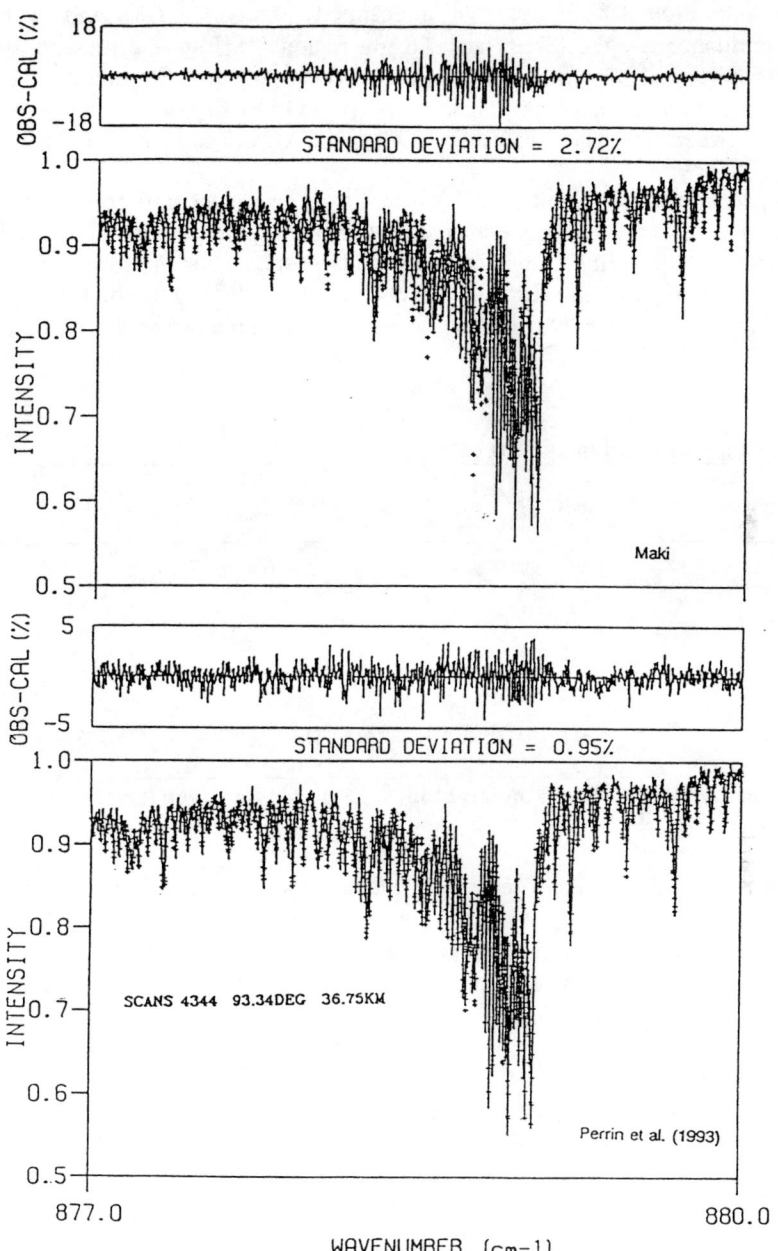

Figure 1. Comparisons of spectral least-squares fitting to ballon-borne solar spectra obtained during D.U. balloon flight of 6 june 1988, with the recent sets of HNO_3 line parameters in the v_5 Q-branch region. Note the improvement brought by the new data of Perrin et al. (8,9) (From ref. (6)).

Reprinted from *J. Q. R. S. T.*, **52** A. Goldman et al., pp 367-377, ©1994 with kind permission from Elsevier Science Ltd., The Boulevard, Langford Lane, Kidlington OX5 1GB, UK.

Michelson interferometer for passive atmospheric sounding (MIPAS) which recorded simultaneously the 11 µm and 7.6 µm region (3). Figure 2 presents two HNO_3 concentration profiles retrieved in either the 11 µm or the 7.6 spectral regions using the spectroscopic data of Perrin et al. (9). One can see the very good agreement between the two profiles, showing the consistency of the spectral parameters.

- Finally, accurate absolute intensities are crucial to derive accurate concentration profiles, an error of x percent in the intensities translating into an error of about the same percentage on the HNO_3 abundance in the atmosphere for linear absorptions (The situation is worse when saturation effects occur). Table 1 gives the total band intensities in the 11 µm and 7.6 µm spectral regions measured either at high or low resolution.

TABLE 1. Comparison of integrated band intensities S

	Ref. (9)	Ref. (12)	Ref. (13)
Temperature	296 K	313 K	273 K
11 µm	S = 502	S = 609	S = 692
7.6 µm	S = 930	S = 1186	S = 1383

The integrated band intensities S are given in $atm^{-1} cm^{-2}$ at the given temperatures.

Even if it is difficult to make a general conclusion, one can say that :

- The ratio of the intensities in the 11 and 7.6 µm regions are consistent for the three references (Let us recall that this ratio is not consistent with the ratio derived from the individual line intensity measurements of ref. (10) for the 11 µm region and of ref. (11) for the 7.6 µm region).

- The intensities of ref. (7) are about 27 % lower than those of refs. (11,12).

As a conclusion, it appears that, if one can be rather confident in the consistency of line intensities i) within the HNO_3 bands at 11 and 7.6 µm and ii) between the bands at 11 and 7.6 µm, subsequent efforts have still to be made in the laboratory in order to derive accurate absolute intensities (For more details, see ref. (5)).

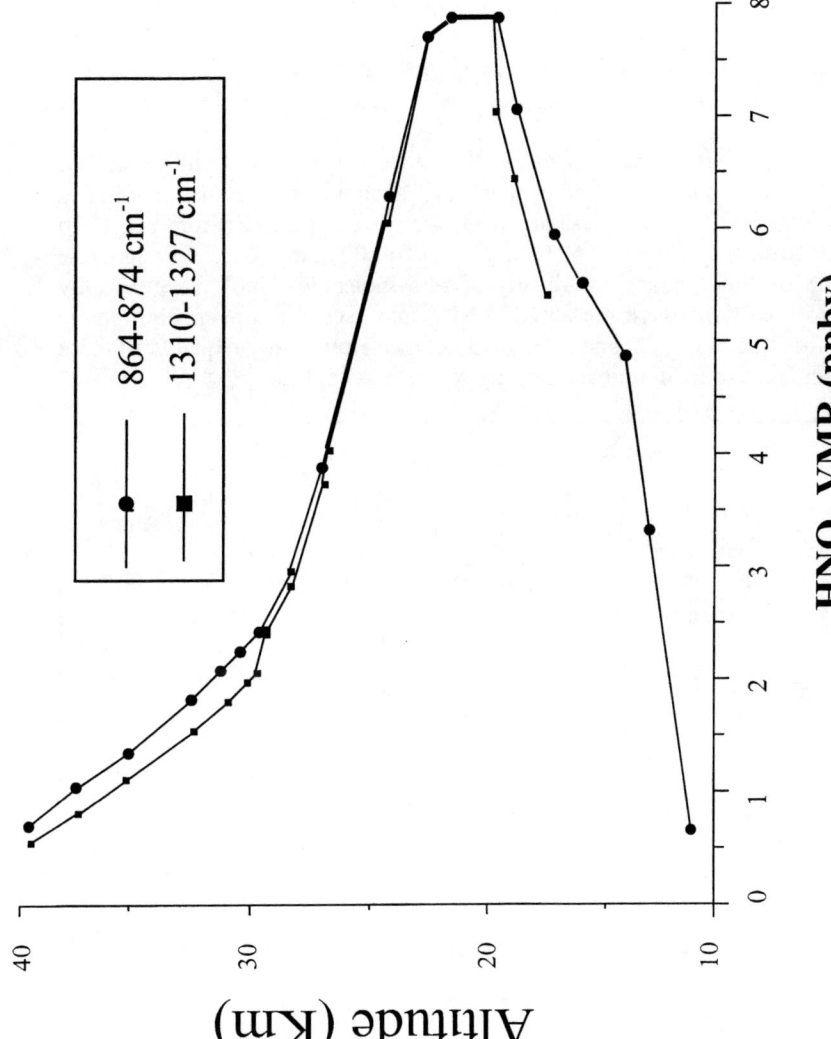

FIGURE 2. HNO$_3$ concentration profiles retrieved using either the 11 μm or 7.6 μm region (From ref. (19))

CℓONO₂ MOLECULE

CℓONO$_2$ is an important reservoir for chlorine and NO$_x$ species and provides a linkage between the chlorine and nitrogen families. It is produced mainly via the reaction

$$CℓO + NO_2 + M \rightarrow CℓONO_2 + M$$

and, in the sunlit atmosphere, is destroyed by photolysis. The situation is different in winter and early spring polar lattitudes where heterogeneous stratospheric chemistry occurs.

From a spectroscopic point of view, CℓONO$_2$ is a heavy molecule with low frequency vibrations and its infrared spectrum at room temperature exhibits a pseudocontinuum with more or less sharp Q branches. Figure 3 (From ref. (14)) shows the absorption spectrum of CℓONO$_2$ between 500 and 1800 cm^{-1} where one can see some of the fundamental bands of this molecule. Until recently, only absorption cross sections were measured (14-16) and used for atmospheric remote sensing (see for example 17-20 and references herein). However, the quality of these analyses suffer from some deficiencies of the spectroscopic data.

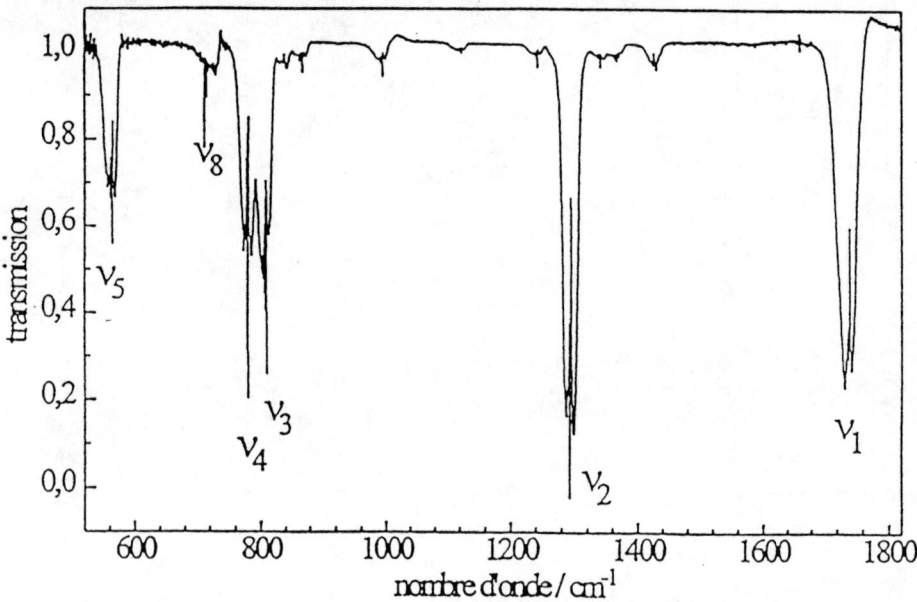

FIGURE 3. Absorption spectrum of CℓONO$_2$ between 20 and 5.5 µm. A number of Q branches from different bands appear clearly (From ref. (14)).

Reproduced with permission.

First, the ClONO$_2$ absorption cross sections are not in excellent agreement. While the absorption cross sections measured in refs. (14,16) agree to within 5 %, they are significantly different from those obtained in ref. (15). Second, they have been measured essentially at two temperatures (220 K and 296 K) and interpolation between these 2 temperatures to span the range of stratospheric temperatures is tricky. Third, no systematic study of pressure effects has been undertaken. Figure 4 (From (19)) shows a fit of the Q branch of the v_4 band near 780 cm^{-1}. It is obvious when looking at the residuals that, when the effect of pressure on the Q branch shape is not considered, it is not possible to achieve a good fit of the observations. An experimental indication of such a pressure shift effect was given in ref. (14) where various laboratory spectra were recorded at 223 K for pressures ranging from 1.8 to 103 Torr. Furthermore, it was observed that the pressure broadening of this Q branch is rather small, probably because of line mixing effects.

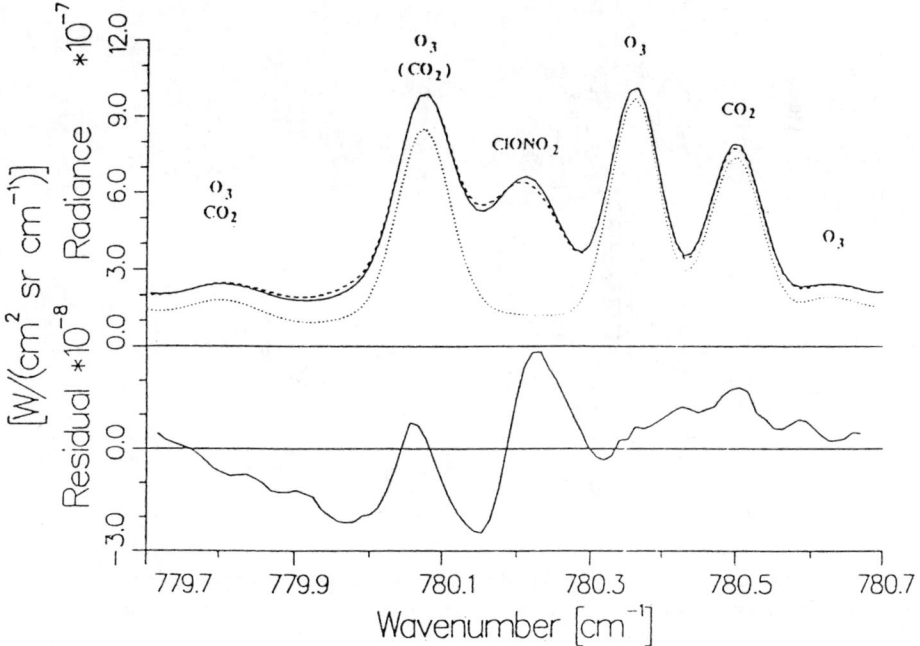

FIGURE 4. Best fit in the ClONO$_2$ region for the 16.43-km tangent altitude. The residual is the difference between the measurement (solid line) and the best fit calculated spectrum (dashed line). The residuum in the ClONO$_2$ Q branch is systematic. The dotted line is a forward calculation without ClONO$_2$ (From ref. (19)).

Recently, in order to cope with these difficulties, a number of studies were performed either in the microwave or the infrared spectral regions using molecular beams (21-25). Given the low temperatures achieved (below 35 K) it is possible to get rid of the absorption of hot bands and to observe the individual lines of the cold bands. Accordingly a line by line analysis can be performed and precise Hamiltonian constants can be derived. Figure 5 (From ref. (22)) presents an example of the accuracy obtained. From these low temperature data, it has been tried to perform rotational contour analysis using also high resolution spectra recorded at higher temperatures (26, 27). However such modeling is very difficult for two main reasons.

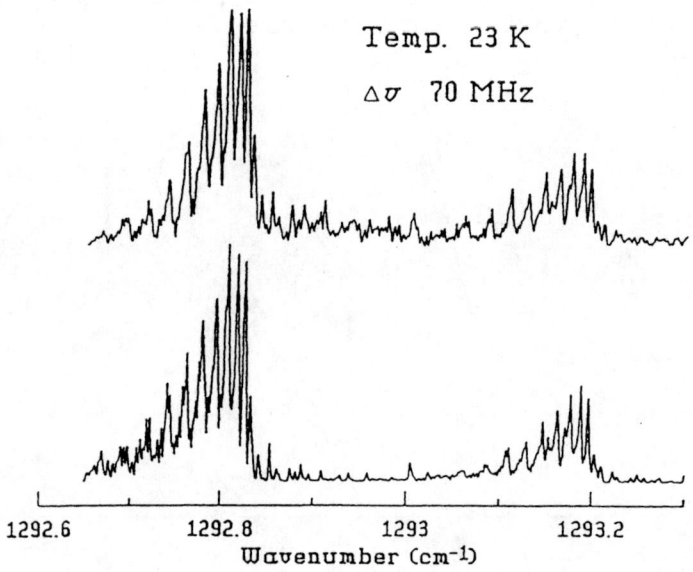

FIGURE 5. The region around the ν_2 band center of chlorine nitrate showing the ^qQ_K subbranch heads. A simulated spectrum calculated from the derived molecular constants is given in the bottom panel. The stronger feature is due to $^{35}C\ell ONO_2$ while the higher frequency series arises from the $^{37}C\ell$ isotopomer. Notice the excellent agreement between observation and calculation (From ref.(22)).

1) In very low temperature spectra, one observes only rather low values of the J and K_a quantum numbers. Extrapolating to high values of J and K_a is delicate and requires, to be efficient, a precise band contour analysis and a determination of higher order Hamiltonian constants.
2) Also, the low temperature spectra do not give information about hot bands and for $C\ell ONO_2$ this is really a serious problem. This molecule exhibits indeed a number of low energy vibrations (ν_9 around 122 cm^{-1}, ν_7 around 273 cm^{-1}, ν_6 around 434 cm^{-1} ...) : at stratospheric or room temperatures these vibrations are involved in hot bands which contribute a lot to the absorption continuum.

A review of all these problems and of the latest efforts performed in order to generate line lists of spectral parameters for some bands of $C\ell ONO_2$ is given in ref. (28).

CONCLUSION

We have tried to show on two different examples, HNO_3 for which line by line calculations are performed and $C\ell ONO_2$ for which absorption cross sections are used, that it is crucial to have good spectral parameters in order to derive precise atmospheric concentration profiles. Also it was our purpose to point out how difficult it is to generate such spectral parameters either because the experiment is delicate or because progress is still to be made concerning theory. In our opinion, given the accuracies required on atmospheric concentration profiles and progress achieved in remote sensing techniques, it is obvious that a lot of efforts have to be devoted in the future to the understanding and the improvement of the spectroscopy of molecules of atmospheric interest.

REFERENCES

1. Chance, K., Traub, W.A., Johnson, D.G., Jucks, K.W., Ciarpollini, P., Stachnik, R.A., Salawitch, R.J., and Michelsen, H.A., *J. Geophys. Res.* **101**, 9031-9043 (1996).
2. Goldman, A., Murcray, F.J., Blatherwick, R.D., Kosters, J.J., Murcray, F.H., Murcray, D.G., and Rinsland, C.P., *J. Geophys. Res.* **94**, 14945-14952 (1989).
3. Oelhaf, H., « Emission Spectroscopy of the Upper Atmosphere », presented at the Fifteenth Colloquium on High Resolution Molecular Spectroscopy, Glasgow, U.K., September 7-11, 1977.
4. Goldman,, A., Murcray, F.J., Blatherwick, R.D., Kosters, J.J., Murcray, D.G., Rinsland, C.P., Flaud, J.-M., and Camy-Peyret, C., *J. Geophys. Res.* **97**, 2561-2567 (1992).
5. Goldman, A., Rinsland, C.P., Perrin, A., and Flaud, J.-M., *J.Q.S.R.T.* (accepted).
6. Goldman, A., Rinsland, C.P., Murcray, F.J., Blatherwick, R.D., and Murcray, D.G., *J.Q.R.S.T.* **52**, 367-377 (1994).
7. Maki, A.G., and Wells, J.S., *J. Mol. Spectrosc.* **152**, 69-79 (1992).
8. Perrin, A., Jaouen, V., Valentin, A., Flaud, J.-M., and Camy-Peyret, C., *J. Mol. Spectrosc.* **157**, 112-121 (1993).
9. Perrin, A., Flaud, J.-M., Camy-Peyret, C., Jaouen, V., Farrenq, R., Guelachvili, G., Kou, Q., Le Roy, F., Morillon-Chapey, M., Orphal, J., Badaoui, M., Mandin, J.Y., and Dana, V., *J. Mol. Spectrosc.* **160**, 525-539 (1993).
10. Brockman P., Bair, C.H., and Allario, F., *Appl. Opt.* **17**, 91-100 (1978).
11. May, R.D., and Webster, C.R., *J. Mol. Spectrosc.* **138**, 383-397 (1989).
12. Goldman, A., Kyle, T.G., and Bonomo, F.S., *Appl. Opt.* **10**, 65-73 (1971).
13. Giver, L.P., Volers, F.P.J., Goorvitch D., and Bonomo, F.S., *J. Opt. Soc. B* **1**, 715-722 (1984).
14. Orphal, J., « Infrared Spectroscopy at High Resolution of Unstable Molecules of Atmospheric Interest : $C\ell ONO_2$, $C\ell NO_2$ and $BrNO_2$ », Thesis n° 3510, Université Paris XI, Orsay, France, 1995 ; Orphal, J., Morillon-Chapey, M., and Guelachvili, G., *J. Geophys. Res.* **99**, 14549-14555 (1994).
15. Davidson, J.D., Cantrell, C.A., Shetter, R.E., Mc Daniel, and Calvert, J.G., *J. Geophys. Res.* **92D**, 10921-10925 (1987). .
16. Ballard, J., Johnston, Gunson, M.R., and Wassel, P.T., *J. Geophys. Res.*, **93D2**, 1659-1665 (1988).
17. Rinsland, C.P., Zander, R., Demoulin, P., and Mahieu, E., *J. Geophys. Res.* **101**, 3891-3899 (1996).
18. Johnson, D.G., Orphal, J., Toon, G.C., Chana, K.V., Traub, W.A., Jucks, K.W., Guelachvili, G., and Morillon-Chapey, M., *Geophys. Res. Lett.* **23**, 1745-1748 (1996).

19. von Clarmann, T., Wetzel, G., Gelhaf, H., Friedl-Vallon, F., Linden, A., Moucher, G., Seefeldner, M., and Trieschmann, O., *J. Geophys. Res.* **102**, 16157-16168 (1997).
20. Zhou, D.K., Bingham, G.E., Rezai, B.K., Anderson, G.P., Smith, D.R., and Nadile, R.M., *J. Geophys. Res.* **D102**, 3559-3573 (1997).
21. Xu, S., Blake, T.A., and Sharpe, S.W., *J. Mol. Spectrosc.* **175**, 303-314 (1996).
22. Domenech, J.L., Flaud, J.-M., Fraser, G.T., Andrews, A.M., Lafferty, W.J., and Watson, P.L., *J. Mol. Spectrosc.* **183**, 228-233 (1997).
23. Suenram, R.D., and Johnson, D.R., *J. Mol. Spectrosc.* **65**, 239-248 (1977).
24. Suenram, R.D., and Lovas, F.J., *J. Mol. Spectrosc.* **105**, 351-359 (1984).
25. Mueller, H.S., Helminger, P., and Young, S.H., *J. Mol. Spectrosc.* **181**, 363-378 (1997).
26. McPheat, R., and Duxbury, G., *J. Geophys. Res.* **101**, 6803-6810 (1996).
27. Goldman, A., « Extended Quantitative Spectroscopy for Analysis of Atmospheric Infrared Spectra », presented at the Fourier Transform Spectroscopy Conference of the Optical Society of America, Santa Fe, New Mexico, February 10-12, 1997.
28. Goldman, A., Rinsland, C.P., Flaud, J.-M., and Orphal, J., *J.Q.S.R.T.* (accepted for publication).

Laboratory Data Needs
and Applications

The Crucial Role of Atomic and Molecular Processes in the Success of Controlled Fusion

Douglass E. Post

International Thermonuclear Experimental Reactor
Garching, Germany and San Diego, USA, Joint Work Sites

Abstract. Atomic processes have played a key role in the success of the present and the next generation of magnetically and inertially confined controlled fusion experiments. Magnetic fusion experiments are beginning to access the plasma regimes needed for fusion reactors. Recent experiments on the TFTR tokamak at Princeton and the JET tokamak at Abingdon, UK, have produced fusion powers of 10-16 MW and temperatures in the 10 to 40 keV range. These achievements were made possible by impurity control and high power auxiliary heating, which both rely upon the successful utilization of atomic processes. Based on these and other experimental successes, in 1988 the US, Europe, Japan and Russia began participation in the International Thermonuclear Experimental Reactor project (ITER), with the goal of designing, constructing and operating a long pulse, ignited tokamak. The engineering design portion of the project will be completed in July 1998, and the ITER partners are now discussing an agreement for construction. Experiments on JT-60 U, JET, DIII-D, Alcator C-Mod, ASDEX Upgrade and other tokamaks together with computational models indicate that atomic processes can be used to reduce the peak heat fluxes on the wall to acceptable levels, control the impurity level, and minimize the impact of plasma disruptions in ITER. Two large stellarators, an alternative to the tokamak, are being built in Japan and Germany and control of atomic processes will be essential for their success. Comparable progress has also been made in Inertial Confinement Fusion. Experiments on NOVA, OMEGA, GEKKO XII and other laser facilities with both direct drive and indirect drive targets produce temperatures in the multi-keV range and capsule compression levels that scale to ignition for the National Ignition Facility (NIF) now under construction at the Lawrence Livermore National Laboratory. Atomic processes play a key role in the pellet compression and heating and are essential for diagnostics.

INTRODUCTION

Atomic processes are important for the success of both magnetic and inertial fusion. This paper reviews the progress in fusion research and gives several illustrative examples of the role of atomic physics in fusion. The detailed role depends on the concept. In magnetic fusion the plasma pressure is balanced by the magnetic field pressure produced by external electromagnets, i.e. $\beta \sim 1$ (Eq. (1))[1]. The ions and electrons gyrate around the field lines—providing containment in two dimensions—but travel freely along the field lines (Fig. 1). Confinement in the third direction is obtained by bending the field lines around into a torus. A tokamak is a toroidal plasma in which a large current is induced around the torus. This provides additional magnetic fields so that the field lines spiral in a helix around the torus and form nested "flux" surfaces. There are major tokamak experiments in almost every developed country (Table 1). The European Union, Japan, Russia and the United States are presently discussing jointly building the International Thermonuclear Experimental Reactor (ITER)[2], a tokamak designed to reach ignition. Although the tokamak has been the most successful magnetic fusion concept, performance improvements and smaller sizes would make a more attractive fusion power reactor and a number of alternative concepts are being studied. Chief among these is the stellarator[3], a toroidal plasma in which the helical field is produced entirely with external coils. Because there are no externally driven currents, stellarators are inherently steady state. They also don't suffer from many of the instabilities involving internal plasma currents that increase the engineering design requirements for tokamak reactors. There are two large stellarators under construction, the Large Helical Device (LHD) at the National Institute for Fusion Studies near Nagoya, Japan[4,5], and the Wendelstein VII-X in Greifwald, Germany[6]. LHD will begin operation in mid-1998 and Wendelstein VII-X construction has just begun. Other alternative concepts being studied on a smaller scale include reversed field pinches, speromaks and mirrors. Although not developed as far as the tokamak, these alternative concepts have either potential physics or engineering advantages compared to tokamaks. The important plasma quantities are:

$$\beta = \frac{3nT}{B^2/8\pi} \qquad \nabla(3nT) = m\dot{\vec{v}} \qquad (1)$$

where n, T, B, m and v are the plasma density, plasma temperature, toroidal field, ion mass and ion velocity, respectively.

TABLE 1 Tokamak Program

Country	Tokamak
Europe (EU)	JET, ASDEX Upgrade, TEXTOR, TORE-Supra, FTU, Compass-D, TCV, START/MAST, ISTOK
Japan	JT-60/U, JFT-2M, JIPPT-2 U, TRIAM
US	TFTR(retired 1997), DIII-D, Alcator C-Mod, NSTX
Russia	T-10, TUMAN-II, T-15
International	International Thermonuclear Experimental Reactor (ITER)*
Other	Korea: KSTAR; small expts.: India, China, Australia, Canada, Iran, Brazil, Mexico

*Being designed by the international ITER team from the European Union, Japan, Russia and the US.

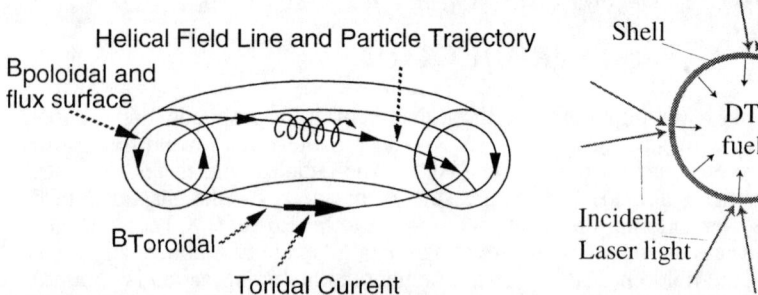

FIGURE 1. Ion spiraling around a magnetic field line in a tokamak.

FIGURE 2. ICF pellet implosion.

The second approach is "inertial confinement fusion" (ICF) in which the plasma pressure is balanced against inertia (Eq.(1), Fig. 2)[7]. This paper gives only a brief discussion of inertial fusion confinement since the topic is covered in detail in a companion paper[8]. Energy—usually high power laser light—is deposited uniformly on the spherical shell of a capsule containing DT gas at high pressures. The ablation of the outer layer of the shell propels the shell inward compressing the DT gas. The gas density needs to be compressed to the point that the alpha particles produced by the fusion reactions slow down in the fuel and further heat it to ignition. The compression also needs to be strong enough to heat the gas to several kilovolts to get a sufficiently large fusion reaction rate. The US inertial fusion program consists of a number of large laser facilities and a light ion facility (Table 2). A laser facility designed to achieve ignition (the National Ignition Facility, NIF) is presently being constructed at the Lawrence Livermore National Laboratory[8,9]. The US inertial fusion program is substantially larger in budget than the US magnetic fusion program but is less prominent abroad (Table 2). Because lasers are relatively inefficient (~ 1% or less of the input energy for Neodymium glass lasers can be converted to energy at the preferred wavelength), most reactor embodiments of inertial fusion are based on heavy ions accelerated to ~ 1 GeV with RF systems because they are potentially more efficient. There are modest heavy ion accelerator programs in the US and abroad[10].

TABLE 2 Inertial Fusion Experiments

Country	Driver	Experiments
US	Laser	NOVA, OMEGA, NIKE, NIF*
	Light Ions	PBFA II
Japan	Laser	Gekko XII

*In construction at the Lawrence Livermore National Laboratory in the US.

PROGRESS IN MAGNETIC FUSION

The progress since the initial magnetic fusion experiments in the 1950's has been impressive. From 1955 to 1998, the measured plasma temperatures have risen from a few eV to ~ forty keV, the densities from the 10^{17} m^{-3} range to the 10^{20} m^{-3} range, and the confinement figure of merit, $nT\tau_E$, from 10^{12}—10^{13} to 9×10^{20} keV m^{-3} s, (n is the plasma density, T the plasma temperature, and τ_E the energy confinement time). ITER will need an $nT\tau_E$ of about 5×10^{21} keV m^{-3} s to meet its goals[11].

The major physics issues include achievement of adequate levels of energy confinement, plasma stability and control, plasma heating, and power and particle control (Fig. 3)[12]. The engineering issues involve the design and implementation of high field, superconducting magnets, long lifetime high heat flux components, neutron shielding and tritium breeding, remote maintenance of the tokamak components, and plasma heating technologies[2]. For energy confinement, recent experiments on JET, TFTR, DIII-D and JT-60 U have demonstrated $nT\tau_E$'s of up to 9×10^{20} keV m^{-3} s, within a factor of 5 of the value needed for long pulse ignited operation (Fig. 4)[13-15]. In addition, there have been major advances in the level of understanding of energy confinement[16]. Highly elongated plasmas are routinely controlled, and high plasma pressures have been obtained[17]. The potentially high local peak heat loads can be reduced by spreading out the power over large areas of the wall by X-ray and ultraviolet radiation from impurities and hydrogen[18,19]. Many plasma technologies (e.g. high energy neutral beams, radio frequency waves, microwaves, etc.) have been successfully developed and used to heat tokamaks to temperatures of up to 40 keV, well above the 15—20 keV needed for a reactor[20]. Perhaps the most impressive progress has been in the detailed understanding of most of the important physical processes that occur in tokamaks. A key part of this is due to the progress in plasma diagnostics such as the measurement of detailed profiles of many of the plasma parameters and properties, including the plasma current profile, impurity distribution, density and temperature fluctuations, radiation losses, impurity line radiation losses, temperature and density profile and heat loads on the plasma facing components[21].

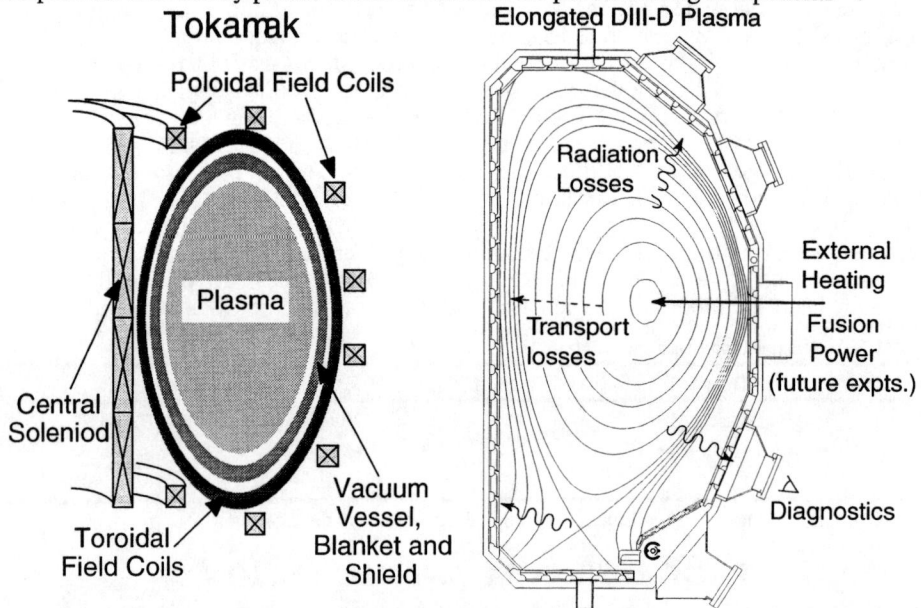

FIGURE 3. Cross section of schematic of a tokamak and a cross section of the DIII-D tokamak illustrating the major physics issues for tokamaks.

The progress in energy confinement consists not only of the achievement of progressively higher $nT\tau_E$'s but of improved understanding of the processes that cause transport[16]. One large improvement has been the discovery that a transport barrier can be formed at the plasma edge when the tokamak is operated with a divertor[22]. Subsequently, it has been determined that similar confinement barriers can be produced in the interior of the plasma by suitable control of the current profile[23-25]. In addition, tremendous progress has been made in controlling the plasma wall interactions through divertors and through conditioning of the walls. This has resulted in substantially reduced impurity levels and lower levels of recycling, which in turn have led to improvements in the plasma performance. The increased understanding has been aided by the development of diagnostic techniques which are now used to measure the properties of the plasma turbulence, and the density, temperature and current density profiles. The wealth of new data, together with steady improvements in computer simulations of plasma turbulence, have resulted in substantial improvements in transport theory, and initial attempts to use the models for projection to new experiments[11].

An international database of tokamak confinement results has been assembled and analyzed using the most modern statistical methods to produce confinement scalings[16]. The database contains the results of about 6000 experiments from 11 different tokamaks. This analysis predicts that a tokamak will need a plasma current of about 20 MA to achieve ignition. A plot of the measured confinement time against the statistical fit shows that the fit is a reasonable representation of the data, but that the extrapolation to an ignition experiment such as ITER is significant (Fig. 5)[16].

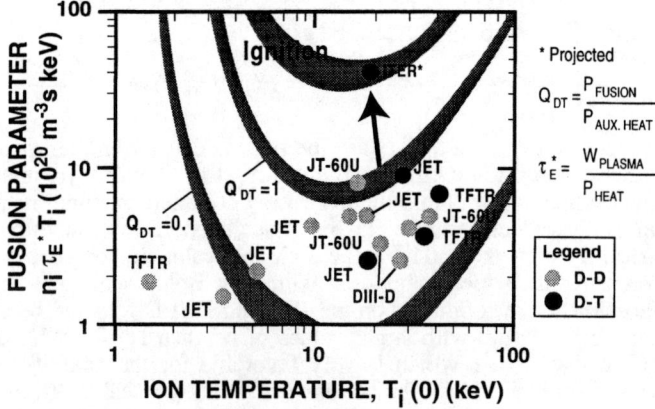

FIGURE 4. $n T \tau_E$ as a function of the central ion temperature for tokamaks[14,26].

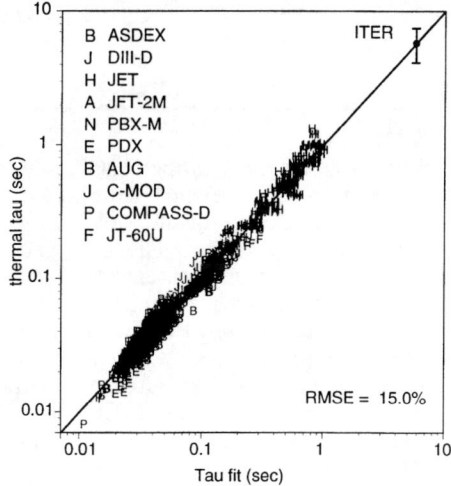

FIGURE 5. The measured thermal confinement time for "ELMy H-mode" operation versus the experimentally measured confinement time using data in the ITER confinement database from 10 different tokamaks[16].

While the statistical approach has been very successful for predicting the performance of the present generation of large tokamaks such as TFTR and JET using data from prior smaller experiments, it lacks the credibility of projections explicitly derived from an experimentally validated theoretical models based on the physics of the transport processes. For this reason, there have been strong efforts to both develop reliable "first principles" models for transport and to conduct experiments which have many of the same dimensionless parameters as the proposed next step experiment. This is similar to the approach used in hydrodynamic turbulence of using smaller, but dimensionlessly similar, experiments to study the airflow and turbulence for candidate aircraft designs. If the transport is due to plasma processes, then it should scale with the variables β, ρ^* and ν^* (Eq. (2))[27]. For experiments with identical β and ν^*, τ_E is only a function of ρ^*. β, ρ^* and ν^* are defined as:

$$\beta \equiv \frac{3nT}{B^2/8\pi}, \nu^* \equiv \frac{\nu}{\omega_{bounce}}, \rho^* \equiv \frac{\rho_{Larmor}}{a} \qquad \tau_E = F(\beta, \nu^*, \rho^*) \propto \frac{F^*(\beta, \nu^*)}{(\rho^*)^{2+\alpha}} \qquad (2)$$

where n, T, B, ν, ω_{bounce}, ρ_{Larmor}, a and τ_E are the plasma density and temperature, the magnetic field, the electron-ion collision frequency, the "bounce frequency", the Larmor radius, the plasma minor radius and the energy confinement time, respectively. This type of analysis has identified two classes of scalings: "Bohm" with $\alpha = 0$ (Eq. (2)) and "Gryo-Bohm" with $\alpha=1$. The earliest global scalings showed a Bohm-like scaling, but the present data is more consistent with gryo-Bohm scaling[16,28]. A set of comparison experiments were conducted on the JET and DIII-D tokamaks with values of β and ν^* similar to ITER, but with larger values of ρ^* than ITER (Table 2). These experiments indicate that $\alpha \approx 1$ which is very favorable for the next generation of tokamak experiments such as ITER. Projections using this ρ^* scaling from JET and from DIII-D predict ignition for the present ITER design (Table 3)[16,29].

TABLE 3 Parameters for two dimensionless parameter scans for JET and DIII-D[16].

Tokamak	β_N	ρ^*/ρ^*_{ITER}	α	δ_α	τ_E ITER (s)*	δ_τ (σ)
DIII-D	2.0	7.3	1.15	±0.44	21	±12
JET	2.2	5.5	0.7	±0.32	6.4	±3

* $\tau_E \approx 6$ s is needed for ignition in ITER.

The third approach for determining the confinement requirements for tokamak reactors is the development and use of detailed transport models in tokamak computational simulations which include all of the relevant processes[30,31]. The transport processes are treated at least two ways. The first, and most successful in terms of matching the experiments, is the use of analytic models developed from theoretical analysis of the relevant transport mechanisms. These models are adjusted to fit a few sample experiments and then applied to the whole range of tokamak data. Models of this type (e.g. the Multi-Mode model[32,33]) have been the most successful in matching present experiments and predict high levels of plasma performance for the next generation of tokamak experiments such as ITER. The second and more ambitious, but so far less successful, approach involves a "first principles" calculation of the turbulent transport processes using large scale computer simulations. Although this second approach is still in its infancy, it should ultimately be the most reliable since it has the potential to be the most firmly based on plasma physics. At the present, two "first principles" methods are being employed, one that approximates the behavior of the ions and electrons as a fluid (gyro-fluid[34]) and one that approximates the plasma as a collection of discrete particles (gyro-kinetic[35]). Initial results from the gyro-fluid approach predicted very large transport levels and very poor confinement for tokamaks beyond the present generation of experiments such as ITER[34]. However, the more accurate gyro-kinetic simulations indicate that these early predictions are much too pessimistic. Furthermore, the initial projections made using the gyro-fluid model used boundary conditions that were much too pessimistic. Projections using the most accurate of the transport models together with the correct boundary conditions ($T_{edge} \sim$ 4—5 keV instead of 100 eV[36-38]) indicate that the present designs for the next generation of tokamaks will achieve ignition[36,38,39].

The second major issue for tokamaks is plasma stability and control[13]. Optimum performance is achieved with elongated plasmas that need dynamic shaping and control (Fig. 3). The required levels of shaping and control are routine elements of the operation of all of the presently operating large tokamaks in the world. Plasma β's of 3 to 4% are needed in ITER to achieve its goals. Present experiments routinely exceed this value, both in absolute magnitude, and in the relevant scaling parameter, β_N ($\beta = \beta_N \frac{I(MA)}{a(m)B(T)}\%$), where I, a and B are the plasma current, minor radius and toroidal field (Fig. 6). Measurements of the plasma pressure and current profiles are routine on many tokamaks and the agreement with the computational models for plasma control and stability is excellent so that the codes can be relied upon as design tools [40]. Nonetheless there are stability issues for the next generation of tokamaks. The collisionality will be low in ITER, and there are instabilities which are damped at higher collisionalities which may provide more stringent stability requirements. In addition, due to slower rotation velocities in larger tokamaks, instabilities due to small error fields and other effects may be more of a concern than in present experiments. Finally, large experiments such as ITER will have significantly larger total stored energies in the plasma and magnetic field than present experiments. If control of the plasma is lost or the plasma becomes unstable, that energy can be deposited on the wall in less than a millisecond during the plasma "disruption", potentially damaging the internal components. The plasma current may also be transferred to the internal components very rapidly, potentially leading to very large mechanical loads [41,42].

Another major issue, and the one with perhaps the closest tie to atomic physics, is power and particle control[18,43]. The power and particle control system for a tokamak experiment must exhaust the heating power without damaging the plasma facing components or introducing unacceptably high levels of impurities into the plasma. In addition, the system must exhaust the He ash from the DT fusion reactions in an ignited experiment. In present experiments, the heat loads on the plasma facing components are high, but are not generally at the level where the peak loads are large enough to damage the internal components. In addition almost all present experiments have pulse lengths shorter than the thermal equilibration time of the components. For the next generation of experiments, the peak heat loads can be unacceptably high, and the pulse lengths are much longer than thermal equilibration times so that cooling needs to be steady-state. The solution that has been successfully developed is to operate with a poloidal divertor which "diverts" the edge plasma to a divertor chamber where the plasma-wall interaction processes can be better controlled. A key part of the strategy is to radiate most of the heating power to the walls and thereby reduce the peak heat loads on the divertor plate to acceptable levels by spreading the heat out over a large surface area. Regimes in which a very cool plasma forms in the divertor, "detached plasmas", with high levels of radiation losses, have been demonstrated on every tokamak with a divertor[18,44]. These experimental results have been used to validate sophisticated two dimensional divertor simulation codes which are then used to design the divertor system for the next generation of tokamak experiments[45]. The radiation losses are produced by a combination of impurities and hydrogen[46]. Gaseous impurities, such as Ne or Ar, are used to trigger detached plasma operation, but the radiation losses are also due to hydrogen and intrinsic impurities such as carbon from the graphite divertor plates. There are also significant radiation losses from the edge of the main plasma (Fig. 7)[47]. Reductions in the peak heat fluxes of 5 to 50 have been observed in the experiments with detached divertor plasmas.

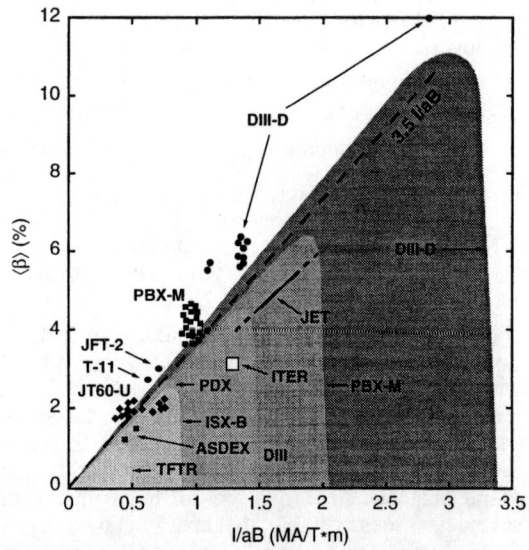

FIGURE 6. Plasma beta's, β, as a function of I/aB, the plasma current divided by the minor radius and the toroidal field, that have been achieved on the present generation of experiments and the beta needed for ITER.

Auxiliary heating is required to produce the high plasma temperatures necessary to achieve ignition[13]. In addition, auxiliary power can be used to extend the pulse length by driving current[11,48]. At least four techniques have been used successfully to heat plasmas and drive currents: high energy neutral beams, radio frequency waves in the 10—100 MHz range at the ion cyclotron frequency, low frequency microwaves (2-5 GHz) at the lower hybrid frequency, and high frequency microwaves (80-150 GHz) at the electron cyclotron frequency. High power neutral beams are the most common technique[49,50]. The present systems have particle energies of 30 to 80 keV/amu and total system powers of up to 40 MW[20]. The systems consist of a source of positive ions, which are then accelerated electrostatically across a potential gap[51]. The beam of high energy ions passes through a gas cell where the ions are neutralized by charge exchange. The high energy neutral atoms then freely cross the confining magnetic field into the plasma, where they are ionized by electron and ion impact or charge exchange. The fast ions then are confined by the magnetic field and heat the plasma by binary coulomb collisions with the background ions and electrons. Neutral beams have proven to be very reliable and successful, and have produced plasma temperatures of up to 40 keV[26]. The larger plasmas required to achieve ignition conditions need beam energies of 400 to 1000 keV/amu for the beams to penetrate to the plasma center. Positive ion systems cannot produce such energies because the cross sections for charge exchange to neutralize the high energy ions are too small for efficient neutralization. The next generation of high energy neutral injectors are therefore based on the acceleration of negative ions, which can be neutralized with good efficiency. 500 keV deuterium injection systems are being tested on the JT-60 U tokamak near Mito, Japan[52], and are being installed on a large stellarator that will begin operation next year at the Japanese National Institute for Fusion Studies near Nagoya, Japan[5] and a 1 MeV system is being designed for ITER.

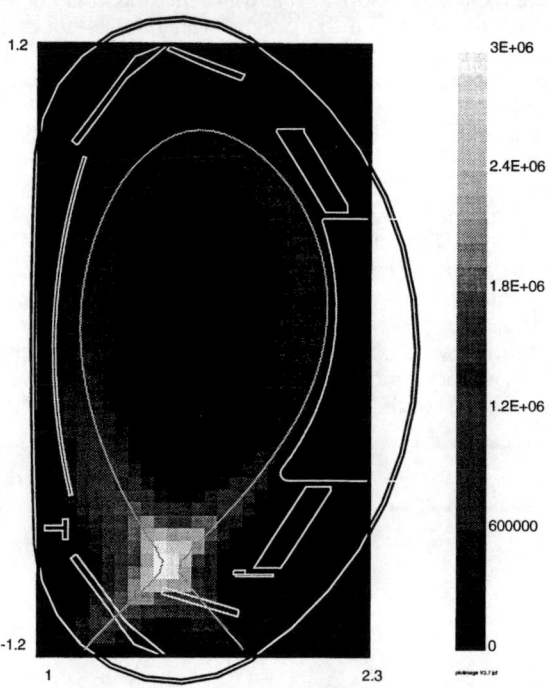

FIGURE 7. Power radiated from different regions of the plasma in ASDEX Upgrade. The greyscale contours show the power radiated in watts m^{-3}. The radiation peaks near the divertor at the bottom, but there is also substantial radiation losses from the plasma edge in the main plasma chamber[47,53].

Radio frequency techniques are also proving to be successful[54,55]. The technology of heating with radio waves (10-100 GHz) at the ion cyclotron resonance frequency (ICRF) is relatively simple, but the physics is complicated. The waves need to couple with the plasma to form standing waves to accelerate ions which then slow down and heat the plasma, or to be absorbed directly by the plasma ions. Systems with powers up to 18 MW are being used on the JET tokamak near Oxford[14], and 4 to 8 MW are being used on Alcator C-Mod at MIT[56]. The chief issues include coupling to the plasma, survival of the antenna in the "hostile" environment near the plasma and achieving a high power density through the antenna and feedthroughs[57].

Microwaves in the 50 to 200 MHz range, at the electron cyclotron resonance frequency (ECRF), have also been successfully used to heat plasmas[57-60]. The physics of ECRF heating is well understood and simpler than the physics of lower frequency systems, and no antenna to couple to the plasma is needed. The major issues include development of long pulse, high power amplifiers for the 140 to 170 GHz frequencies needed for the next tokamak generation. In addition, insulating windows are needed which can withstand high powers. Due to their high thermal conductivity, diamond windows offer a lot of promise and are being fabricated and tested.

Microwaves in the 1 to 10 GHz range at the lower hybrid resonance frequency (LHRF) have also been used for plasma heating and current drive[48,61]. Although this technique has been the most successful current drive method, it is not very efficient for heating. The issues include control of the plasma-wall interactions at the antenna and poor penetration to the plasma center[57].

An important development during the last ten years has been experimental verification that the plasma pressure gradient can drive appreciable plasma currents (up to 80% of the total current) for the appropriate conditions[62]. Studies indicate that it should be possible to make very long pulse or steady state tokamaks with feasible current drive systems[1].

The dominant heating method in the next generation of tokamaks will be alpha particle heating. In 1994, TFTR achieved a fusion power of 10.7 MW for a few 100 milliseconds, and this year JET achieved 16 MW for about half a second[14,26]. Although the alpha particle heating was on the order of 0.1 to 0.2 of the auxiliary heating, there was nonetheless measurable plasma heating from the alpha particles. Although some instabilities were observed (Toroidal Alfven Eigenmodes), they were of the level predicted by theory, and didn't affect the heating[63]. These instabilities are predicted to not be a problem for the next generation of experiments[64].

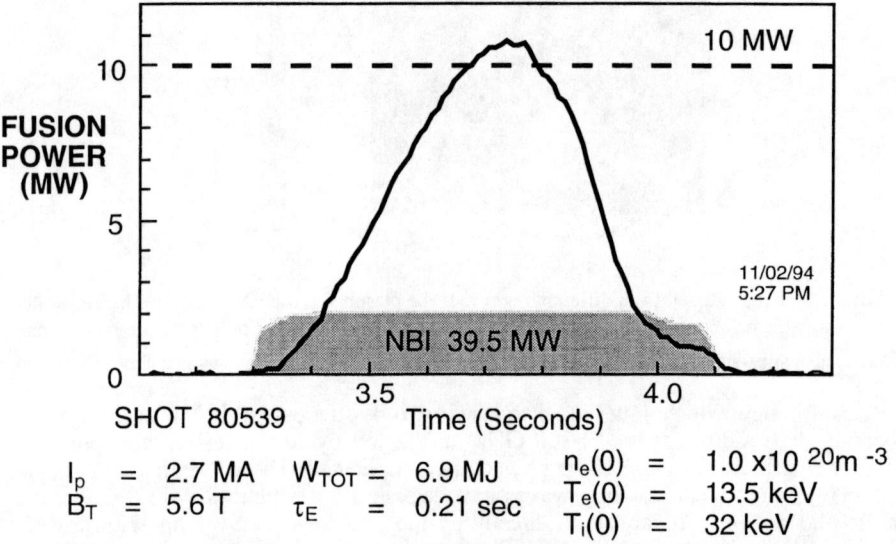

FIGURE 8. Highest fusion power shot for TFTR. The shot parameters are shown. I_p, B_T, W_{Tot}, τ_E, $n_e(0)$, $T_e(0)$ and $T_i(0)$ are the plasma current, the toroidal field, the total stored thermal energy, the energy confinement time and the central plasma density, electron temperature and ion temperature. The neutral injection heating power (NBI) was 39.5 MW[20].

Among the areas where the progress has been the greatest is plasma diagnostics[65-67]. First, the quality and detail of the standardly available tokamak measurements have dramatically improved. Part of this is due to improved instrumentation and computers, and part is due to improvements in the knowledge of the atomic processes and other effects. In addition, new diagnostic techniques have been developed to measure quantities that are important for developing a quantitative understanding of tokamak physics. Both have encouraged and facilitated the development of much better models and theories for tokamak plasmas.

One can define three classes of tokamaks: "conventional", "advanced" and "small aspect ratio". Conventional tokamaks have been described above. The performance is good, but a tokamak that is designed to provide some margin for ignition with conventional physics will require a plasma current of about 20 MA and will be fairly large and costly[2]. By active control of the plasma and density profile, every tokamak has been able to achieve parameters which, under optimum conditions, exceed those which typify the conventional tokamak[14,24,25]. Extrapolated to the next generation of tokamaks, these conditions would result in improved performance (or reduced machine parameters for the same performance[11].) As promising as these results are, no tokamak experiment has yet been able to reliably achieve these results simultaneously with the other necessary parameters for a pulse duration that would allow one to rely upon these regimes as a basis for a next step experiment. Thus intensive research that has the promise of producing improved operating regimes is a major focal point of the international fusion program, and the flexibility to study and exploit those regimes has been incorporated into the ITER design[11]. Small aspect ratio tokamaks, with R/a ~ 1.5 (ratio of the major radius to minor radius) compared to ~ 3 for conventional tokamaks, have the promise of better stability and confinement[68]. Several small experiments to test this concept are in operation or under construction (START now operating in Culham, UK; MAST and NSTX in construction at Princeton and Culham)[69].

Stellarators have two advantages compared to tokamaks: they are steady state because they do not require externally driven currents, and they are significantly less likely to have problems with plasma disruptions because there is no externally sustained current to drive plasma instabilities. Stellarators supply the required magnetic field helicity with an external helical coil which can be superconducting and therefore steady state. The external helical field leads to toroidal asymmetries, however, which potentially introduce additional plasma transport[3].

Two major stellarator experiments have operated recently, the Heliotron E at the U. of Kyoto in Japan[58], and Wendelstein VII-AS at Garching, Germany[6,70]. Wendelstein VII-AS has produced plasma parameters close to those of similar sized tokamaks (T ~ 2—4 keV, τ_E ~ 55 ms, β ~ 1.8 %,...)[6,71]. These two groups are presently building new, larger stellarators. The National Institute for Fusion Studies near Nagoya, Japan, will complete construction of the Large Helical Device (LHD) early in 1998 and begin operation shortly thereafter[5,72]. LHD has a major radius of 4 m, a minor radius of 60 cm, and cost ~ $1 B. The Max Planck Institute for Plasma Physics at Garching, Germany is beginning the construction of a different type of stellarator, Wendelstein VII-X at Greifswald, Germany (just north of Berlin on the Baltic Sea). Wendelstein VII-X will have a major radius of 5.5 m and a minor radius of 55 cm and cost about $600 M. It will begin operation in ~2004. Both LHD and Wendelstein VII-X have superconducting coils[72].

Other alternative concepts being pursued include reversed field pinches[73,74] and spheromaks[75]. These are small, compact, high β plasmas which use internal currents to provide almost all of confining magnetic field. They have the potential to be smaller and cheaper than tokamaks, but need to demonstrate that they can achieve the same high levels of plasma performance as tokamaks and can produce long pulse discharges. Just as in tokamaks, internal currents tend to decay resistively, and may need to be sustained with external means.

During the last ten years, the international tokamak community has been developing a design for a tokamak "test reactor" that can operate with long pulse ignited conditions, the International Tokamak Experimental Reactor (ITER) which builds on the exciting JET and TFTR results (Fig. 4)[14,20]. The project began in 1988 as an international collaboration with equal participation of the European Union, Japan, Russia and the US[76]. ITER is being carried out in three phases: The Conceptual Design Activity (CDA) from 1988-1990, Engineering Design Activity (EDA) (1991-1998) and a Construction Activity (1999-2010). The CDA successfully concluded in 1990 with a design concept[76]. The purpose of the EDA is to produce a detailed engineering design for the tokamak and facility and to complete the physics and technological Research and Development necessary to support a construction start in 1999. The EDA is almost complete and the four parties are discussing conditions for construction[2]. The design features include superconducting magnets, a blanket and shield to protect the magnets and vacuum vessel from the intense neutron flux, the capability to remotely maintain and service the device, the ability to handle large quantities of tritium, a cryostat to provide the thermal insulation needed for the superconducting coils, a poloidal divertor system mounted in divertor cassettes for ease of maintenance, auxiliary heating systems, power supply systems, heat exchange systems, and a tritium and gas handling system (Fig. 9). The device parameters are listed in Table 4. The total capital cost will be $ 10 billion (in 1997 $). A 20 year experimental program is planned from 2010 to 2030.

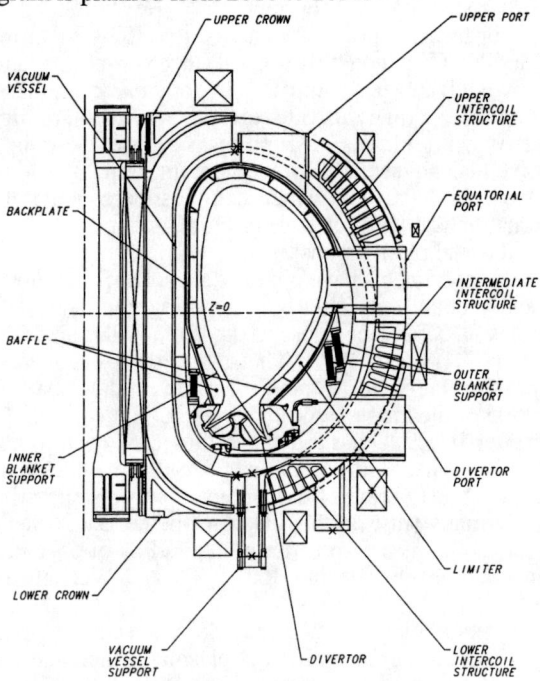

FIGURE 9. ITER device and plasma cross-section

ITER is a pioneering international scientific and technological collaboration. Each of the partners has been an equal participant in the technical, political and managerial decisions and work. The technical and managerial staff at all levels is comprised of members from the four ITER "countries". The CDA was carried out in Garching, Germany. The EDA has been carried out in three Co-centers located at San Diego, California; Garching, Germany; and Naka, Japan. These design sites are linked electronically, and the design activity goes on around the clock. The sun never sets on

ITER. The experiment will be operated in the same fashion. Remote control rooms for ITER are planned for all of the major fusion institutions in each participating country. These institutions will participate in real time in the experimental planning, operation, data exchange and data analysis. This type of remote operation minimizes personnel transfer and costs, and maximizes the scientific benefit to the participants.

TABLE 4. ITER design parameters[2].

Parameter	Value
Major/minor radius	8.14 m / 2.80 m
Plasma configuration	Single null divertor
Plasma volume	~ 2000 m^3
Plasma surface area	~ 1200 m^2
Nominal plasma current	21 MA
Toroidal field	5.68 T (at R = 8.14 m)
Fusion power (nominal)	1.5 GW
Burn duration (ignited, inductive drive)	≥ 1000 s
Auxiliary heating power	100 MW

INERTIAL CONFINEMENT FUSION—APPROACHES AND RESULTS

There are two approaches to compressing inertial fusion capsules to the levels needed to achieve ignition: direct drive[77] and indirect drive[7]. The energy for direct drive is focused directly on the capsule shell (Fig. 10). The outer layer of the shell ablates and drives the shell toward the center producing a ~ 1000 fold compression through convergence. To achieve symmetric convergence, the deposition of energy on the surface of the shell must be very uniform. The indirect drive approach (Fig. 10) attempts to ensure uniformity by depositing the driver energy on the walls of a cavity. This heats the cavity walls which then radiate large fluxes of thermal X-rays to illuminate the laser capsule relatively uniformly, potentially leading to a more symmetric convergence of the capsule.

The efficiency of the coupling of the driver energy to the target is important, since the glass lasers used for experiments have electrical efficiencies of about 1 %. The compression must be symmetric since $R_{initial}/R_{final}$ ~ 20—30, and significant asymmetries would destroy the high compression ratio needed to achieve the high densities and temperatures required for ignition conditions[7]. The thin shell must be stable during compression since $R/\Delta R$ is ~ 20—30. The driver beams must be sufficiently uniform to ensure uniform illumination of target. The beams must have sufficient power density that multiple beams can be focused down onto the capsule with enough energy to compress the capsule. The major drivers now being used in experiments are large neodymium glass lasers which produce light in the near infrared (1 micron). Driving the capsule with 1 micron light produces a hot electron tail which preheats the interior of the capsule and reduces the compression efficiency. To avoid this, the 1 micron light is frequency tripled to 1/3 micron before illumination of the target. The major facilities are the NOVA laser at the Lawrence Livermore National Laboratory[78], the OMEGA laser at the University of Rochester[79], the NIKE at the Naval Research Laboratory[77] and the GEKKO XII at the University of Osaka in Japan[80]. In addition, there are smaller programs in the UK, France and Germany[7].

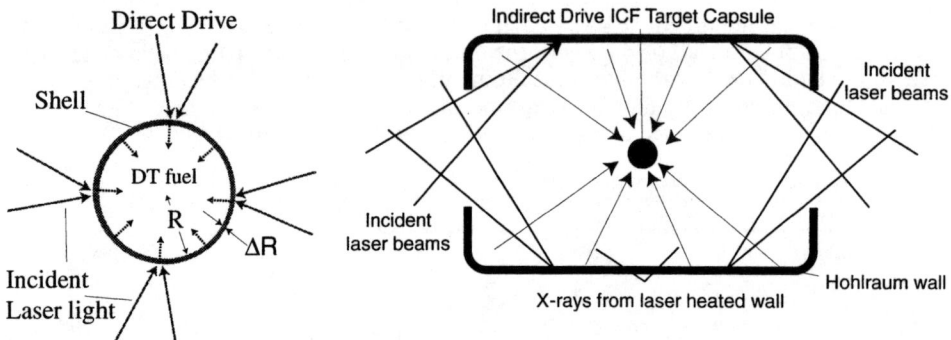

FIGURE 10. Direct and indirect drive approaches to inertial confinement fusion.

The progress in ICF is indicated by the degree of understanding embodied in the computer simulation codes used to analyze ICF experiments and predict the performance of the experiments. The degree of agreement between the models and experiments (Table 5) and the level of understanding of the detailed processes is very good[7]. The convergence achieved in NOVA is adequate for ignition in NIF, and LASNEX calculations indicate that the NIF parameters should be adequate to achieve ignition with an indirect drive hohlraum[7].

TABLE 5. Comparison of NOVA experiments, LASNEX calculations, and Ignition requirements[7]

	Observed (NOVA)	Calculated (LASNEX)	Ignition requirements (NIF)
Convergence	23.6 ± 1.8	20.7	20-35
Post-compression Density (g/cm^3)	19.1 ± 4.4	14	45—75
Neutron yield (10^7)	4.2 ± 0.4	3.8	

The Lawrence Livermore National Laboratory has just begun construction of the National Ignition Facility (NIF) which is designed to compress capsules to ignition. It will have a total energy of 1.8 MJ with a peak power of 500 TW at 0.35 microns. The estimated cost for the facility is $ 1 B[8].

EXAMPLES OF THE ROLE OF ATOMIC PROCESSES IN FUSION

The success of fusion experiments, both magnetic and inertial, has been possible because we have learned to control the role of atomic processes. Early magnetic fusion experiments were plagued by energy losses due to impurity radiation and collisions of the plasma with recycling neutral atoms and molecules. By use of divertors and appropriate wall materials, development of techniques for cleaning loosely bound impurities from the wall and control of plasma recycling, these energy losses have been brought under control. Now, specific properties of atomic processes are being used to solve important design problems. In addition, many plasma diagnostics rely upon the measurement and interpretation of atomic processes and diagnostics advances have been essential to the progress in developing an understanding the physics of tokamaks and other magnetic fusion experiments[21,67]. The understanding and control of atomic processes has also been essential for inertial fusion[8]. The physics of laser light absorption, transport of thermal x-rays, equation of state for the materials during compression, and diagnostic techniques all involve atomic processes[7].

Two specific examples of the use of atomic processes to solve tokamak problems are the use of radiation losses to reduce the peak heat loads on the divertor plates[18] and the use of radiation to rapidly quench a disrupting plasma, thereby minimizing the damage to the internal components from the disruption[81]. As discussed earlier, a major problem for the next generation of tokamaks with high levels of fusion power is the high peak loads on the plasma facing components. The strategy now employed in the design of the power and particle control systems for these experiments involves using radiation losses from the plasma edge and divertor to transfer most of the heating power from the plasma to the tokamak walls before it reaches the divertor plates. The walls have a much larger surface area than the divertor plates, and so the peak heat loads would be reduced to manageable levels. Losses from both the divertor and the plasma edge are important in this strategy (e.g. Fig. 7). This approach has been successfully demonstrated on every divertor tokamak(e.g.[82]). Even more important is that the plasma behavior can be successfully simulated with sophisticated two dimensional divertor modeling codes[45]. When the radiation losses become comparable to the heating power, there is not sufficient energy to sustain the plasma ionization losses near the divertor plate, and the plasma "detaches" from the plate and the plasma flux on the plate drops by an order of magnitude or more. This leads to very long component lifetimes. The radiation losses for detached plasmas in present experiments consist of hydrogen line radiation, impurity radiation from intrinsic impurities such as carbon from the divertor plates and radiation from externally added gaseous impurities such as N, Ne or Ar (Fig. 12)[46]. Pressure balance is maintained along the field lines by charge exchange and elastic collisions between the ions and recycling neutral atoms and molecules which transfer the momentum from the plasma to the wall. The temperature of the detached plasma can drop to the 1 eV range where volume recombination can contribute significantly to the reduction of the plasma flux on the divertor plates and to the momentum loss from the plasma (Fig. 13)[83,84].

This complicated set of physics involves the transport of plasma and neutral atoms and molecules, their interaction with the plasma facing materials and the transport of multiple charge states of each impurity species including ionization due to electron impact collisions; recombination due to dielectronic, charge transfer, three-body and two body processes; and electron excitation; all in a complex geometry (e.g. Figs. 7, 11 and 12)[85]. This physics has been included in the divertor modeling codes which have first been validated with experimental data[45] and then used to predict the performance of divertors for the next generation of tokamaks (Fig. 14)[86].

Given the strong role of volume recombination in detached plasmas and the high neutral atom densities ($\sim 10^{20}$ m^{-3}), radiation trapping can potentially be important[85,87,88]. For neutral densities in this range, the divertor can be many optical depths in width for the dominant Lyman alpha line of hydrogen (Fig. 15)[89]. Preliminary calculations indicate that radiation trapping effects can change the ionization balance, as well as reduce the energy losses.

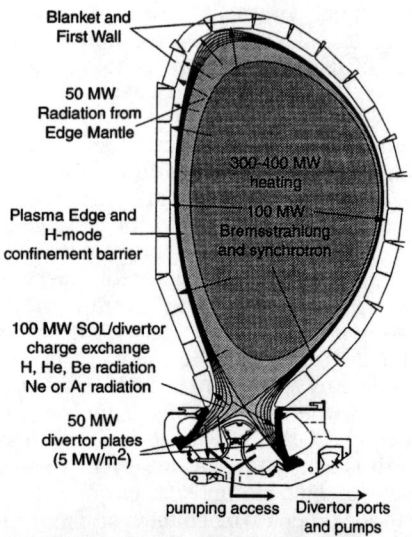

FIGURE 11. Schematic power losses and power balance for ITER.

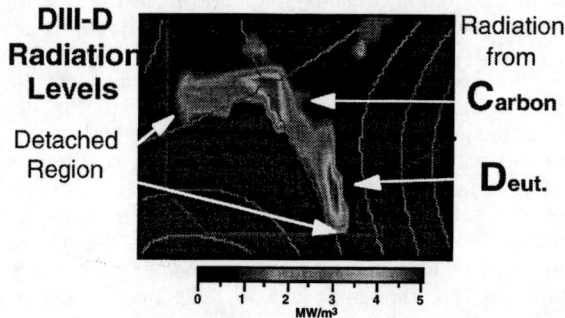

FIGURE 12. Radiation losses in the DIII-D divertor due to carbon and deuterium line excitation[82].

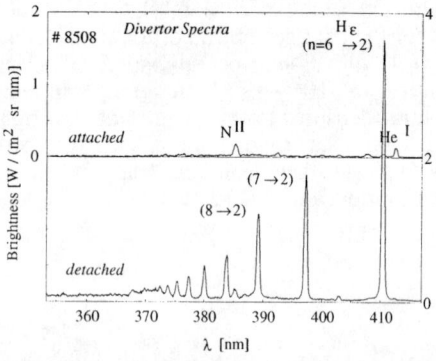

FIGURE 13. Comparison of Hydrogen line spectra for attached and detached plasmas on ASDEX Upgrade. The detached spectrum is characteristic of a recombining plasma and attached plasma is consistent with excitation from the ground state[84].

Molecular processes may also be important in divertor detachment and recombination[85]. Desorption and molecular recombination from the wall and ion and electron collisions can produce vibrationally excited molecules. These vibrationally excited molecules have the potential to increase the effective ion-electron recombination rate for hydrogen ions by several orders of magnitude through the reactions involving vibrationally excited molecules (Eq. (3)[85,90-92]):

$$\begin{aligned}&1{:}\quad H_2(v)+e^- \to H^o + H^- \;\vdots\; H^+ + H^- \to H^o + H^o(n=3)\\&2{:}\quad H_2(v)+H^+ \to H^o + H_2^+(v) \;\vdots\; e^- + H_2^+(v) \to H^o + H^o(n=3)\end{aligned} \quad (3)$$

The rates for these reactions are much larger than the rates for two and three body recombination. Calculations of these rates have been performed[92], and there is some evidence that they are important in cold linear plasma devices. Experimental investigation of these effects is just beginning on tokamaks.

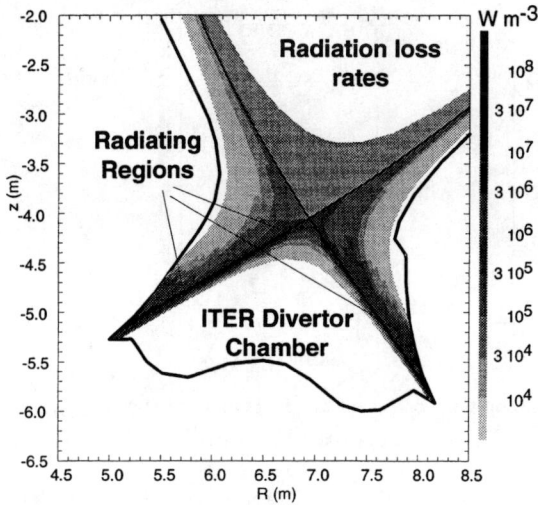

FIGURE 14. B2-EIRENE calculation of the radiation emission rates for the ITER divertor. The contours are in watts/m^3.[19]

A second application where atomic processes and line transfer effects in particular are important is the use of "killer" pellets to quench tokamak plasmas which have developed conditions which will likely lead to a plasma disruption. If control of the plasma position is lost in ITER, the plasma can move and come into contact with the wall, and could then deposit up to 60 MJ/m^2 on the contact point. In addition, a portion of the plasma current would be transferred to the wall. Part of this current (up to 50% of the initial current) could flow poloidally, leading to very large $\vec{j} \times \vec{B}$ forces acting on the first wall components (Fig. 16)[41,42].

A proposed solution for avoiding these problems is to inject "killer" pellets to increase the radiation losses by factor of 10 to 100 or more[81]. The plasma would then cool rapidly before it moved and the energy would be spread out over the walls by radiation, thus avoiding high peak heat loads. The current would decay resistively in the plasma before the plasma had time to move to the wall, and the large mechanical forces would be avoided. A major difficulty is that the rapid cooling can lead to the formation of multi-MeV runaway electrons. As the plasma cools, the resistivity, η, increases rapidly. Since $\eta \propto T^{-1.5}$, when the central temperature drops from several keV to ~1—3 eV, η increases by a factor of about 10^5. Since the time is not sufficient for the current to change, the toroidal electric field could increase by a similar factor. During normal operation, $E_{tor} \approx \eta I_p \approx 0.002$ V. E_{tor} could thus increase to 20 V/m or more following killer pellet injection. When $E_{tor} > E_{crit}$, runaway electrons can form. E_{crit} for ITER is about 0.1 $n_e(10^{20}$ m^{-3}) V / m (0.1 V/m for $n_e = 10^{20}$ m^{-3}), much less than 20 V/m. There is no advantage for "killer pellets" if the pellet ends up replacing 21 MA of plasma current with 10 to 20 MA of runaway electrons[81,93].

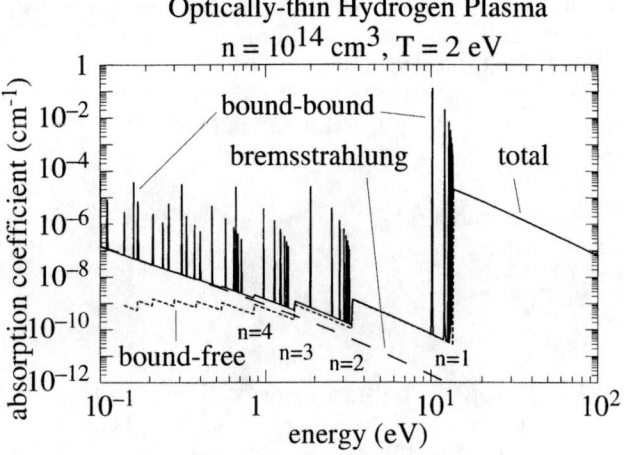

FIGURE 15. Photon absorption coefficient as a function of photon energy for a 2 eV, 10^{20} m^{-3} hydrogen density plasma[89].

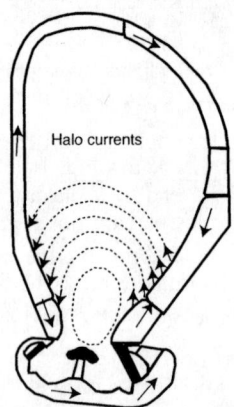

FIGURE 16. Halo currents flowing in the plasma and first wall following the motion of a plasma when control of the vertical position has been lost[41].

A technique for rapidly quenching the plasma without forming runaways is therefore needed. Since E_{crit} is proportional to the density, the criteria to avoid runaway production is $E_{tor}/E_{crit} \leq 1$. $E_{tor}/E_{crit} \sim n_e^{-1} T^{-1.5}$—which implies that one wants to maximize T_e and n_e. Initial analysis of high Z impurities such as Xe indicates that with such pellets, the plasma cools to 1 eV, and more than 40% of the plasma current is converted into runaways (Fig. 17)[81]. Almost all impurities behave similarly, but lower Z impurities could have higher densities (less line radiation per ion) and therefore higher E_{crit}'s. Further analysis indicates that hydrogen offers the most promise. It can have the largest increase in electron density, and potentially can have only bremsstrahlung and radiative recombination losses from the center if the central temperature remains above ~ 5 eV. Calculations of the runaway production for hydrogen injection consistent with a 100-fold increase in the plasma density indicate that, while the final temperature is higher than for any other pellet material, it is still sufficiently low (~ 1-3 eV) that runaways are formed. Line radiation losses contribute too much cooling from the recombining outer layer to avoid runaway production. The edge cools rapidly and then a cold front quickly propagates to the center, cooling the whole plasma to 1-3 eV.

However, if line transport effects are included, the penetration time of the recombination front lengthens to the point that few runaways are formed (Fig. 18)[89]. The line radiation escapes slowly, and the discharge quietly terminates with no runaways[81]. The Lyman alpha mean free path (the dominant line for radiation losses) in the recombined layer at the edge is of order 10^{-6} m, much smaller than the width of the recombining layer (several cm) or the main plasma ($r \approx 3$ m)[89]. The central plasma cools slowly by bremsstrahlung and radiative recombination, the central temperature remains at about 6 eV and runaways are avoided (Fig. 19). The equilibrium profiles are very broad for the cases with radiation transport. The neutral hydrogen fraction in the center remains low so that cooling from Lyman alpha lines is minimal. However, the calculation (Fig. 19) doesn't treat the current decay and assumes a constant loop voltage. New calculations are being done which include the current evolution.

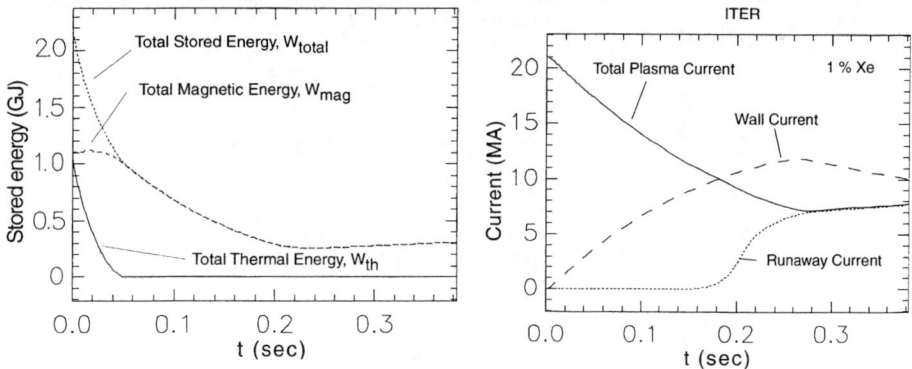

FIGURE 17. Calculated time traces for the stored energy and current distribution ion ITER following the injection of a Xe pellet sufficient to cause a 1% Xe concentration[81].

Many plasma diagnostic techniques are based on atomic processes[66,67]. These include spectroscopy to measure impurity radiation losses and the impurity density, Doppler broadening of impurity emission lines to measure the ion temperature, visible bremsstrahlung Z_{eff} measurements, excitation of neutral atoms in a probe beam to measure density fluctuations and measurement of the energy distribution of escaping fast neutral hydrogen and helium atoms to determine the energy distribution of the confined ions. Two examples of new measurements are the use of the Motional Stark Effect (MSE) to measure the plasma current profile[94,95] and Thomson scattering measurements of the electron density and temperature profiles in the divertor[96]. The plasma current profile is one of the main determinants of plasma stability. It is difficult to choose between competing models for stability without knowing the current profile. The motional stark effect diagnostic is based on the measurement of the polarization of the Doppler shifted H_α line which is split by the Stark effect from the $\mathbf{E}=\mathbf{v}\times\mathbf{B}$ motion of a high energy neutral beam in the plasma. The change in the polarization of a particular Stark component is a measure of B_p/B_T, the poloidal magnetic field to the toroidal magnetic field, which is inversely proportional to $q(r) = r\,B_T / (R\,B_p)$, the MHD safety factor (Fig. 1). Once q(r) is known, the plasma current distribution can be determined. The MSE diagnostic has literally revolutionized stability studies and led to a much better theoretical understanding of plasma stability.

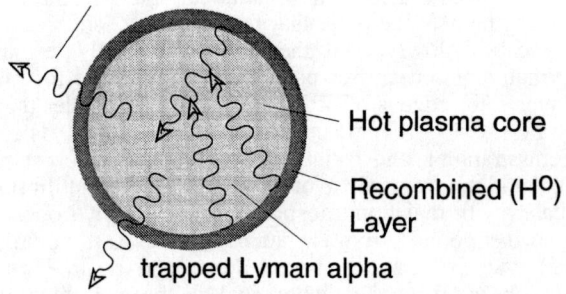

FIGURE 18. Schematic illustration of trapping of the Lyman alpha in a tokamak with a high hydrogen density.

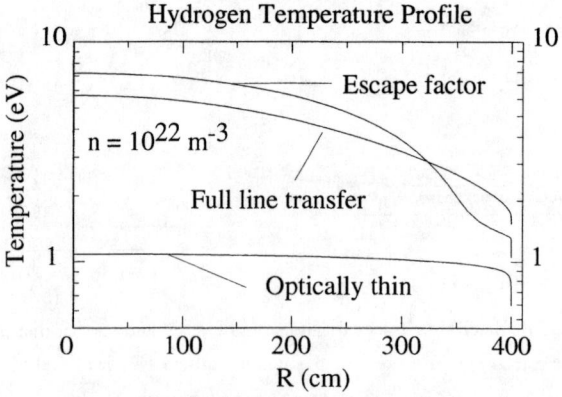

FIGURE 19. Final temperature profile for a hydrogen ITER plasma with $n=10^{22}$ m^{-3} for three cases: optically thin with coronal equilibrium radiation losses, full line transfer (including absorption and re-emission) of Lyman alpha, and transport using an escape factor formalism for Lyman alpha[89].

Detailed measurements of the divertor temperature and density with Thomson scattering have resulted in the identification of the importance of volume recombination in detached plasmas. Before these measurements, due to kinetic effects[97] the temperatures measured with Langmuir probes appeared to be 2 to 5 eV so that volume recombination would be a small effect. The measurements indicated the temperature was 1 eV or less[96], conditions for which volume recombination is important and subsequently, experimental signatures of volume recombination were observed (Fig. 13)[84,98]. In addition, measurements of the two dimensional temperature and density distribution have allowed the divertor simulation codes to test their predictions and improve the physics in the codes[45].

Atomic processes play an essential role in inertial fusion[7,99]. The compression of the capsule involves hydrodynamics and an equation of state is needed to relate the pressure, the density and the temperature, and the energy required to ionize the plasma (Eq. (4)). The pressure, p, is defined as:

$$p = p(\rho,T) = N(f_o T_o + f_i T_i + f_e T_e) = \rho \frac{N_o}{A}(f_o T_o + f_i T_i + f_e T_e) \quad \vdots \quad f_x = \frac{n_x}{N} \quad (4)$$

where N is the number density of nuclei, f is the fraction of neutrals, ions or electrons, ρ is the mass density, N_o is Avagadro's number and A is the atomic weight of the element. Computing the fraction of neutrals, ions and electrons and the energy for ionization and recombination of the material as a function of temperature and density requires detailed atomic structure calculations[99]. Radiation transport also is important in the compression of ICF capsules, especially for the indirect drive approach in which the laser energy is converted to x-rays which then are used to drive the capsule implosion. Solution of the radiative transfer equation requires the opacity, which requires a detailed knowledge of the atomic structure and ionization, recombination, and excitation processes involving ion, electron and photon collisions with partially ionized plasmas in high density conditions[99,100].

Atomic processes also play a key role in diagnostics for laser produced plasmas[99,101]. Spectroscopic lines characteristic of particular charge states can be used to identify the temperature of a region. Line ratios for tracer impurities can be used to determine the temperature as well, particularly when linked to powerful modeling codes such as LASNEX. Ti/Cr tracers were used to seed a hohlraum (Fig. 10) without a capsule to measure the temperature that could be obtained in the hohlraum irradiated in the NOVA laser facility. The measured line intensities indicate that the temperature was about 3 keV (Fig. 20). These measurements were among the large amount of data used to validate the LASNEX code, which also indicated that the hohlraum temperature would be expected to be in the 3 keV range (Fig. 21) for these line ratios.

Atomic data would be much more useful to the general plasma physicist if it were cast into a form that was immediately useful and could be conveniently accessed. Most data need to be extensively processed before the plasma physicist can use it. For instance, plasma modellers need rates, not cross sections. They need global energy loss rate coefficients, not oscillator strengths for single lines. In addition, many plasma physicists do not have a detailed understanding of atomic physics and are therefore unable to interpret the general atomic physics literature and extract and process the data they need and understand the uncertainties and caveats associated with the use of the data. Data for the plasma physics community therefore needs to be processed into a convenient form and be easily accessible to potential users. Computer based databases and database centers that are accessible over the internet offer the potential for this access, and are a large step in the right direction.

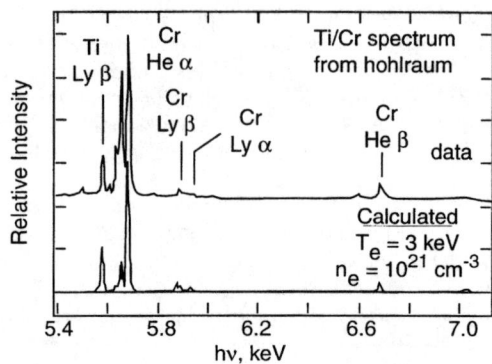

FIGURE 20. Measured intensities of x-ray lines for a Ti/Cr tracer seeded hohlraum on NOVA. The line ratios are consistent with an electron temperature of 3 keV and an electron density of 10^{21} cm^{-3}.[7]

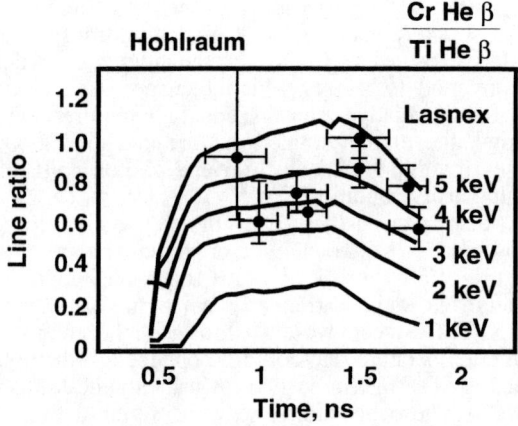

FIGURE 21. LASNEX calculations of the line ratio expected for hohlraum temperatures of 1, 2, 3, 4 and 5 keV for NOVA experiments with a hohlraum with no capsule. The NOVA data is consistent with the LASNEX calculation of 3 keV.[7]

DATA NEEDS

The areas where new and improved data for atomic processes are needed in magnetic fusion include impurity rate data for low temperature divertor plasmas, data for hydrogen and impurity molecular collisions and data for elastic scattering[66,85,91,102,103]. Accurate data is needed to compute the radiation losses in divertor plasmas with $10^{19} \leq n_e \leq 10^{21}$ m^{-3}, $0.5 \leq T_e \leq 50$ eV. These data should include collisional radiative effects. The elements that need special consideration include H, He, Be, C, N, Ne, Ar, Kr, Mo and W. The collisional radiative effects need to be projected into a model that considers a single "effective" density for each charge state. An improved characterization is needed for processes involving molecular assisted recombination (e.g. Eq. (3)), including isotope effects. Chemical erosion of graphite is a major issue, and detailed knowledge is needed to calculate the transport and break-up of the various hydrocarbon molecules that are formed at the graphite divertor plates and broken up in the plasma. The gas pressure in the divertor chamber is sufficiently high that elastic collisions are important for calculating the neutral gas transport, so that accurate cross sections are needed for atomic and molecular collision speeds of 10^3 to 10^5 m/s. These needs are summarized in a number of reviews (e.g. 85,102,103) and the publications of the IAEA Atomic Fusion Data Center[104-109]. With regard to ICF, improved opacity and equation of state data is needed for the temperatures and densities expected to be achievable in NIF with $10^{23} \leq n_e \leq 10^{26}$ m^{-3}, $1 \leq T_e \leq 5000$ eV [99]. Data is needed not only for pure materials (Au, Eu, Fe, C,...) but for mixtures (Au-C, Pb-Li,..). Detailed data is needed for heavy ion beam stopping, including the stopping power for any partially stripped ion in the $0.1 \leq E/A \leq 100$ MeV energy range[99]. A more complete list of data needs for ICF is given in Deutsch[99].

SUMMARY

Results from the present generation of fusion experiments (MFE: TFTR, JET, JT-60/U,...; ICF: NOVA, OMEGA, GEKKO XII, NIKE,...) strongly indicate that the fusion program is ready to proceed with the construction and operation of tokamak and ICF ignition experiments. The tokamak ignition experiment, ITER, is nearing the end of the design phase prior to beginning construction, and construction of NIF has begun. Atomic processes have played a key role in both programs through being a major factor in the energy and particle balance in tokamaks as well as contributing to power and particle control and disruption avoidance, through equation of state and opacity data for laser driven ICF capsules and through diagnostics for both approaches. The success of these programs has hinged on accurate data for atomic processes. The codes used to analyze present experiments and predict the performance of the next generation of experiments rely upon the quality and completeness of the data. There is a continuing need for improved data, especially for new regimes such as low temperature plasmas in tokamak divertors and the very high energy densities expected to be achieved in the NIF facility. Convenient access to the data over the internet is potentially very important, and the development of such databases and data centers is strongly encouraged.

ACKNOWLEDGMENTS

The author gratefully acknowledges contributions from S. Allen, M. Bell, N. Fujiwara, G. Cordey, R. Hawryluk, Yu. Igitkhanov, G. Janeschitz, R. Janev, A. Kukushkin, J. Lindl, D. Meade, K. McGuire, R. More, V. Mukhovatov, H. D. Pacher, R. Parker, F. Perkins, S. Putvinski, D. Reiter, M. Rosenbluth, N. Sauthoff, H. Scott, D. Stotler, M. Sugihara, F. Wagner, A. Wan, H. Wanner and J. Wesley. This paper has been prepared through the ITER Publications Office as an account of work performed under the Agreement among the European Atomic Energy Community, the Government of Japan, the Government of the Russian Federation, and the Government of the United States of America on cooperation in the Engineering Design Activities for the International Thermonuclear Experimental Reactor ("ITER EDA Agreement") under the auspices of the International Atomic Energy Agency(IAEA).

REFERENCES

1. J. Wesson, *Tokamaks* (Clarendon Press, Oxford, 1997).
2. R. Aymar, V. Chuyanov, M. Huguet et al., IAEA Fusion Energy-1996 **1**, 3–18 (1997).
3. A. Boozer, Physics of Plasmas, to appear (1998).
4. A. Iiyoshi and et al, Fusion Technology **17**, 169 (1990).
5. M. Fujiwara, Transactions of Fusion Technology **27**, 58 (1995).
6. F. Wagner and W7-AS group, Physics of Plasmas, to appear (1998).
7. J. Lindl, Physics of Plasmas **2** (11), 3933-4024 (1995).
8. D. Crandall, International Conference on Atomic and Molecular Data and Their Applicatons, Gaithersburg, Maryland, 1997, this volume.
9. D. Crandall, Fusion Technology **30**, 391 (1996).
10. R. Bangerter, A. Friedman, and W. Herrmannsfeldt, Plasma Physics and Controlled Nuclear Fusion Research-1994 **2**, 701 (1995).
11. D. Boucher, Y. Barano, B. Fischer et al., IAEA Fusion Energy–1996 **2**, 945-953 (1997).
12. D. Post, K. Borrass, J.D. Callen et al., *ITER Physics* (International Atomic Energy Agency, Vienna, 1991).
13. J. C. Wesley and ITER Joint Central Team, Physics of Plasmas **4** (7), 2642 (1997).
14. A. Gibson and JET Team, Physics of Plasmas, to appear (1998).
15. S. Ishida, T. Fujita, H. Akasaka et al., Physical Review Letters **79**, 3917 (1997).
16. J. G. Cordey, Plasma Physics and Controlled Fusion **39**, B115 (1997).
17. O. Sauter, R. LaHaye, Z. Chang et al., Physics of Plasmas **4**, 1654-1664 (1997).
18. D. Post, G. Janeschitz, A. Kukushkin et al., Physics of Plasmas **4**, 2631 (1997).
19. A. Kukushkin, V. Abramov, M. Baelmans et al., IAEA Fusion Energy 1996 **2**, 987-994 (1996).
20. K. McGuire, C. Barnes, S. Batha et al., Fusion Energy 1996 **1**, 19-36 (1997).
21. P. Stott, G. Dorini, and E. Sindoni, *Diagnostics for Experimental Thermonuclear Fusion Reactors* (Plenum Press, New York, 1996).
22. F. Wagner, M. Keilhacker, and ASDEX team, Journal of Nuclear Materials **121**, 103-113 (1984).
23. P. Breger, A. Cherubini, S. Davies et al., 24th EPS Conference on Plasma Physics and Controlled Fusion, Berchtesgaden, Germany, 1997 1, pp. 69-72.
24. R. Bell and TFTR Group, Physics of Plasmas, to appear (1997).
25. C. Rettig and DIII-D group, Physics of Plasmas, to appear (1997).
26. R. Hawryluk and TFTR Team, Physics of Plasmas, to appear (1998).
27. B. B. Kadomstev, Soviet Journal of Plasma Physics **1**, 295 (1975).
28. P. Yushmanov, T. Takizuka, K. Riedel et al., Nuclear Fusion **30**, 1999-2006 (1990).
29. J. Cordey and JET Team, Fusion Energy-1996 **1**, 603-610 (1996).
30. J. Connor, M. Alexander, S. Attenberger et al., Fusion Energy-1966 **2**, 935-944 (1997).
31. J. Connor, Plasma Physics and Controlled Fusion **37**, A119 (1995).
32. G. Bateman and et al, Physica Scripta **51**, 597 (1995).
33. J. Kinsey and et al, Physics of Plasmas **2**, 811 (1995).
34. M. Kotscheneruther, W. Dorland, and et al, Fusion Energy-1996 **2**, 371-384 (1996).
35. A. Dimits, T. Williams, J. Byers et al., Physical Review Letters **77**, 71-74 (1996).
36. G. Janeschitz, A. Hubbard, Yu. Igitkhanov et al., 24th European Physical Society Conference on Controlled Fusion and Plasma Physics, Berchesgarten, Germany, 1997 3, pp. 993-996.
37. A. Loarte and et al., Contributions to Plasma Physics, to appear (1998).
38. R. Groebner, Physics of Plasmas, to appear (1998).
39. A. Hubbard and Alcator C-Mod Group, Physics of Plasmas, to appear (1998).
40. F. Perkins, D. Post, M. Rosenbluth et al., IAEA Fusion Energy 1996 **2**, 963-969 (1997).
41. J. Wesley, N. Fujisawa, S. Ortolani et al., IAEA Fusion Energy 1996 **2**, 971-978 (1997).
42. R. Parker, S. Chiocchio, F. Federici et al., Journal of Nuclear Materials **241-243**, 1 (1996).
43. G. Janeschitz, A. Antipenkov, S. Chiocchio et al., 15th International Conference on Plasma

Physics and Controlled Nuclear Fusion Research **2**, 549-560 (1994).
44. G. F. Matthews, Journal of Nuclear Materials **220-222**, 104-116 (1995).
45. A. Loarte, Journal of Nuclear Materials **241-243**, 118 (1997).
46. R. Isler, R. Wood, C. Klepper *et al.*, Physics of Plasmas **4**, 355 (1997).
47. J. Neuhauser, M. Alexander, G. Becker *et al.*, Plasma Physics and Controlled Fusion **37**, A37-A51 (1995).
48. N. J. Fisch, Reviews of Modern Physics **59**, 175 (1987).
49. R. Hemsworth, E. Pietro, M. Hanada *et al.*, IAEA Fusion Energy 1996 **2**, 927-933 (1997).
50. D. Stork, Fusion Engineering and Design **14**, 111 (1991).
51. D. E. Post and R. Pyle, "Neutral Particle Beam Production and Injection," in *Atomic and Molecular Physics Controlled Thermonuclear Fusion*, edited by C. Joachain and D. E. Post (Plenum, New York, NY, 1983), Vol. B101, pp. 477-518.
52. M. Kikuchi, Fusion Technology **30**, 660-668 (1996).
53. O. Gruber, A. Kallenbach, M. Kaufmann *et al.*, Physical Review Letters **74** (21), 4217—4220 (1995).
54. D. Hwang and J. Wilson, Proceedings of the IEEE **69**, 1030 (1981).
55. R. Cairns, *Radio frequency heating of plasmas* (Adam Hilger, Bristol, 1991).
56. Y. Takase, R. Boivin, F. Bombarda *et al.*, Physics of Plasmas **4**, 1647 (1997).
57. G. Bosia, R. Elio, M. Makowski *et al.*, IAEA Fusion Energy 1996 **2**, 917-925 (1997).
58. T. Obiki, F. Sano, K. Kondo *et al.*, Fusion Energy 1996 **2**, 13 (1997).
59. R. Prater, Journal of Fusion Energy **9**, 19 (1990).
60. M. Bornatici, R. Cano, O. DeBarbieri *et al.*, Nuclear Fusion **23**, 1153 (1983).
61. P. Bonoli, "Linear theory of lower hybrid waves in tokamak plasmas," in *Wave heating and current drive in plasmas*, edited by V. Granatstein and P. Colestock (Gordon and Breach, New York, 1985), pp. 175.
62. M. Kikuchi and M. Azumi, Plasma Phsyics and Controlled Fusion **37**, 1215 (1995).
63. R. Nazikian and TFTR Group, Physics of Plasmas , to appear (1998).
64. C. Cheng, H. Berk, D. Borba *et al.*, Fusion Energy 1996 **2**, 953-962 (1997).
65. H. P. Summers and M. von Hellermann, "Atomic and Molecular Data Exploitation for Spectroscopic Diagnostics of Fusion Plasmas," in *Atomic and Plasma-Material Interaction Processes in Controlled Thermonuclear Fusion*, edited by R. Janev and H. Drawin (Elsevier, Amsterdam, 1993), pp. 87-118.
66. H. Summers, International Conference on Atomic and Molecular Data and Their Applicatons, Gaithersburg, Maryland, 1997, this volume.
67. I. H. Hutchinson, *Principles of plasma diagnostics* (Cambridge University Press, London, 1987).
68. M. O'Brien, R. Akers, R. Bamford *et al.*, IAEA Fusion Energy-1996 **2**, 57-69 (1997).
69. A. Sykes, R. Akers, L. Appel *et al.*, Plasma Physics and Controlled Fusion **39**, B247 (1997).
70. M. Kick, J. Baldzuhn, J. Geiger *et al.*, Fusion Energy-1996 **2**, 27 (1997).
71. E. Suvorov, E. Halzhauer, W. Kasparek *et al.*, Plasma Physics and Controlled Fusion **39**, B337 (1997).
72. A. Iiyoshi, Fusion Energy 1996, Montreal, Canada, 1997 1, pp. 113-128.
73. Y. Hirano, Y. Maejima, T. Shimada *et al.*, IAEA Fusion Energy 1996 **2**, 95-104 (1997).
74. J. Drake, H. Bergsaker, P. Brunsell *et al.*, IAEA Fusion Energy 1996 **2**, 193-199 (1997).
75. M. Yamada, H. Ji, T. Carter *et al.*, IAEA Fusion Energy 1996 **2**, 253-261 (1997).
76. K. Tomabechi, J. Gilleland, Yu. Sokolow *et al.*, Nuclear Fusion **31**, 1135 (1991).
77. S. Bodner, Physics of Plasmas , to appear (1998).
78. K. Budil, Physics of Plasmas , to appear (1998).
79. T. Murphy, Physics of Plasmas , to appear (1998).
80. S. Nakai, M. Nakatsuka, H. Fujita *et al.*, Fusion Technology **30**, 625-633 (1996).
81. S. Putvinski, N. Fujisawa, D. Post *et al.*, Journal of Nuclear Materials **241-243**, 316 (1997).
82. A. Leonard, S. Allen, M. Fenstermacher *et al.*, 22nd European Physical Society Conference on

Controlled Fusion and Plasma Physics, Bournemouth, UK, 1995 19C, Part III, pp. 301-304.
83. D. Lumma, J. Terry, and B. Lipschultz, Physics of Plasmas **4**, 2555-2566 (1997).
84. V. Mertens, A. Herrmann, A. Kallenbach *et al.*, IAEA Fusion Energy-1996 **1**, 413-424 (1997).
85. D. E. Post, Journal of Nuclear Materials **220-222**, 143-157 (1995).
86. A. Kukushkin, H. D. Pacher, M. Baelmans *et al.*, Journal of Nuclear Materials **241-243** (268) (1997).
87. S. Krasheninnikov and A. Pigarov, Plasma Physics and Controlled Nuclear Fusion Reseach **3**, 387 (1986).
88. H. E. Dalhed, A. S. Wan, H. A. Scott *et al.*, Journal of Nuclear Materials **220-222**, 1102-1106 (1995).
89. H. Scott and D. Post, Plasma Physics, to appear (1997).
90. R. K. Janev, D. E. Post, W. D. Langer *et al.*, J. Nucl. Mater. **121** (May 1984), 10-16 (1984).
91. D. Reiter, International Conference on Atomic and Molecular Data and Their Applicatons, Gaithersburg, Maryland, 1997, invited talk.
92. S. Krasheninnikov, A. Pigarov, D. Knoll *et al.*, Physics of Plasmas **4**, 1638 (1997).
93. M. Rosenbluth, P. Parks, D. Post *et al.*, Fusion Energy 1996 **2** (1997).
94. F. Levinton, "The Motional Stark Effect Diagnostic on TFTR," in *Tenth APS Topical Conference on Atomic Processes in Plasmas*, edited by A. Osterheld and W. Goldstein (AIP Press, San Francisco, California, 1996), Vol. 381, pp. 143-150.
95. F. Levinton and et al., Physical Review Letters **63**, 2060 (1989).
96. S. Allen, D. Hill, T. Carlstrom *et al.*, Journal of Nuclear Materials **241-243**, 595 (1997).
97. O. Batishchev, S. Krasheninnikov, P. Catto *et al.*, Physics of Plasmas **4**, 1672-1680 (1997).
98. J. Terry and Alcator C-Mod Group, Physics of Plasmas, to appear (1998).
99. C. Deutsch, "Atomic Data Needs for Inertial Confinement Fusion," in *Atomic and Plasma-Material Interaction Processes in Controlled Thermonuclear Fusion*, edited by R. Janev and H. Drawin (Elsevier, North-Holland, Amsterdam, The Netherlands, 1993), pp. 327-354.
100. M. Klapisch, Physics of Plasmas, to appear (1998).
101. P. Amendt, S. Glendinning, B. Hammel *et al.*, Physics of Plasmas **4**, 1862 (1997).
102. R. Janev, *Atomic and Molecular Processes in Fusion Edge Plasmas* (Plenum Press, New York, 1995).
103. R. K. Janev and H. W. Drawin, (Elsevier, Amsterdam, 1993), pp. 484.
104. R. Janev, Nuclear Fusion, Supplement on Atomic and Plasma-Wall Interaction Data for Fusion **1**, 1-138 (1991).
105. R. Janev, Nuclear Fusion, Supplement on Atomic and Plasma-Wall Interaction Data for Fusion **2**, 1-133 (1992).
106. R. Janev, Nuclear Fusion, Supplement on Atomic and Plasma-Wall Interaction Data for Fusion **3**, 1-125 (1992).
107. R. Janev, Nuclear Fusion, Supplement on Atomic and Plasma-Wall Interaction Data for Fusion **4**, 1-180 (1993).
108. R. Janev, Nuclear Fusion, Supplement on Atomic and Plasma-Wall Interaction Data for Fusion **5**, 1-265 (1994).
109. R. Janev, Nuclear Fusion, Supplement on Atomic and Plasma-Wall Interaction Data for Fusion **6**, 1-264 (1995).

The Use of Atomic and Molecular Data in Fusion Plasma Diagnostics

H P Summers[†], H Anderson[†], N R Badnell[†], F W Bliek[§], D C Griffin[♭], M von Hellermann[‡], R Hoekstra[§], A Howman[‡], L D Horton[‡], R Konig[‡], G M McCracken[‡], C F Maggi[‡], M G O'Mullane[†], M S Pindzola[♮], R E Olson[♯] and M F Stamp[‡]

[†] Department of Physics and Applied Physics, University of Strathclyde, Glasgow G4 0NG, UK
[‡] JET Joint Undertaking, Abingdon, Oxon. OX14 3EA, UK
[§] KVI, Atomic Physics, Zernikelaan 25, 9747 AA Groningen, Netherlands
[♭] Department of Physics, Rollins College, Winter Park, Florida 32789, USA
[♯] Department of Physics, University of Missouri, Rolla, Missouri 65401, USA
[♮] Department of Physics, Auburn University, Auburn, Alabama 36849, USA

Abstract.
Considerable attention in modern large fusion experiments is focused on neutral beam penetrated plasma and on divertor plasma. In the former, ion/atom reactions drive the population dynamics while in the latter, although electron processes predominate, there are significant ion/atom and atom/atom influences. Neutral atoms in beams, impurities in beam penetrated plasma and complex partially ionised impurities in divertor plasma find themselves in the collisional radiative regime where the simple stationary excitation/cascade picture of the coronal model is invalidated by secondary collisions. Thus atomic data needed in such fusion applications are not a few reaction cross-sections but large complete sets which support full collisional-radiative models. Also, experiment analysis does not make use directly of fundamental atomic data but rather effective coefficients deduced from such models.

In the paper, we describe two cases. The first is neutral deuterium beam attenuation and beam emission and its linking to active diagnostic spectroscopy using beams. We describe how the diagnostic analysis and experimental data reduction are carried out efficiently using derived effective stopping and effective emission coefficients. Then we describe the organisation of the fundamental ion/atom data and how collisional-radiative modelling generates the effective coefficients from them. This is an area in which long-term, intensive collaborative effort on both the fundamental data and the detailed spectral reduction is now bearing fruit. The second case is the more general task of modelling ionisation state, power and impurity line emission for interpretation of divertor observations. We explain the matching of effective collisional-radiative coefficients to the divertor plasma models and of photon emission coefficients to the post-processing for particular spectrometers and lines-of-sight. Then, we illustrate how just one part of the substantial fundamental atomic data requirement - dielectronic recombination - must be prepared for the application. Finally, the generalised collisional-radiative model, which prepares the effective coefficients, is described and we explain how it gives proper attention to the dynamic nature and non-negligible density of the divertor plasma and why resolution of metastable states is a requirement. The Atomic Data and Analysis Structure, ADAS and the JET Joint Undertaking Experiment are used in illustration.

I INTRODUCTION

This paper is concerned with how atomic and molecular data are used in fusion research. Rather than attempting an exhaustive review of applications we have chosen to focus on just two important areas. These are (a) charge exchange / beam emission spectroscopy with neutral deuterium beams and (b) modelling / spectral diagnosis of divertor plasma. Both are of very considerable interest at the present time and are expected to remain so into the future. Also, both use large quantities of atomic data and this use is quite well developed. Thus the problems of access to data, its general management and its provision at the point of diagnostic or modelling need in appropriate forms have been examined and solved. They will also illustrate a secondary theme of the paper, namely, that they are derived atomic/molecular data produced by reaction kinetic (collisional-radiative) models in plasmas rather than fundamental data which are actually used in application. Such models use and maneouver so much fundamental data that the latter must be carefully structured and organised from the beginning.

As is well known, the introduction of neutral deuterium heating in fusion plasmas initiated a complete new approach to the spectroscopy of the core plasma. This followed because electron transfer from the deuterium beam atoms to bare nuclei in the plasma core allowed the latter to radiate. Since such 'charge transfer' is generally to excited states, consequential spectral line radiation is observable in the visible. Quantitative study of this radiation requires knowledge of the local neutral deuterium beam atom density in the viewed volume. This can be deduced in principle provided the beam particle energies and fluxes are known at the point where the beam enters the plasma and the beam attenuation process can be modelled. A number of the key attenuation reaction processes have been measured and very large efforts have been made over the last decade to expand and improve the theoretical data collection so that the effects of all light impurities likely to be present in the plasma can be included. Nonetheless the demand of quality on such data is very high since analysis using calculated attenuation to the observed volume strongly amplifies the influence of error in the fundamental reactions [14]. The more recent observations of the emission by excited deuterium in the

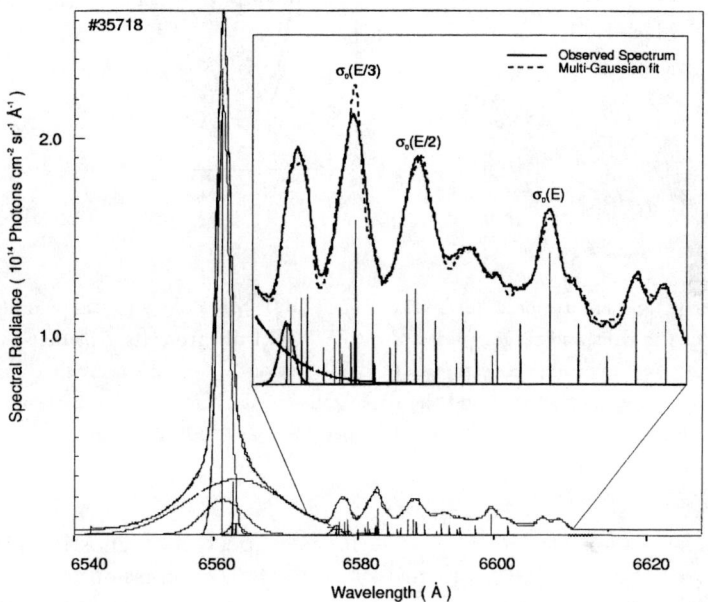

FIGURE 1. Observed motional Stark D_α spectral feature. The emission feature was recorded during the JET pulse 35718 using track 4 of the multichord visible spectroscopy system. The tangential bank was active with a primary energy of 140keV. (cf. von Hellermann (1993) [9] for a specification of the geometry). The constrained multigaussian spectral fit to the interval is also shown. Particular attention should be brought to the annotation of the full, half and one-third σ_0 Stark components. Due to the existence of three fractional energy components in the beam this gives rise to three Stark multiplets. Each Stark multiplet is Doppler shifted according to the velocity of the relevant beam neutrals and as a result the overal picture is an overlap of each Stark feature. A somewhat more complicated situation arises when the radial bank is also on as this results in the overlap of 6 Stark multiplets.

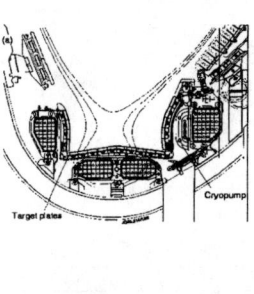
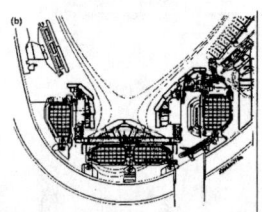
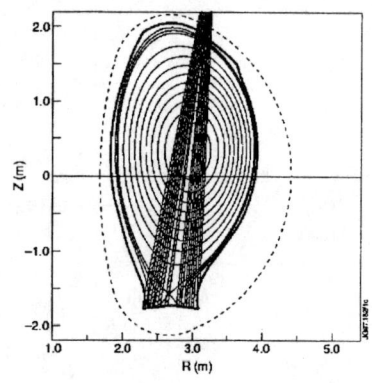

FIGURE 2. (a) Schematic poloidal section of the JET torus showing the geometrical arrangement of the coils and target plates of the JET Mk1 divertor. (b) Schematic of the Mk2A divertor. (c) Typical reconstructed poloidal section of the magnetic flux surfaces showing the last closed flux surface and the strike zones of the scrape-off-layer plasma with the target plates. Viewing lines used for visible spectroscopic studies are superimposed for illustration.

beams themselves [4] introduced a powerful new spectral diagnostic built on the subtlety of the motional Stark distortion of the D_α emission feature and the substantial Doppler shifts of the features in oblique viewing lines. It also promised an improvement over the beam attenuation calculation for the local neutral density since the density could instead be deduced directly from the localised D_α intensity. In practice, this has proved difficult to achieve for several reasons. The basic observed beam emission feature is a Doppler shifted D_α motional Stark multiplet. The observed D_α Stark multiplet is in fact a superposition of such features from different injectors and from different energy fractions within each individual injector. The appearance of the 6540Å - 6620Å spectral interval is shown in figure 1. Such observations together with visible charge exchange spectroscopic signals from impurities allow the local deposition of beam energy and particles, local impurity densities and related parameters to be deduced and this must be done self-consistently. At JET Joint Undertaking, the matching of the various observations and their merging with additional JET diagnostic signals, including radial profiles of electron temperature and density, are conducted iteratively in a complex experimental analysis package called CHEAP ('CHarge Exchange Analysis Package') [9].

Current thinking on control of the interaction of the particle and energy

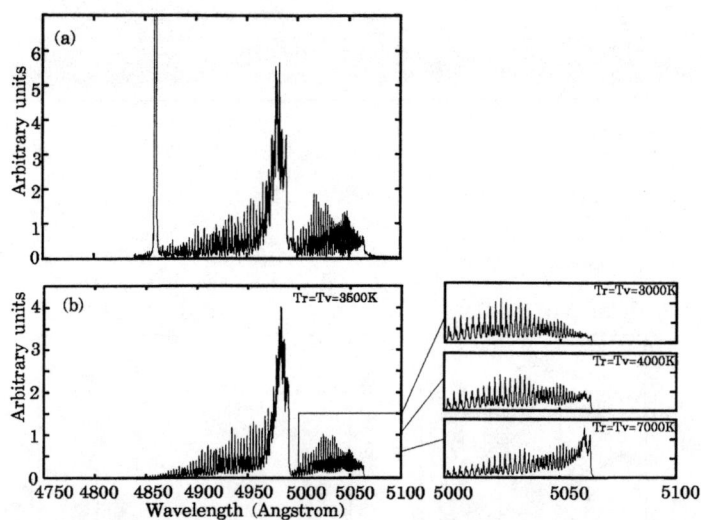

FIGURE 3. (a) Observed spectrum of the 4850-5100Å region (JET Pulse #35687 at 14.5s) showing BeD ($A\ ^2\Pi - X\ ^2\Sigma$). (b) Simulated 0-0, 1-1 and 2-2 bands at T_r=3500K and $T_v = 3500$K. The insert shows the variation of the P branch with temperature. [7]

flux flowing from the fusion plasma with material surfaces indicates that a cool divertor region remote from the main plasma is a necessary part of fusion reactor design. Over the last eight years implementation of the axi-symmetric poloidal divertor concept on the JET machine has led to extensive redesigning of the vessel interior and magnetic field coils to produce appropriate magnetic null (X-point) positions, connection lengths and divertor target tiles for the plasma flowing down the scrape-off-layer. Experimental campaigns have investigated the properties of the Mk1 and Mk2A designs. The next stage in this evolution is the Mk2GB (gas box) design due for installation at the end of 1997. Many diagnostic systems were upgraded and optimised prior to plasma operation with the Mk1 divertor [5]. As a result of the improved divertor spectroscopy the MK1 divertor campaign has given the opportunity for detailed sets of observations of impurity atoms and ions and in some cases their molecular precursors in JET. Spectral observations of the divertor are rich. Figure 3 of a visible wavelength segment and figure 4 in the VUV show some examples. The geometrical complexity of the divertor plasma region, the open magnetic field line structure, the localisation of power deposition flowing in the scrape-off-layer at the strike zones and divertor operational strategies such as the creation of detached radiating plasma present a diagnostic challenge. The divertor plasma is neither homogeneous nor stationary as the time

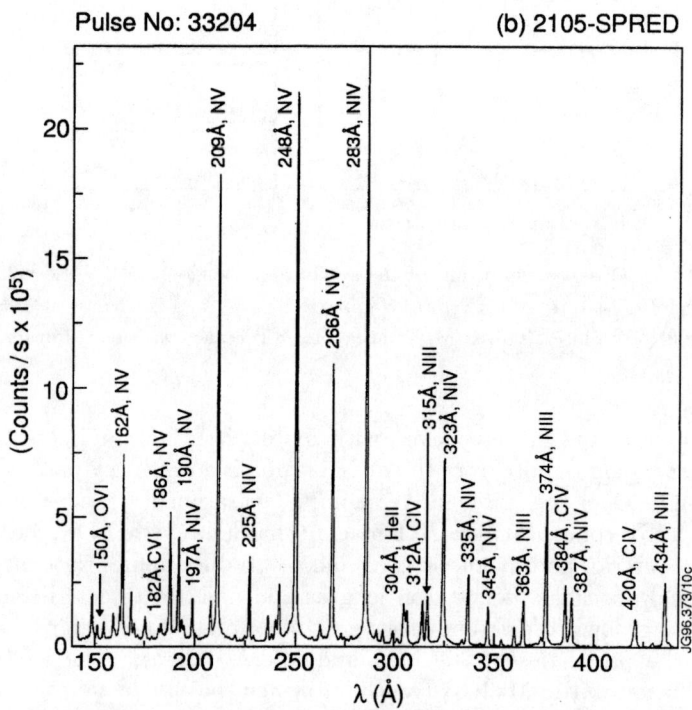

FIGURE 4. VUV spectra of the divertor from the 2105-SPRED during a nitrogen seeded radiative divertor discharge (Pulse 33204) [11]

constants for parallel movement of low ionisation stages of impurities from the strike zones and for high stages of ionisation from the bulk plasma into the divertor can be comparable to ionisation and recombination times. Thus the usual equilibrium ionisation vehicle for spectral interpretation is not sustainable. Instead, it is necessary to link the spectral observations, dynamic divertor plasma models and the atomic modelling of the radiating plasma closely from the beginning. Multiple probes along the divertor target, tomographic bolometric reconstructions together with comprehensive wavelength spectroscopy along multiple lines of sight through the divertor provide the experimental pulse data. Such signals are simulated in an interconnected set of modelling and post-processing steps and then the simulations and experiments compared.

In its approach to the study of these two areas, JET is fairly typical of other fusion experiments such as ASDEX-upgrade, ALCATOR-Cmod, JT60 and DIIID.

II THE ATOMIC DATA AND ANALYSIS STRUCTURE, ADAS

At JET Joint Undertaking, we took the view that atomic data should be a responsibility of the experimental, spectroscopy division (Experimental Division II) since it was there that the most testing use and scrutiny would take place. A tightly defined, integrated computational approach was taken to the provision of appropriate derived atomic data for diagnosis and modelling. This is known as the Atomic Data and Analysis Structure, ADAS. Several fusion laboratories have worked with JET in the shared development of ADAS into its present guise as a UNIX workstation code and data package. In the discussion of beam and divertor atomic physics data and modelling, we shall from time to time refer to ADAS [15]. ADAS data, both fundamental and derived, are currently assigned to one of twenty-six data format classes, called $adf's$, such as $adf01$ and $adf02$. ADAS codes are allocated to six series each containing about ten codes. Thus the ninth code of series 2 is called ADAS209. In the schematics of data and program flows of the later parts of the paper, we use these names when we have an ADAS implementation.

III DEUTERIUM BEAM STOPPING, BEAM EMISSION AND CHARGE EXCHANGE SPECTROSCOPY

Deuterium atoms in beams penetrate the confining magnetic field of the plasma as neutrals until their point of ionisation. The rate of beam energy deposition at any point along the beam path is obtained therefore from the

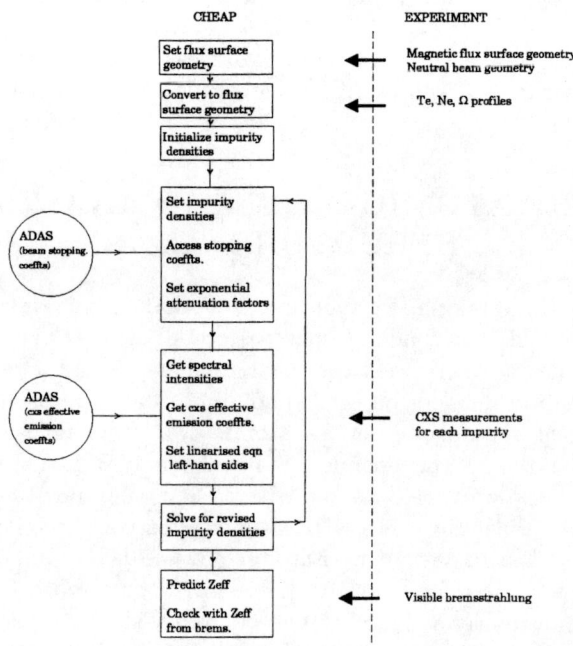

FIGURE 5. Organisation of the CHEAP analysis showing the entry points of derived atomic data

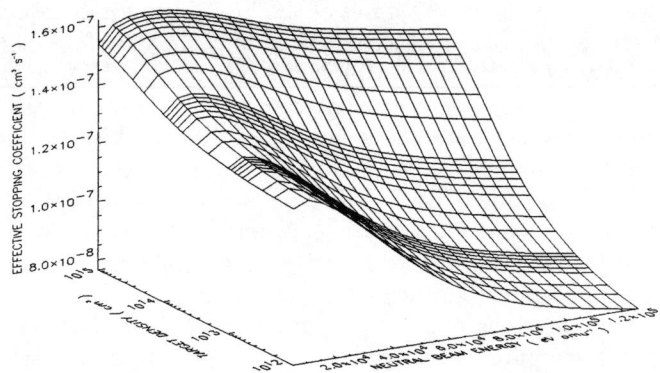

FIGURE 6. Variation of the effective deuterium beam stopping coefficient with beam energy and electron density. Pure deuterium plasma at $T_i = 2.0 * 10^3$ eV.

ionisation rate, R, or from the effective ionisation rate coefficient, $S_B^{(e)}$ with $R = N_e N_{D,beam} S_B^{(e)}$ if the electron density N_e is known. $S_B^{(e)}$ is usually known as the stopping coefficient. The neutral deuterium density in the beam, $N_{D,beam}$ at any point of the beam path can be obtained by using $N_e S_B^{(e)}$ to calculate the attenuation from where the beam enters the plasma. $S_B^{(e)}$ is written with reference to the electron density in the plasma, but it is primarily collisions with thermal deuteron and thermal impurity ions (fully ionised) which cause the ionisation. So $S_B^{(e)}$ depends on the impurity concentrations. Unfortunately, the concentrations, N_{D+}/N_e, and $\{N_{Z_i}/N_e : i = 1,...\}$ in the plasma are not initially known except in so far as their mean Z_{eff} can be inferred from bremsstrahlung measurements. However, a part of the ionisation of the beam atoms takes place via a charge transfer reaction to the impurity ions in the plasma which then emit measurable spectral line radiation. Thus, the charge exchange spectroscopic line-of-sight intensity of radiation, $I_{Z_i,n\to n'}$ in a spectrum line $n \to n'$, which is localised at the intersection of the beam/ spectrometer viewing line intersection, L, may be used to infer N_{Z_i}, from $I_{Z_i,n\to n'} = \int_L N_{D,beam} N_{Z_i} q_{Z_i,n\to n'} dl$, if the effective emission coefficient $q_{Z_i,n\to n'}$ is known. We are therefore led to a circular iterative analysis based on plasma measurements and spectroscopy. At JET, the implementation is called 'CHEAP' and a schematic is shown in figure 5.

A The derived data

The beam stopping coefficient depends on density. This is because there are stepwise losses through excited states of the beam deuterium atoms at fusion plasma densities. At the precision required of beam stopping coefficients for calculating attenuation to the centre of the plasma ($\leq 10\%$) such effects matter. Thus the beam stopping coefficient is an effective coefficient including the influence of the excited states. It is properly called a collisional-radiative coefficient. The behaviour of the coefficient is shown in figure 6. The secondary collisions which cause this step-wise ionisation are principally due to ion collisions (deuterons and impurity nuclei) and so the effective stopping coefficient is *itself* a function of the impurity concentrations. The *ab initio* re-evaluation of the stopping coefficient for varying impurity mixtures in the iterative CHEAP analysis is quite impractical. At JET we have evaluated the beam stopping for pure impurity plasmas, that is with only a single impurity and with the electron number density being that which comes from charge balance. Then a linear combination of coefficients for a mixed impurity plasma can be produced. The error of this step is assessed by comparison with exact calculation for a mixed plasma [1]. The error is below (5%). Thus a tabulation of the derived stopping coefficient is made for every light impurity $H, He, Li, Be, B, C, N, O, F$ and Ne. The effective coefficients are also functions of T_i, N_i and E_{beam}. The rapid extraction of the stopping coefficient from tabulations, required for inter-pulse analysis, does not permit interpolation through a multi-dimensional grid. The primary dependence of the stopping coefficient is on beam energy and ion density. We define reference conditions for the plasma parameters and beam energy relevant to JET and establish the stopping coefficient at a two-dimensional grid in E_{beam}/N_i at the reference condition of T_i. Then a one-dimensional scan in T_i is made at the reference values of E_{beam} and N_i. Such a mixed two- and one-dimensional scan structure of the tabulations delivers the required precision in the final interpolated coefficient and the required look-up speed. The prescribed structure follows ADAS format *adf*21. Revision of *adf*21 data takes place at intervals when there are improvements of the fundamental database from which it is built. The most recent reprocessing was carried out in June 1997 and made available through the ADAS network.

Charge exchange spectroscopy and beam deposition studies in the current CHEAP analysis calculate the attenuation of the beam from the point of entry to find the local neutral beam density at points along the beam line. The exponential attenuation causes a corresponding increase in the error bounds of this local density arising from the error in the stopping coefficient. Since the beam emission is observable (cf. figure 1) it might be thought that the local neutral density would be best deduced from this local observation. Such a deduction requires knowledge of the theoretical effective emission coefficient for Balmer alpha emission. The influences on this coefficient are the same as for the beam stopping coefficient with emphasis on losses from excited states of the beam deuterium atoms. Figure 7 shows the behaviour of the effective beam emission coefficient. The two methods of deduction have historically

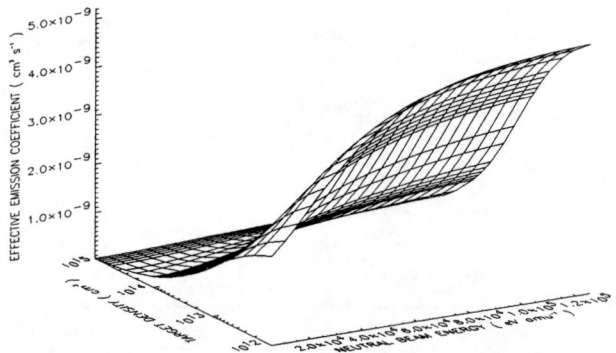

FIGURE 7. Variation of the effective Balmer alpha beam emission coefficient with beam energy and electron density. Pure deuterium plasma at $T_i = 2.0 * 10^3$ eV.

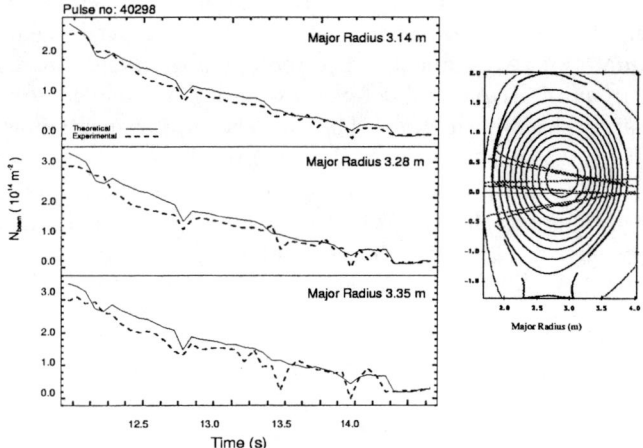

FIGURE 8. Time evolution of the line-integrated neutral beam density at three different radial positions. The solid curve is the line integral neutral deuterium number density from the attenuation calculation. The dashed curve is the line integral number density obtained from the measured Balmer alpha intensity. The latter also uses the collisional-radiative model to convert from excited population to ground population. The insert shows the paths of the individual contributing beam lines in flux surface geometry.

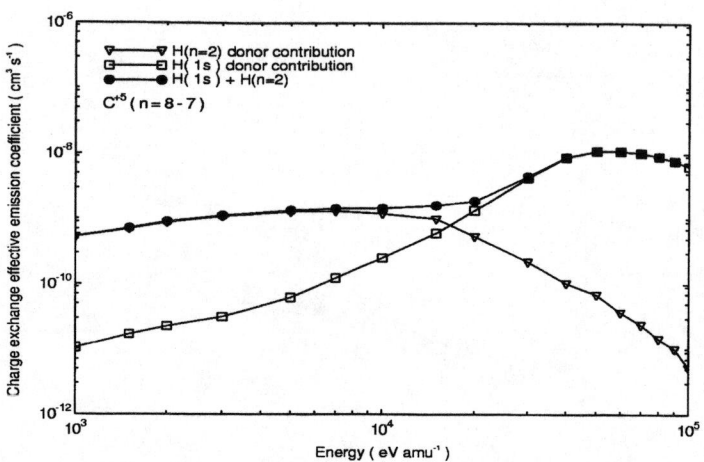

FIGURE 9. Behaviour of the charge exchange effective emission coefficient for the $n = 8 - 7$ line of CVI. $Z_{eff} = 2, T_i = 8.0 * 10^3 \text{eV}, N_i = 5.0 * 10^{13} \text{cm}^{-3}$

given results which differed typically by up to factors of three. This situation now seems to have been corrected. We have conducted a detailed review of the fundamental atomic data, error analysis of the derived coefficients, step by step validation of experimental spectroscopic, beam and plasma data entry and flow through CHEAP and detailed examination of our fitting of the motional Stark split observed feature. The various improvements now yield an accuracy and consistency of $\sim 20\%$. Present results are shown in figure 8. We now believe the beam emission can be used in the manner originally envisaged. It is to be noted that the structure used for access and storage of beam stopping coefficients are immediately usable again for beam emission. We use an identical ADAS data organisation but called *adf*22 rather than *adf*21.

The improved precision in our knowledge of the excited populations in the beam allows progress in charge exchange spectroscopy. For relatively fast beams ($\sim 70 \text{keV/amu}$) with modest fractional energy components, charge exchange spectroscopy in the visible spectral region is dominated by donation from the ground state of the deuterium beam atom. This stuation changes as we move to lower energies where donation from excited states matters. We have improved the state selective charge exchange database for donation from $D(n = 2)$ to light impurity nuclei [3]. These new data are now being combined with the beam population results to provide a low beam energy correction to the charge exchange effective emission coefficients. Figure 9 illustrates the role of the excited beam donor in the CVI (n=8-7) line emission.

B The fundamental data

The derived collisional-radiative effective beam coefficients used and described above depend on high quality collision-cross-section data for positive ion/hydrogen reactions. These include ionisation, charge transfer and excitation reactions with all fully ionised light impurity nucei from hydrogen to neon. Electron impact cross-sections for ionisation and excitation are also needed. The effective coefficients for charge exchange spectroscopy of plasma impurities in turn require very large collections of state selective charge exchange cross-sections for neutral hydrogen (ground and excited state) with fully stripped light impurity ions. These data are required over extended energy ranges and resolved to the nl levels of the receiver. Considerable effort has been put into creating and assembling such data over the years. There are some substantial new additions. These are described in the accompanying paper by Hoekstra [10].

C Collisional-radiative modelling and the computational implementation

We have found it advantageous to embed the calculation of beam stopping and beam emission for hydrogen in a more general picture of neutral hydrogen as a radiating, ionising and recombining species in the fusion plasma. This is because an initial fast neutral beam hydrogen atom can lead to a high temperature thermal hydrogen atom in a beam halo and ultimately to a lower temperature thermal hydrogen atom at the plasma periphery. Also with the variation of primary beam energies which occur in fusion experiments (40 keV/amu - 70 keV/amu typically for deuterium) and the inevitable fractional energy components (35 keV/amu and 23 keV/amu for the JET primary 70 keV/amu beams) there is a wide range of energies of hydrogen donors present which influence observed charge exchange impurity spectra and contribute to the overall composite $D\alpha$ feature. For fast beams, ground state donors primarily drive the visible impurity charge exchange spectra while at thermal energies it is almost entirely excited donors which do this. From the point of view of the hydrogen atom populations, the most important type of collision alters also from being ion impact at beam energies to electron impact in the lower temperature thermal regimes. In the thermal regime, recombination of hydrogen nuclei by both free electron capture and charge exchange capture balances ionisation while in the beams, the hydrogen atom is strictly ionising and a charge exchange donor. From this viewpoint, the beam atom is distinguished simply as one with a high fixed translational velocity which enters into the collisionality together possibly with enhanced state mixing from a motional Stark electric field. Because of the strong l-shell mixing, we work in the bundle-n picture. Consistent modelling of the type sought here should

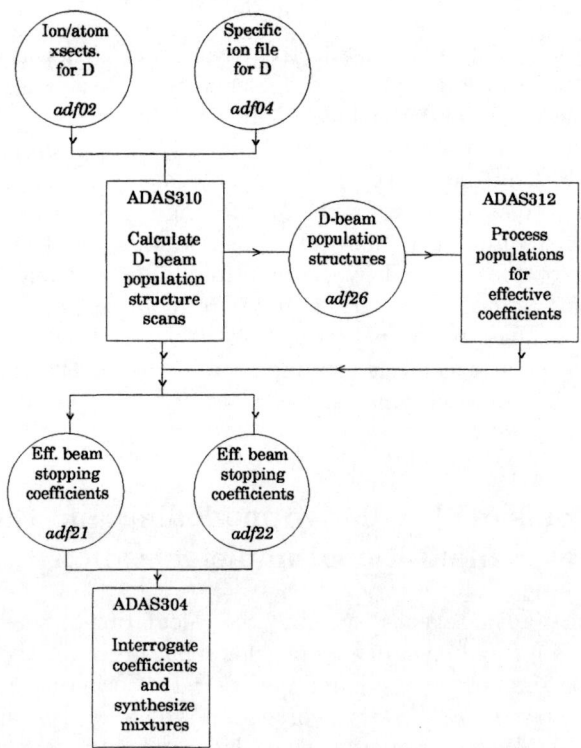

FIGURE 10. Schematic of ADAS codes and data sets used in neutral deuterium beam studies. There is an equivalent schematic for charge exchange spectroscopy with deuterium beams which is not shown here.

work smoothly between these different regimes and use the same fundamental data. It is not a difficult task to generalise computationally in this manner and it has the added merit of allowing stringent testing of Saha-Boltzmann limits, zero-density ionising extremes etc.

Three ADAS codes and five data formats provide the computational and data archiving functionality for the beam analysis as shown in the schematic of figure 10. *adf*02 and *adf*04 are structured assemblies of fundamental reaction cross-sections and rates. *adf*21, *adf*22, and *adf*26 are derived data generated by the collisional radiative codes.

- ADAS304 - interrogates effective beam stopping and emission coefficients and assembles effective coefficients for impurity mixtures

- ADAS310 - calculates beam atom excited populations and stopping coefficients in the bundle-n approximation

- ADAS312 - post-processes population data from ADAS304 to assemble effective coefficient collections suited to rapid experiment analysis

IV MODELLING AND SPECTRAL DIAGNOSIS OF THE DIVERTOR

The highly ionising environment near the divertor strike zones, from the point of view of neutral and near neutral impurities emerging from the surfaces, has become familar in recent years. Atomic modellers calculate the derived 'ionisation per photon' coefficients for quick estimation of impurity fluxes in these circumstances [8]. However, the divertor also has recombining environments and it is found that the passage of ions through them and the power which they radiate there matters for the overall divertor power balance. Also, most observations have lines-of-sight which cross these different, usually geometrically complex regions. The 'experimental' approach at JET to diagnostic study of the divertor is that of modelling observed signals and then comparison with the actual observed signals [12]. Such an approach is almost mandatory for other than the simplest reductions of observed data. This has the disadvantage of requiring the use of the complete (or nearly complete) theoretical computational models of divertor fluid and particle transport (and their slow execution times) but with the advantage of allowing manipulation of relevant physical parameters in seeking the simulation/observational match. This means that derived collisional-radiative atomic data enters the picture both in the form of source terms for the fluid and particle transport codes and as effective emissivity and radiant power data for the post-processing of the modelled ion distributions into simulated line-of- sight signals. A schematic of the JET approach is shown in figure 11. The entry points, indicated in the figure as ADAS, show the derived atomic data needed for the JET plasma modelling and diagnosis. Such derivative data are the actual targets of fundamental atomic data creation and the subsequent atomic modelling.

A The derived data

Collisional-radiative theory allows us to identify the appropriate types of derived data needed for each impurity ion from a consideration of time constants. Slowly relaxing state populations are large and must enter the plasma dynamic equations explicitly. So for those we need effective recombination and ionisation coefficients and related power coefficients. Fast relaxing state populations are small and they reach a local quasi-static equilibrium with respect to the slowly relaxing populations. For those we need effective emission coefficients which relate emission in any line to the large populations. In certain situations in the divertor, the large populations include metastable as well as

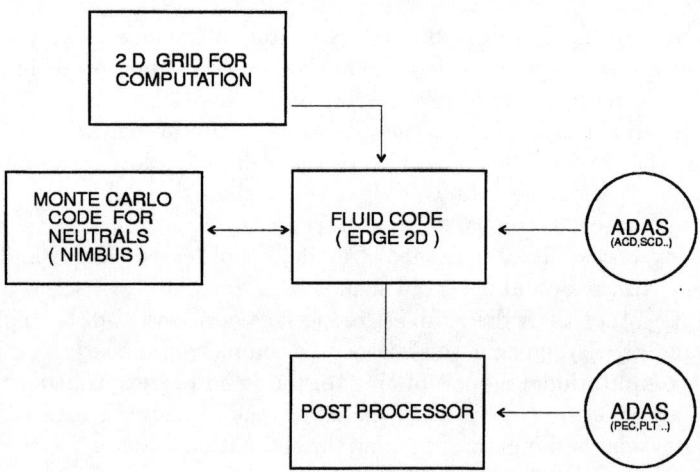

FIGURE 11. Schematic of the plasma modelling codes and post-processing used in simulating and analysing JET divertor signals. *2D grid*: from reconstruction of the magnetic equilibrium, generates a 2D field aligned net on which the computation is carried out. *Fluid code*: solve the 2D fluid equations for each species: electron, plasma and impurity ions (every charge state) are treated as species. *Monte Carlo code for neutrals*: particle, momentum, energy sources and losses for the plasma, neutral hydrogen isotope distributions, neutral impurity distributions. *Post − processor*: after the solution is obtained, performs integrations along diagnostic lines of sight of line emissivities, radiation profiles etc. *ADAS*: points of insertion of collisional-radiative atomic data.

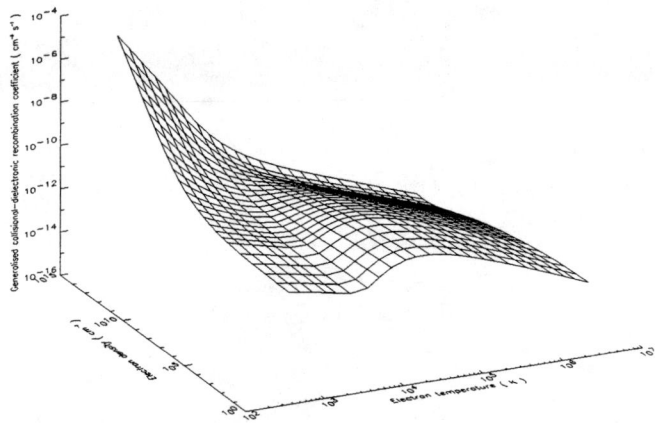

FIGURE 12. Generalised collisional-radiative recombination coefficient for the process $C^{+2}_{2s^2}\ {}_1S + e \to C^{+1}_{2s^2 2p}\ {}_2P$. This is the ground state to ground state part of the metastable resolved data. Note the high density three-body regime at low temperatures and the suppression of the high temperature dielectronic part at higher densities before three-body recombination takes over.

ground state populations. At JET and in ADAS, we have taken the approach that a complete set of metastables is the best starting collection of 'large' (usually called 'dynamic') populations [16]. Such a collection can be simplified (by grouping) in less dynamic regions or kept separate in more dynamic regions. In fusion plasma modelling, it is unnecessary to treat excited state populations as dynamic. We use the terms 'generalised collisional-radiative modelling' and 'metastable resolved treatment' when we distinguish metastables. Metastable resolution provides a sound basis for calculating accurate direct dielectronic recombination coefficients and ionisation coefficients even in static conditions.

In summary, for modelling and spectral analysis of the divertor, we prepare a number of types of derived datasets. These fall into two classes, namely those required to establish the distribution of impurity ions in the divertor. These data are used by the impurity transport models. Then there is the class of derived data required to interpret the radiation losses from the plasma. These include total radiated power at the grossed-up end of the scale and emissivities of specific lines at the other. Also, for quick inferences of impurity influx and emission shell shapes, one or two other forms of derived data are usually created. Such data can allow either reduction of the calibrated observed photon counts to physical quantities or can enter simulation of the spectrometer signals for matching with observation. These datasets are summarised in Table 1.

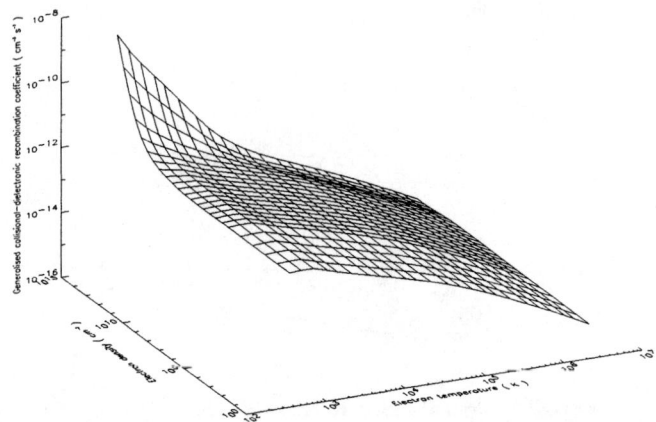

FIGURE 13. Generalised collisional-radiative recombination coefficient for the process $C^{+2}_{2s2p\ ^3P} + e \to C^{+1}_{2s^22p\ ^2P}$. This is the metastable to ground state part of the metastable resolved data. Note that for the metastable parent dielectronic recombination is suppressed through secondary autoionisation.

TABLE 1. Definition of ADAS derived datasets used in divertor modelling

adf type	Subcode	Content
adf11	acd	generalised collisional-radiative recombination coefficient
adf11	scd	generalised collisional-radiative ionisation coefficient
adf11	ccd	generalised collisional-radiative CX recomb. coefficient
adf11	qcd	generalised collisional-radiative cross-coupling coefficient
adf11	xcd	generalised collisional-radiative parent cross-coupling coefficient
adf11	prb	generalised collisional-radiative recomb/bremss. power coefficient
adf11	plt	generalised collisional-radiative low-level line coefficient
adf11	prc	generalised collisional-radiative CX recom. power coefficient
adf11	pls	generalised collisional-radiative specific line power coefficient
adf11	met	generalised collisional-radiative metastable fractions
adf13	sxb	ionisation per photon coefficients
adf15	pec	photon emissivity coefficients

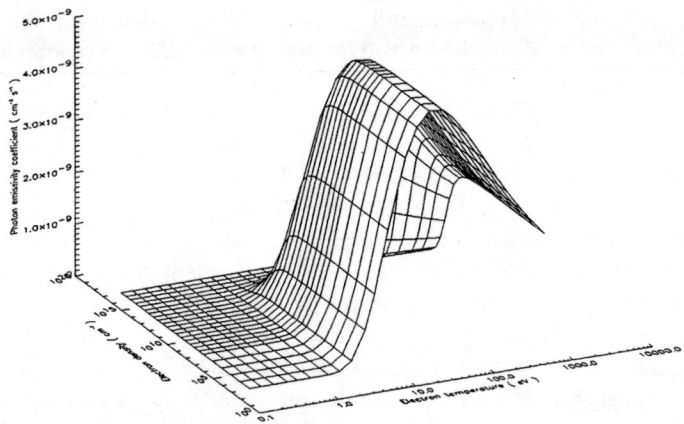

FIGURE 15. Generalised collisional-radiative emission coefficient for the transition $CII(2s^2 2p\ ^2P - 2s^2 3s\ ^2S)$ at 858.4Å. This coefficient is driven by excitation from the $C^{+1}_{2s^2 2p\ ^2P}$ ground state as evident from the exponential rise at low temperature. Note the suppression at high density as the collision limit moves down towards the upper level of the transition.

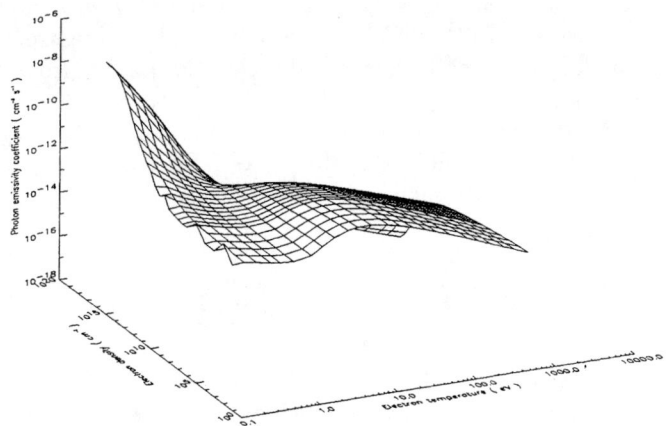

FIGURE 14. Generalised collisional-radiative emission coefficient for the transition $CII(2s^2 2p\ ^2P - 2s^2 3s\ ^2S)$ at 858.4Å. This coefficient is driven by recombination from the $C^{+2}_{2s^2\ ^1S}$ state. Both direct radiative and dielectronic recombination as well as cascade are influential on populating this level at low density. At high density, the role of collisional-radiative redistribution is evident. The slight ripple at the lowest temperature is a numerical artifact.

The various derived data sets summarised above are each functions of plasma electron temperature and plasma electron density. Also they are metastable resolved. Thus the effective recombination coefficient for $C^{+2} \to C^{+1}$ has separate parts for $2s^2\ {}^1S \to 2s^22p\ {}^2P$, $2s2p\ {}^3P \to 2s^22p\ {}^2P$ and $2s2p\ {}^3P \to 2s2p^2\ {}^4P$. In our progressive refinement of this type of data, we have now moved to tabulations at the fairly dense grid of temperature and densities shown in figure 12 for the $2s^2\ {}^1S \to 2s^22p\ {}^2P$ effective recombination coefficient for $C^{+2} \to C^{+1}$. Note that in practice the tabulation of data for members of the same isoelectronic sequence is made at z-scaled temperatures and densities. In contrast, the $2s2p\ {}^3P \to 2s^22p\ {}^2P$ coefficient representing recombination from the C^{+2} metastable has quite a different character because of secondary autoionisation.

In fusion application, we have become accustomed to emission coefficients for spectral transitions of an ion being driven principally by electron impact excitation from the ground state of the ion. In the generalised collisional-radiative picture, we must also include separate driving from distinguished metastables of the ion. With the focus of recent years being principally on influx of ions under ionising conditions, the above was usually sufficient. However, the recombining plasmas which now matter indicate that we must include the effective emission coefficients driven from the metastables of the next higher ion by recombination. These coefficients, formally present in generalised collisional-radiative modelling, were usually ignored previously except for less common lines emitted from higher n-shell states. Figure 14 shows the behaviour of this coefficient for a doublet line in the CII spectrum driven from the $2s^2\ {}^1S$ ground of C^{+2} and is to be contrasted with figure 15 which is the coefficient for the same line, driven by electron impact from the C^{+1} ground state $2s^22p\ {}^2P$. The divertor plasma at the time and point of detachment from the target is currently of interest and shows clear signatures of recombination. Figure 16 illustrates this in JET for deuterium itself. The analysis by McCracken draws upon collisional-radiative emissivity coefficients from the ADAS database.

V THE FUNDAMENTAL DATA

In this section, we wish to discuss briefly the steps involved in building collections of fundamental atomic data which are in some sense complete for the divertor application. In ADAS, these are called *adf*04 files. An *adf*04 file relates to a set of low levels of an ion - usually complete up to some n-shell - which spans the observational spectroscopy. The electron impact excitation part has been discussed on many occasions and again at this meeting. Here we wish to touch on the recombination part.

We seek to add free electron capture rate coefficients to each energy level of the set adopted for the ion. There are three parts to this, namely radiative re-

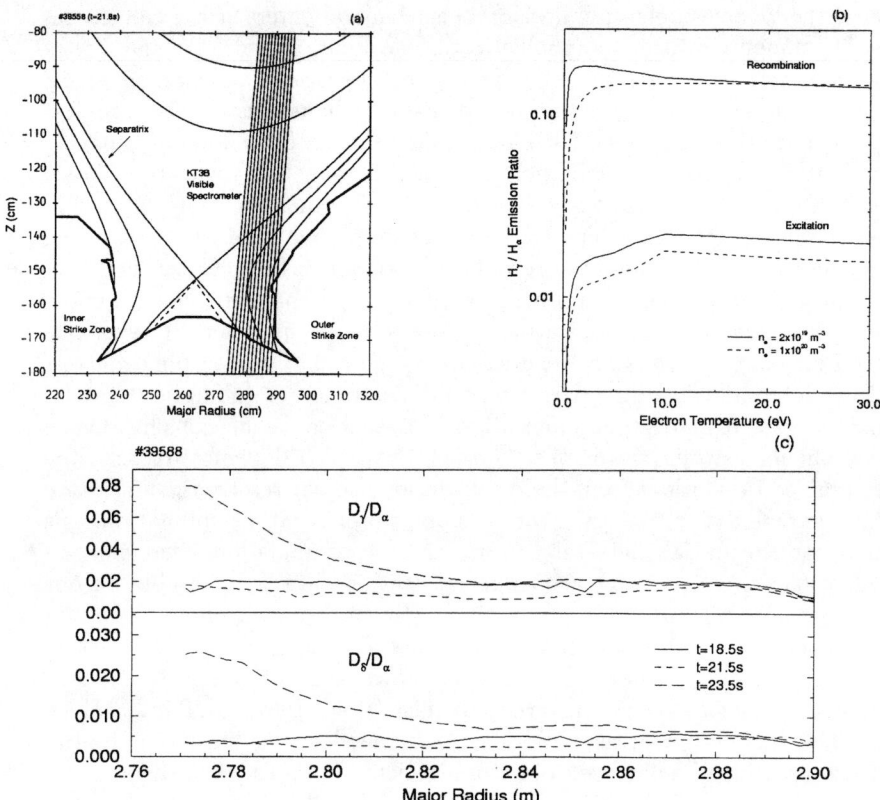

FIGURE 16. (a) Twelve lines of sight of the KT3B visible spectrometer directed at the outer divertor target; (b) Theoretical D_γ/D_α emissivity ratios drawn from the ADAS adf15 emissivity coefficient database; (c) Observed spatial variation of the D_γ/D_α and D_δ/D_α line ratios from the different KT3B lines of sight for JET discharge 39588 at times 18.5, 21.5 and 23.5 secs [D_α is a flux camera measurement]. From (c) at 18.5 sec., the D_γ/D_α ratio is ~ 0.02 over the range $R = 2.75 - 2.8$m consistent with an electron impact excitation mechanism for the lines at $N_e \sim 2*10^{19} \text{m}^{-3}$. At 21.5 sec, all three line intensities have increased but D_α has peaked at ~ 2.80m resulting in a decrease of the D_γ/D_α and D_δ/D_α line ratios. This indicates a density increase but the line formation mechanism remains excitation. At 23.5 sec, D_γ/D_α changes markedly at small major radius - increasing to ~ 0.10. From (b), this is consistent with a change in line formation mechanism to recombination. Comparison with adf15 emissivity coefficients from ADAS indicates that both line ratios are consistent with $T_e = 0.7 - 0.9$ eV and $N_e \sim 2*10^{20}\text{m}^{-3}$

combination, dielectronic recombination and three-body recombination. The latter is the inverse of electron impact ionisation and enters at the collisional-radiative modelling stage. In a sophisticated formulation, radiative and dielectronic recombination occur as parts of one general process and can in principle interfere with each other. In practise, at least for light ions this is unimportant and radiative and dielectronic recombination can be treated as separate processes. Radiative recombination is obtained from the associated photo-ionisation cross-section via the Milne relation. If R-matrix photo-ionisation cross-sections [6] are used which include resonances then implicitly part of the dielectronic capture is added to the radiative recombination. This added part is not quite perfect since radiation damping of the resonances is not normally handled in R-matrix photo-ionisation cross-section calculations. However this is not a serious problem since regimes where the radiation damping matters can be isolated and treated separately as a dielectronic recombination calculation. Extended R-matrix photoionisation cross-section calculations have been carried out for astrophysics in the 'Opacity Project' [13]. Unfortunately, the only archived tabulations from these calculations do not resolve the final state of the photoionisation process! From the point-of-view of recombination, this means that the initial metastable from which recombination takes place is not identifiable. Thus the Opacity photoionisation data is not usable for our present purpose. Some recomputation of Opacity R-matrix photo-ionisation cross-section data with resolution of final state and then preparation of state selective radiative recombination coefficients has taken place for Be-like ions $\rightarrow$ B-like ions at Queens's University, Belfast (Reid, 1997 - JET internal report). This has been contrasted with simpler one-electron, effective potential calculations using the observed quantum defects. In cases when dielectronic recombination can be ignored agreement is very good (error $\leq$ 20 %). It is our view that for fusion applications the latter method is mostly adequate. In ADAS, we archive state selective radiative recombination data in the data format *adf*08. This format is written to from the external refined calculation and also from an internal ADAS code operating the simpler approximation. Note that at moderate to high temperatures, the predominant part of the overall radiative recombinations takes place to resolved low levels spanned by our spectroscopic interest. At very low temperatures, the contributions to radiative recombination from capture to higher n-shells increase. Since purely hydrogenic rate coefficients are accurate to within a few per cent, the latter are used for high n-shells and generated internally in collisional-radiative codes.

Dielectronic recombination presents a more substantial problem. There are an abundance of dielectronic recombination results in the literature but these are virtually all summed over final (recombined) states and are only for recombination from the ground state of the recombining ion. Also, dielectronic recombination coefficients are required not only to low levels in a resolved picture but also to very many higher n-shells. These data must be archived unsummed so that the modifications of the effective recombination

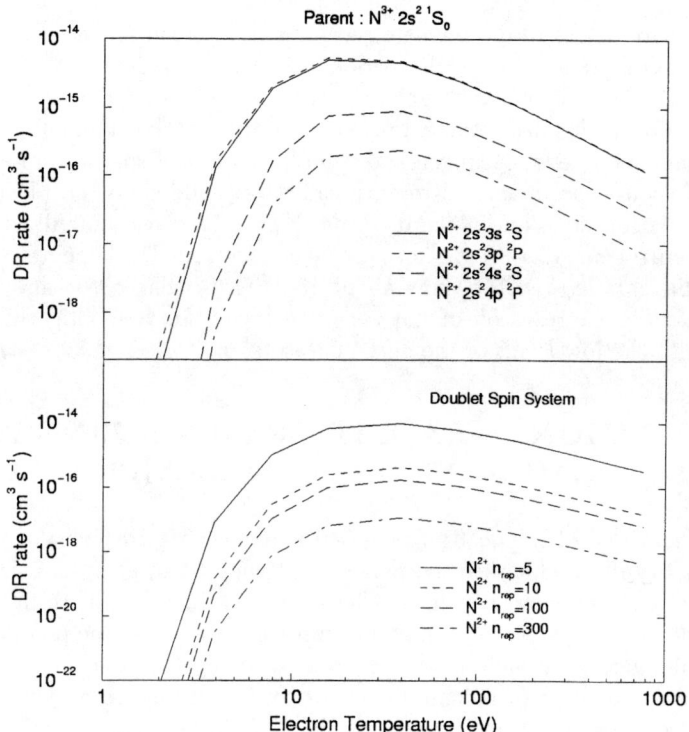

FIGURE 17. State selective dielectronic recombination coefficients to LS coupled terms and nS spin system shells of N^{+2} from the $N^{+3}_{2s^2\ ^1S}$ state occuring via $\Delta n = 0$ parent transitions. LS-resolved data is archived up to n=7. nS shell sums are archived for approximately 50 representative principal quantum numbers spanning from 2-1000. Note secondary auto-ionisation restricts the range of data built on metastable parents.

due to redistribution and reionisation can be evaluated properly. It must be stressed that zero-density dielectronic calculations of total effective recombination, even though accurately done, are quite inappropriate for light ions at the typical densities of divertor plasma. With such large data flows involved, we have found it a great advantage to interact directly with those able to generate such dielectronic data. This is done so that a detailed prescription of the data layout can be agreed to which fulfills the fusion need prior to substantive calculations. Within ADAS, this corresponds to another data format, *adf*09. Extensive state selective dielectronic recombination calculations, structured for ADAS were carried out in the period 1991-92 by Badnell (cf. Badnell, 1986 [2]). These data, for key light impurities of concern for fusion at that time, have now been substantial extended by O'Mullane and Badnell to cover all light ions up to neon. Nitrogen and neon have been of special importance in recent fusion experiments as impurities actively added to the plasma to enhance radiation emission in the divertor. Figure 17 gives some illustrative results for nitrogen. It is worth noting that the current fundamental dielectronic coefficients in ADAS occupy about 15 Mbytes. The *adf*08 and *adf*09 datasets provide the reservoir of state selective recombination data which are mapped onto the low levels of the *adf*04 datasets.

VI COLLISIONAL-RADIATIVE MODELLING AND THE COMPUTATIONAL IMPLEMENTATION

Clearly data sets of the *adf*04 type described above are the focus of all the highest quality data which can be mustered in support of analysis using the ion in question. As such they may reflect very considerable effort by many people. The datasets however remain incomplete for study of the populations in a plasma since they include only processes among the designated levels. From the point of view of recombination when dielectronic recombination is active, a very substantial downward cascade of electrons from higher n-shells occurs which markedly alters populations and indeed contributes the bulk of the effective recombination in the collisional-radiative sense. Conversely, for an ion in a low stage of ionisation in the relatively high densities and low temperatures of a divertor, populations of excited levels, for example in the $n = 3$ shell of N^{+2}, experience ionising collisions. Also this ionisation is not necessarily direct but may occur in a stepwise fashion through a series of higher excited n-shells. Finally even at low density, excitation from the ground level to higher n-shells followed by cascade gives a significant correction to that from direct excitation. The effects of higher n-shells must be included therefore to allow the precision obtainable in principle from the quality of the lower n-shell data. The estimation of the influence of higher n-shells may be calculated in simpler approximation since for such levels sub-shell mixing within an n-shell is very strong. At JET, we conduct such calculations within a parent and spin system resolved 'bundle-n' model and generate from it so-called 'condensed projection matrices'. These are archived in the database and need not change as refinement of the low level data proceeds with time. Thus for the complete treatment of an ion, collisional-radiative population codes draw both upon the detailed specific ion file of type *adf*04 in ADAS and also upon the projection matrices (called *adf*17 in ADAS). Together, these datasets and the subsequent population calculations allow us to realise all the derived collisional-radiative data actually used in analysis and modelling. For us, such datasets are the structures into which we seek to meld disparate fundamental data from many sources.

The above considerations indicate that our derived atomic calculations must be in the collisional-radiative framework. That is, as well as dealing with the mixture of collisional and radiative processes affecting population lifetimes, we must contrast atomic and plasma transport timescales. Collisional-radiative theory provides the prescription for separating populations into those which are short lived, reflecting local conditions only and those which are long-lived, reflecting their past history. The former are small populations (the true excited state populations) which can be treated as quasi-static yet provide the spectral emission, while the latter are large and must enter the plasma transport equations explicitly. Generically, we call these large populations metastables. They include the ground populations of ionisation stages. Most plasma transport models, in fact, only include the ground state populations (or effectively the ionisation stage populations). The earlier illustrations in this paper indicate that the ground and low lying metastables are the 'correct' set of 'metastables' for a comprehensive picture. The production target of our collisional-radiative atomic modelling should be derived data relative to the complete set of metastables in the first instance, although it is perfectly reasonable to simplify further later according to more specific conditions or available resources. From a practical point of view, the collisional radiative separation (we use the name 'generalised collisional-radiative' when we include the complete set of metastables) provides effective ionisation coefficients, effective recombination coefficients etc. which link metastable population to metastable population and then effective emission coefficients which give the spectral emission possible from the quasi-static populations once the metastable populations are known. Figure 18 illustrates the implementation of these ideas in ADAS as used for JET divertor modelling and analysis.

VII DISCUSSION AND CONCLUSIONS

The level of complexity involved in metastable resolution certainly matters for ionisation stages which are targetted for spectroscopic diagnostic use, but may be excessive for less important ionisation stages and simpler modelling. This is especially true for heavy species with many ionisation stages.

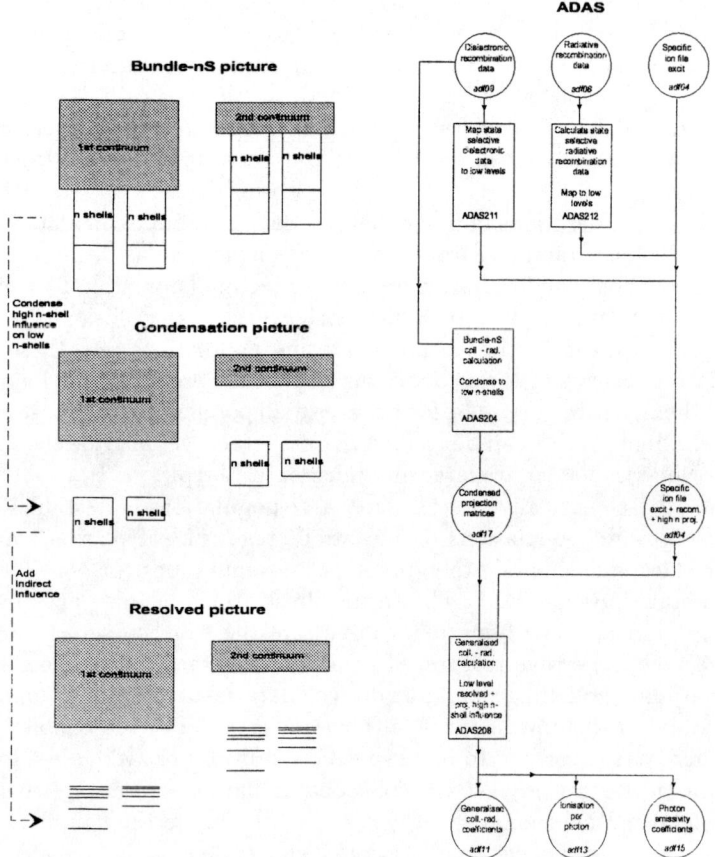

FIGURE 18. Schematic of mapping state-selective radiative and dielectronic recombination coefficients to specific ion files of type *adf*04 for low level resolved population calculations and to bundle-nS population models for the very many n-shell analysis. The lower section of the schematic shows the condensation of the results of the bundle-nS model onto low levels via condensed projection matrices (*adf*17) and then their mapping into the final high-level-influence-supplemented, resolved, low level, collisional-radiative calculation in which all the required metastable-resolved generalised derived data of table 1 are generated.

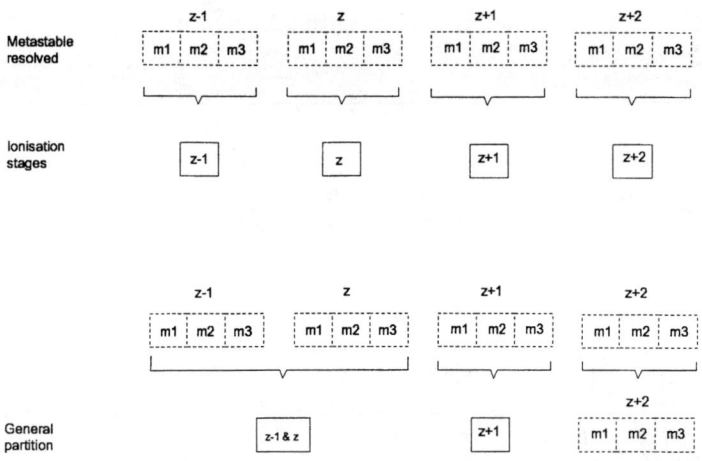

FIGURE 19. Schematic of flexible partitioning and condensing of metastable resolved collisional-radiative coefficient data as appropriate for the modelling and analysis.

To allow practical computations with reasonable execution times and storage requirements, it is helpful to reduce the effective number of populations (stages, grounds or metastables) to be handled. We have found it useful to arrange our primary storage of metastable resolved collisional-radiative data by isoelectronic sequence as indicated above. A separate computational step then gathers together those ions from the iso-electronic sequence archives to assemble an iso-nuclear sequence (that is the data for a particular element). This gathering step can simultaneously implement a condensation of the data in a manner tuned to the intended use. A schematic is shown in figure 19. We define a partition of the complete list of metastable states which specifies a grouping. The metastable resolved equilibrium ionisation balance can be used to decide the proportions to be used in generating a new condensed set of collisional-radiative data linking partition member to partition member. The simplest condensing partition is to combine metastables belonging to the same ionisation stage. Such a condensation restores the usual stage to stage data but with a sounder pedigree. At JET for light ions we have worked mostly with the fully metastable resolved partition and the stage partition.

It is satisfying to observe in the illustrations given here and studies elsewhere that extensive and sophisticated fundamental atomic data are being brought into play in analysis of fusion plasma diagnostic measurements. However, this does require exploitation of the collisional-radiative method, a structured approach to fundamental data and the provision of appropriate derived data at the point of analysis. At JET, we have found that ADAS has helped in these matters. Through its shared use and development with associated laboratories operating different machines, it has strengthened interpretation.

REFERENCES

1. Anderson, H. *in preparation* (1997).
2. Badnell N. R. *J. Phys. B* **19**, 3827 (1986).
3. Bliek F. W., Hoekstra R., Olson R. E. et al. *J. Phys. B* - in preparation (1997)
4. Boileau A., von Hellermann M., Mandl W. et al. 1989 *J. Phys. B* **22**, L145 (1989).
5. Breger P. and Vlases G. *JET Joint Undertaking Report* JET-TN(91)04 (1991).
6. Burke P. G. and Berrington K. A. *Atomic and Molecular Processes: An R-matrix Approach* (IOP Publishing Ltd.) (1993).
7. Duxbury G., Summers H. P. and Stamp M. *JET Joint Undertaking Report*, JET-P(97)18 (1997).
8. Griffin D. C., Pindzola M. S., Shaw J. A., Badnell N. R., Summers H. .P and O'Mullane M. *J. Phys. B* **30**, 3543 (1997).
9. von Hellermann M. *Atomic and Plasma-Material Interaction Processes in Controlled Thermonuclear Fusion* - ed. R K Janev and H W Drawin, p135 (Elsevier, Amsterdam) (1993).
10. Hoekstra R., et al. *these proceedings* (1997).
11. Maggi C. F. M. *Ph.D. Thesis* University of Strathclyde (1997).
12. Maggi C. F., Elder J. D., Fundamenski W. et al. *J. Nucl. Mat.* **241-243**, 414 (1997).
13. Seaton M .J. *J. Phys. B* **20**, 6363 (1987).
14. Summers H. P. *Adv. At. Mol. and Opt. Physics* **33**, 275 (1994).
15. Summers H. P. *Atomic Data and Analysis Structure* JET Joint (1994). Undertaking Report JET-IR(94)06; $http://patiala.phys.strath.ac.uk/adas/$
16. Summers H. P. and Dickson W. *Recombination of Atomic Ions* - ed. W G Graham, p31 (Plenum, New York) (1992).

The National Ignition Facility and Atomic Data

David H. Crandall

Office of Inertial Fusion and the NIF, Defense Programs
US Department of Energy, Washington DC, 20585

Abstract: The National Ignition Facility (NIF) is under construction, capping over 25 years of development of the inertial confinement fusion concept by providing the facility to obtain fusion ignition in the laboratory for the first time. The NIF is a 192 beam glass laser to provide energy controlled in space and time so that a millimeter-scale capsule containing deuterium and tritium can be compressed to fusion conditions. Light transport, conversion of light in frequency, interaction of light with matter in solid and plasma forms, and diagnostics of extreme material conditions on small scale all use atomic data in preparing for use of the NIF. The NIF will provide opportunity to make measurements of atomic data in extreme physical environments related to fusion energy, nuclear weapon detonation, and astrophysics. The first laser beams of NIF should be operational in 2001 and the full facility completed at the end of 2003. NIF is to provide 1.8 megajoule of blue light on fusion targets and is intended to achieve fusion ignition by about the end of 2007. Today's inertial fusion development activities use atomic data to design and predict fusion capsule performance and in non-fusion applications to analyze radiation transport and radiation effects on matter. Conditions investigated involve radiation temperature of hundreds of eV, pressures up to gigabars and time scales of femptoseconds.

STATUS OF THE NIF

The idea of using lasers to drive small amounts of deuterium and tritium mixtures to fusion conditions was born in the 1960s and began development in the 1970s. The basic concept of radiation drive of matter to attain extreme pressure and temperature had been applied, through the leadership of Edward Teller, to make two-stage thermonuclear weapons. Ideas for laboratory inertial confinement fusion combined knowledge of the performance of the secondary portion of such weapons with the conceptual ability of the laser to concentrate energy in space and time. Consequently, in the US, the development of inertial confinement fusion has been under the sponsorship of Defense Programs at the Department of Energy since the early 1970s.

Today, the National Ignition Facility (NIF) and the inertial confinement fusion program are a cornerstone of the nation's Stockpile Stewardship program. The President and the Congress have mandated a Stewardship program to preserve the technical and intellectual capabilities of the US in nuclear weapon science while using this science basis to maintain the safety and reliability of our nuclear weapon stockpile in the absence of nuclear weapon testing. The mission of the NIF further explicitly includes contributions to fusion energy, to high energy density science (including astrophysics), to testing the effects of radiation on strategic systems, and to technology development for national economic strength. The NIF has been designed to direct high intensity radiation onto various material targets in flexible ways that address these multiple missions.

The NIF is a product of the Department of Energy's inertial confinement fusion program with technical construction capability centered at Lawrence Livermore National Laboratory under the direction of Jeff Paisner. The NIF is expected to cost $1.2 billion with funding appropriated on an annual basis by Congress. Through fiscal year 1998, 42% of this funding has been provided and the current fiscal year 1999 request of $291 million would raise the appropriated amount to 66% of the total. Construction is within cost and on schedule to be completed at the end of fiscal year 2003. Included in the project team are the Department of Energy's Oakland Operations Office, Sandia National Laboratories, the Los Alamos National Laboratory, the University of Rochester's Laboratory for Laser Energetics, several engineering companies, a number of construction firms, and companies that produce laser and optical components.

Figure 1 gives the engineering layout of the stadium-sized NIF and an aerial photograph of the construction site on January 13, 1998. The 192 lasers are in bundles of 4 high by 2 wide, additionally packaged into two groups of 6 such bundles in each of two laser bays. Each laser beam will have a 40 cm by 40 cm square cross section during the primary amplification by multiple passes through 16 large neodymium-doped glass slabs. After transport to the switchyard near the target chamber, 4 beams are combined into one so that a total of 48 laser beam lines are brought to the target. The beams can be directed to the target as shown from top and bottom or reconfigured to be brought to the target in a spherically uniform arrangement.

For most experiments the red laser light (1ω) is tripled in frequency to blue light (3ω) as it enters the chamber; provision can also be made for bringing in red light or frequency doubled green light. Figure 2 shows the full energy and power that could be delivered by the NIF as blue light (3ω). Of course the power is just the energy divided by the pulse duration. The intent is to compress the energy in space and time to achieve the high energy density needed for fusion and other nuclear weapon physics.

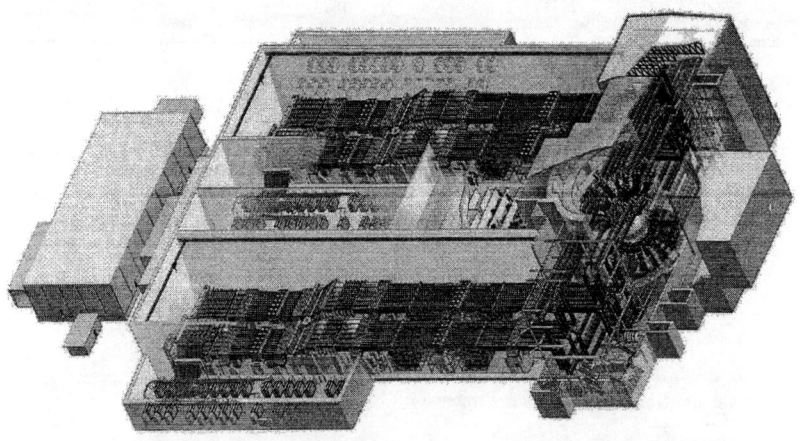

FIGURE 1. The engineering layout of the NIF and an aerial view of the construction site from optics assembly area at the left to target area at right with laser bays in the center.

For this audience, greatest interest is in the underlying features of the high energy density science to be enabled by the NIF and the details of inertial fusion approaches. The first edition of the Facility Use Plan of the NIF (1) provides a general background on these subjects and the review article by Lindl (2) provides the scientific description of the baseline indirect-drive inertial fusion approach. Other technical papers give underlying technical concepts for additional uses of the NIF, for example for direct drive inertial fusion (3) and for radiation effects tests (4).

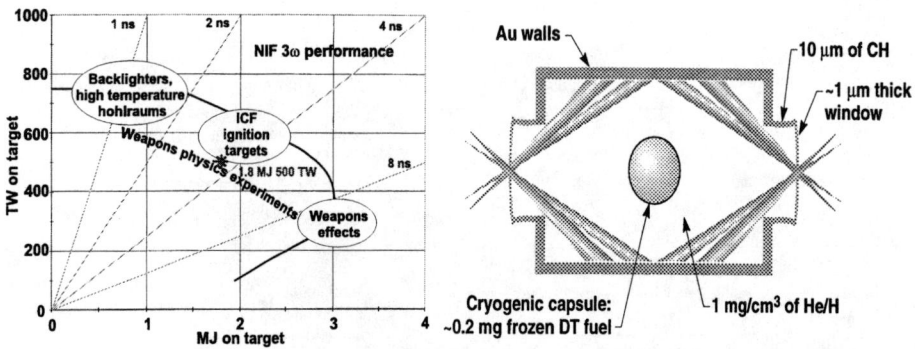

FIGURE 2. The energy and power to be produced by the NIF and the types of experiments anticipated. A typical ignition target is also shown with central spherical capsule having 1 mm diameter.

NIF AND ATOMIC PHYSICS

Atomic physics plays important roles in all of the high energy density conditions to be obtained in the NIF; only a few examples can be cited here. The transport and conversion of radiation in various colors or temperatures of light is the heart of the use of the NIF. Atomic collisions are also important for some of the applications anticipated, but only examples of atomic data involving light will be shown here. Opacity (the impedance to flow of light) is a critical property of high energy density systems or physically extensive systems such as found in astrophysics.

For indirect drive inertial fusion the laser light will enter a small cylinder (called a hohlraum) in the manner depicted in Figure 2. The inside walls of this cylinder are typically made of heavy material such as gold which is to be heated by the NIF laser light to radiation temperatures of a few hundred eV. High opacity results in efficient conversion of the laser light to x-rays which in turn fill the cylinder providing the environment for inertial fusion capsule compression and for many of the other weapon physics experiments. Heavy materials are needed for the walls so that the opacity is high enough to efficiently convert the laser radiation to a high temperature radiating "wall". It is the x-rays from this wall at radiation temperature about 300 eV that uniformly illuminate the center capsule and cause the ablation that in turn compresses this capsule to the conditions of high density and central temperature needed to initiate fusion burn.

Gold has been the baseline material for the wall surface, in part because of its opacity. However, adding small amount of other materials could increase the opacity allowing some increase in x-ray temperature in the hohlraum for a given incident laser energy. Figure 3 shows an example calculated opacity for a "cocktail" wall of mixed gold and gadolinium. The trick is to mix materials so that low opacity in one is matched by high opacity in the other. This approach to a design change for ignition assumes that the input atomic data and calculated opacities are accurate spectrally so that highs and lows are matched as shown in the figure. The calculations suggest a Rosseland mean opacity for gold of 452 cm²/g, for gadolinium 177, and for a 50:50 mixture 808. The gold/gadolinium mixture proved qualitatively successful on the Nova laser in increasing the x-ray temperature (5) and subsequent calculations have suggested other candidates that could be even better (6). These designed experiments, and others involving opacity of the hohlraum walls or of sample materials exposed to hohlraum radiation or direct laser illumination, all rely on predictions of opacity that in turn rely on detailed atomic calculations for a wide range of heavy materials.

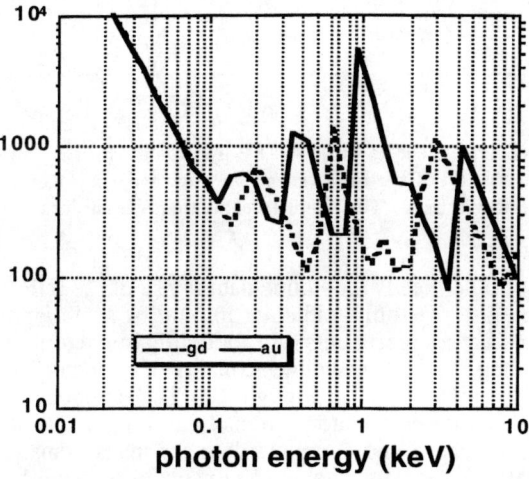

Figure 3. Opacity of gold (Au) and gadolinium (Gd) at electron temperature of 300 eV as a function of photon energy.

The physical environments produced by large lasers can also be used to measure opacity features in detail. Figure 4 shows an example absorption measurement on iron accomplished using the Nova laser (7). This measurement was conducted by placing a tamped sample of iron on the inside of a hohlraum where it is heated by the x-ray bath to uniform temperature and density. Spectrally resolved light transmission through the iron plasma is measured during a laser shot. For the observed spectra, the iron was a plasma at about 50 eV electron temperature and 0.001 g/cm line density. The absorption is calculated relying on correctly predicting the opacity of iron at these conditions. The large absorption above 60 eV photon energy is not correctly calculated unless 3 to 3, $\Delta n=0$ transitions of iron are included as was done in the Opal code to obtain the reasonable agreement shown in the figure for the calculated absorption. Opacity in this region of conditions for iron is important in astrophysics.

The NIF can be used to make such measurements in a larger range of materials and plasma conditions. For a given input laser energy, larger hohlraums give lower radiation temperatures: smaller ones give higher. The NIF should be able to make such measurements like those shown in Figure 4 even for uranium at conditions of interest for analyzing nuclear weapon performance.

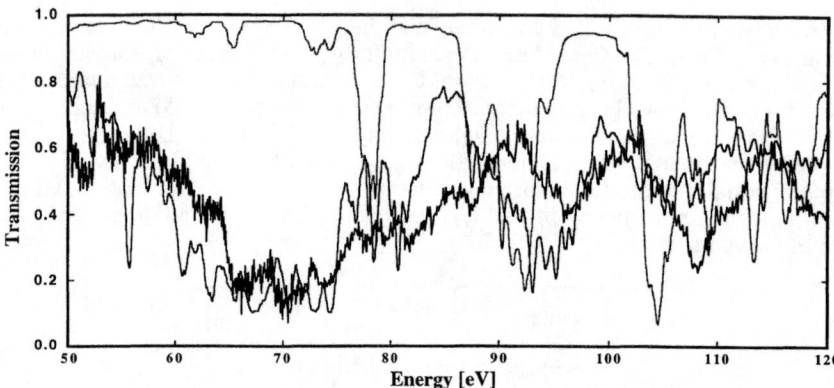

Figure 4. Absorption of light of energy 50 to 120 eV by iron at electron temperature of 50 eV as measured in the Nova laser (7). Experiment is the heavy line and calculation with the Opal code is the lighter line close to the data. The simpler calculation shown does not give reasonable agreement.

There are a number of other needs for atomic data in use of the NIF. As one further example, consider the use of atomic radiation line ratios as a diagnostic of fusion capsule behavior. Today's experiments and modeling suggest that subtle effects such as mix of imploding materials and dimensionality of evolving instabilities can be measured if line ratios are known well enough (8). These applications demand the best possible atomic data for selected atomic transitions in particular atomic species in contrast to the need for broad and complete data for many transitions in opacity predictions. Thus atomic physicists should consult closely with inertial fusion scientists if they plan to produce or improve atomic data of particular value in inertial fusion.

SUMMARY

The National Ignition Facility is under construction at Lawrence Livermore National Laboratory. This $1.2 billion facility will deliver 50 times more energy to targets than any existing laser and will be a cornerstone of the US Stockpile Stewardship program. Through concentration of light energy in space and time and consequent radiation transfer and material responses, the NIF is to produce the highest combinations of pressure and temperature in matter ever available for laboratory experiments. These conditions are expected to be sufficient to achieve ignition of nuclear fusion in the laboratory for the first time.

Detailed atomic data for a range of materials will be used to design ignition experiments for the NIF. Experiments in the NIF will be able to test the validity of atomic data used in analyzing a range of physical environments. By spanning the range of conditions needed for tests of fusion energy, astrophysics and nuclear weapon detonations, the NIF will address long-term missions in national security and economic development for the US.

REFERENCES

1. Allan A. Hauer, Robert Kauffman, Ann J. Satsangi, Thomas Haill, Robert Cauble, and Theodore T. Saito, "Facility Use Plan of the National Ignition Facility" prepared for Defense Programs, Department of Energy by the Los Alamos National Laboratory, LALP-97-7 (1997).

2. J. Lindl, Physics of Plasmas, $\underline{2}$ (11), 3933 (1995).

3. Stephen E. Bodner, Dennis G. Colombant, John H. Gardner, Robert H. Lemberg, Stephen P. Obenschain, Lee Phillips, Andrew J. Schmitt, John D. Sethian, Robert L. McCrory, Wolf Seka, Charles P. Verdon, James P. Knauer, Bedros B. Afeyan, and Howard T. Powell, "Direct-drive Laser Fusion: Status and Prospects", to be in Physics of Plasmas, (May 1998)

4. Christina A. Back, Christopher D. Decker, Gregory J. Deepest, Michael Gerassimenko, Robert A. Managan, Franklin J.D. Serduke, Gregory F. Simonson, and Laurance J. Suter, "High Power Laser Source Evaluation", Lawrence Livermore National Laboratory, UCRL-ID-129096 (1997).

5. T. J. Orzechowski, M. D. Rosen, H. N. Kornblum. J. L. Porter, L. J. Suter, A. R. Thiessen, and R. J. Wallace, Phys. Rev. Lett., $\underline{77}$, 3545 (1996).

6. D. Columbaut, M. Klapisch, and A. Bar-Shabur, Phys. Rev. E, $\underline{57}$, 3411, (1998).

7. P. T. Springer, D. J. Fields, B. G. Wilson, J. K. Nash, W. H. Goldstein, C. A. Iglesias, F. J. Rogers, J. K. Swenson, M. H. Chen, A. Bar-Shabur and R. E Stewart, Phys. Rev. Lett., $\underline{69}$, 3755 (1992).

8. Steven H. Langer, Howard A. Scott, Christopher J. Keane, Otto L. Landen and Michael M. Marinak, J. Quant. Spectrosc. Radiat. Transfer, $\underline{58}$, 709 (1997).

Atomic Data and Methods for Low-Pressure Discharge Lamps

Timothy J. Sommerer

Corporate Research and Development, General Electric Company
P. O. Box 8, Schenectady, New York 12301 USA

Abstract. The development of a mercury-free replacement for conventional fluorescent lamps is used to illustrate the role of and need for atomic and molecular data in developing and improving light sources.

INTRODUCTION

General-purpose lighting consumes about one-quarter of all electricity produced.[1] Hence there is an interest in improved lighting technologies, both by individual consumers motivated by aesthetics, economics, and safety, as well as the world in general, where the motivations might be reduced greenhouse gas emission and increased worker productivity, to name a few. The term "general purpose" light source is intended to recall lighting systems used to illuminate areas in homes, offices, factories, shops, stadiums, and roadways, as distinct from specialized light sources such as industrial lasers and ultraviolet water treatment systems.

This purpose of this paper is first to briefly introduce the technologies contained in existing general-purpose light sources and to describe the ways in which the lighting industry measures lamp performance. We will see that all high-efficiency light sources are presently based on some form of electric discharge plasma, either "low-pressure" such as fluorescent lamps or "high-intensity discharge" such as roadway and stadium lighting. I will then focus on a recent project to develop a completely new fluorescent lamp, work which forced us to revisit the entire process of developing a new light source. This story will serve as an example to illustrate the thinking process which goes into developing new light sources, and to highlight the important role of available, high-quality atomic and molecular data.

I want to clarify a few points before proceeding much further. First, the background information I will present is common knowledge to those associated with lamp development, and information specific to the development of a mercury-free fluorescent lamp is in the existing archival literature. This paper is intended to inform technically skilled people in atomic and molecular physics and allied fields about the issues encountered when developing new light sources, and to provide some usable entry points into the literature. Second, we in the business tend to use the word "lamp" to refer to the light-emitting bulb itself, and then use the word "light" to refer to the emitted visible spectrum of the lamp. Finally, I will try to avoid getting lost in details, while noting that it is difficult at times to write informative generalizations

---

[1] This is a US statistic.

without glossing over significant exceptions and details. For a more complete introduction to the topic I will refer you to a recently revised book (1) for a general overview of all aspects of lighting, from the technology behind lamp operation to electronic control gear, fixture optics, methods of characterizing lamp output and color, and information about human visual perception. Another introductory source is an article which appeared in *Physics Today* on the 100th anniversary of Edison's demonstration of the carbonized thread lamp (2).

Types of lamps for general-purpose illumination

Lamp developers tend to think of lamps in the following categories: (i) incandescent, (ii) low-pressure discharge, and (iii) high-intensity discharge or HID. The incandescent lamp is Edison's lamp, where a filament of material is resistively heated to incandescence by the passage of electrical current. The incandescent lamp is known for its pleasing appearance to most human eyes and its low cost, but also for its relatively short life (1000 hours) and low efficiency (15 lumens/watt, to be defined below). Incandescent lamps typically convert only five percent of input power into visible light. Over the years there are have been variations and improvements on the theme such as "halogen" and "krypton" lamps, but the light-producing mechanism is unchanged. Atmospheric molecular chemistry actually has significant overlap with the gas-phase processes occurring in a halogen lamp, but I will not be covering this topic here; if you are further interested in this topic, Ref. (3) is a recent starting point in the literature.

All lamps for general purpose illumination, other than incandescent lamps, are based on some sort of electric discharge plasma, and hence are referred to as "discharge lamps." Discharge lamps are then further divided into "low-pressure" discharge lamps and "high-intensity-discharge" or HID lamps (4). Low-pressure discharge lamps are the topic of interest here, with the most familiar example being the fluorescent lamp.[2] Low-pressure discharge lamps are characterized by relatively large size, low surface brightness, and an operating temperature not too far above the ambient air. The discharge plasma itself is a weakly ionized, nonequilibrium positive column which converts input electrical power into atomic resonance radiation. Fluorescent lamps are known for their high efficiency (approaching 100 lumens/watt), long life (10,000–20,000 hours, or 1–2.5 years of continuous operation), and pleasing color.[3]

---

[2] The second type of low-pressure discharge lamp in general use is the low-pressure sodium lamp, used exclusively for roadway lighting and similar applications primarily in the United Kingdom, Belgium, and some other countries. The so-called SOX lamp emits nearly monochromatic yellow sodium D-line resonance radiation near 590 nm. For the present purpose the SOX lamp discharge is fundamentally no different than the fluorescent lamp mercury discharge, which emits nearly monochromatic mercury resonance radiation at 254 nm. SOX lamps are the most efficient general-purpose light sources, around 200 lumens per watt, about double the efficiency of fluorescent lamps.

[3] The color of a fluorescent lamp is determined not by the discharge itself, but rather by the phosphor. The most familiar fluorescent lamps contain a phosphor material which produces a pale bluish light from the luminescence of antimony and manganese ions; while perhaps "acceptable," the color is not

Since most people rarely see high-intensity-discharge (HID) lamps up close, we tend to think of HID lamps in terms of their application: streetlamps, indoor and outdoor stadium lighting, and the ceiling lamps in many factories and "big box" retail stores. HID lamps include high-pressure mercury (the blue streetlights), high-pressure sodium (the gold-yellow streetlights), and metal halide (the blue-white stadium lights). The discharges in these lamps operate at relatively high pressure, say 0.5–20 atmospheres, and hence are high-pressure arcs near local thermodynamic equilibrium. The key outward feature of these lamps is a very compact, high-brightness light source which operates in a quartz[4] or ceramic[5] envelope at temperatures of 600–1200°C. Besides compactness and brightness, HID lamps can be even more energy efficient and longer-lived than fluorescent lamps. HID lamps are finding increased use indoors as improvements are made to their color and as they are successfully scaled down to smaller "lumen packages" comparable to 100–300 watt incandescent lamps.

Figures of merit for lamps

Lamp properites of greatest interest to users are (i) total power consumption, (ii) total light output, (iii) lamp color, and (iv) color rendering. I will briefly describe each of these properties in turn in this section.

Total power consumed is probably the most familiar measure of a lamp, as we regularly refer to lamps as "a 100 watt incandescent" or "a 40 watt fluorescent." Note that it is not a measure of light output; a 40 watt fluorescent lamp emits nearly twice the light of a 100 watt incandescent lamp. The power stamped on a lamp is the power consumed by the bulb itself, and does not include the power consumed by any requisite electric control gear. For incandescent lamps there are no additional electrical components and a 100 watt lamp truly consumes 100 watts of electrical power from the wall plug.[6] Discharge lamps universally require some electrical circuitry, commonly known as a "ballast," for stable operation when connected to a constant voltage source. A ballast is required because all discharge lamps have a negative differential electrical impedance, and if powered directly from a wall socket, would draw a rapidly increasing current until limited by a fuse, circuit breaker, or some catastrophic failure. The other key requirement of the ballast is that it provide the starting sequence for the lamp.[7] Ballasts may be either "electromagnetic," where the current is limited by an inductor in series with the lamp across the line, or "electronic," usually consisting of a resonant power converter to operate the lamp at

---

sufficiently pleasing for many people to put such lamps in the living areas of their homes, and hence they are relegated to the basement or garage. Much improved light is possible through the use of modern phosphors based on rare-earth ions where white light is made by mixing red, green, and blue phosphors. Such "tri-color" phosphors are the ones used in retail stores and in "compact fluorescent lamps" which fit into conventional incandescent lamp sockets.

[4] The material is commonly referred to as quartz, but is actually vitreous silica and not crystalline.
[5] The only ceramic in wide use is densely sintered polycrystalline alumina.
[6] The lamp will consume its rated power only if connected to a source delivering the rated voltage.
[7] The starting sequence usually consists of some sort of higher-voltage transient. Starting requirements can account for a significant portion of the overall cost and complexity of a ballast.

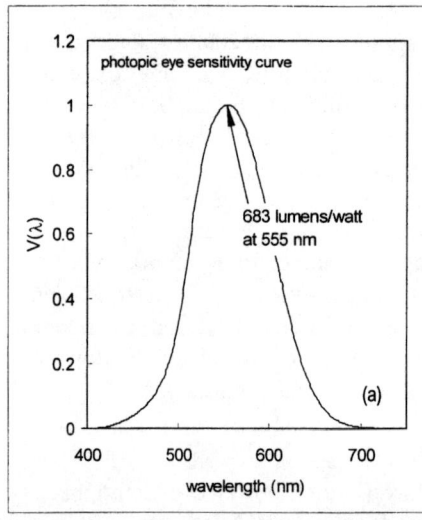 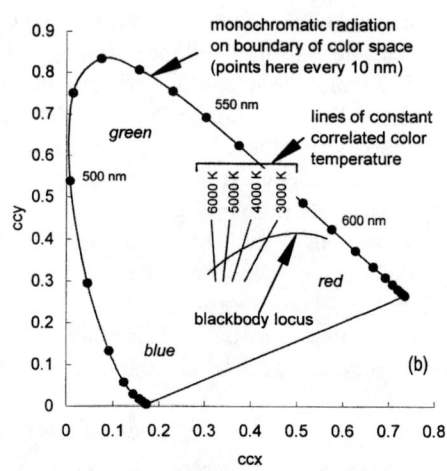

FIGURE 1. (a) Relative sensitivity of the human eye under bright illumination. (b) Chromaticity (color) space. The area inside the enclosed bullet-shaped outline corresponds to physically allowed colors.

higher frequency.[8] Ballasts ordinarily consume 10–20 percent of the lamp power, in addition to the power consumed by the lamp.

Total light output is measured in units of lumens; a 100 watt incandescent lamp emits about 1700 lumens. The lumen itself is one of the SI units, defined as 1/683 watt of monochromatic green light at a frequency of 540×10^{12} Hz. This frequency corresponds to a wavelength of about 555 nm, which is where the human eye is most sensitive under photopic (bright) illumination.[9] We then rely on psycho-physical testing of human subjects to determine the relative sensitivity of the human eye at other wavelengths. This weighting function is shown in Fig. 1(a), and has remained basically unchanged since it was first put forth in 1924.[10] Lamp efficiency is then defined as lamp output (lumens) per unit power input, and is commonly known as efficacy.

---

[8] For fluorescent lamps "higher" frequency usually means 20–100 kHz. Electronic ballasts can also provide all manner of specialized power control, starting sequences, and control functionality such as dimming.

[9] From this definition you can see that the most efficient possible light source would be one converting electrical power into monochromatic 555 nm light with unit efficiency, yielding an efficacy of 683 lumens/watt. The maximum efficacy possible from a "white" light source depends on how stringent one is in defining "white," but is usually quoted to be in the range 200–250 lumens/watt. The maximum efficacy of a perfectly efficient blackbody source is about 100 lumens/watt.

[10] There are other weighting functions appropriate for night (scotopic) vision, plant photosynthesis, and so forth. Unless otherwise specified, output (lumens) and efficacy (lumens/watt) are computed using the photopic curve characteristic of human response to high levels of illumination.

Lamp color can be described completely and compactly by a pair of scalars using a very large body of work based on psycho-physical knowledge of human color vision (5). The retina of the human eye contains four types of light-sensitive receptors: the rods, which give us our monochromatic night vision, and three types of cones, which give us our trichromatic color vision under bright illumination. It should therefore not come as a complete surprise that colors as perceived by the human eye can be described by three scalars. One of the scalars can then be dropped because perceived color is independent of intensity.[11] The result is a two-coordinate color space, for which a common manifestation is the x-y space shown in Fig. 1(b). The x-y pair is known as the "chromaticity" of a color. There is a special locus of x-y points in color space known as the blackbody line; as you might expect, these points correspond to the spectrum emitted by a perfect blackbody radiator at various temperatures. From this we get the concept of a "color temperature," by which we mean that the color of a source (lamp) is perceived to be identical to the perceived emission from a blackbody source at a given temperature. Color temperature can also be computed off the blackbody locus, in which case it properly goes by the name of "correlated color temperature" or CCT; however, the usefulness of the concept of a color temperature tends to diminish as you move away from the continuum emission of a blackbody radiator toward the monochromatic radiation at the edges of color space. Lamps for general purpose illumination are designed to fall near the blackbody locus.

Finally, color rendering is a measure of the accuracy of the perceived color of a selected set of 8 or 14 color samples when illuminated by a test light source, compared to the perceived color of the same samples under blackbody illumination. Note that it is the first measure I have listed which describes how an object, and not the lamp, will appear. It is entirely possible, for example, that a light source could have exactly the chromaticity of the sun, yet objects illuminated by this light source would appear distinctly different (and because of our prejudices, we would likely say "worse").

Lamp measurements are typically performed with the lamp placed at the center of an integrating sphere 0.5–3 m in diameter, the interior of which has a spectrally flat reflectance.[12] The spectrum can then be sampled through a hole in the sphere which is baffled to prevent direct viewing of the lamp. All of the lamp quantities discussed here (lumen output, chromaticity, color temperature, and color rendering) can be computed using weighting functions and straightforward procedures given the spectral output of a light source (*e.g.*, watts/nm) as a function of wavelength over the visible range 390–750 nm.

LOW-PRESSURE DISCHARGE LAMPS

If you were to remove the phosphor from a fluorescent lamp you would see the structure illustrated in Fig. 2. The lamp consists of a soda-lime glass tube with electrode structures hermetically sealed into the ends of the tube. The tube is filled

---

[11] There is ongoing work to quantify the effect of intensity on color vision, particularly under "mesopic" brightness levels where both the rods and cones are active.
[12] Barium sulfate is the usual choice to obtain a spectrally flat reflectance.

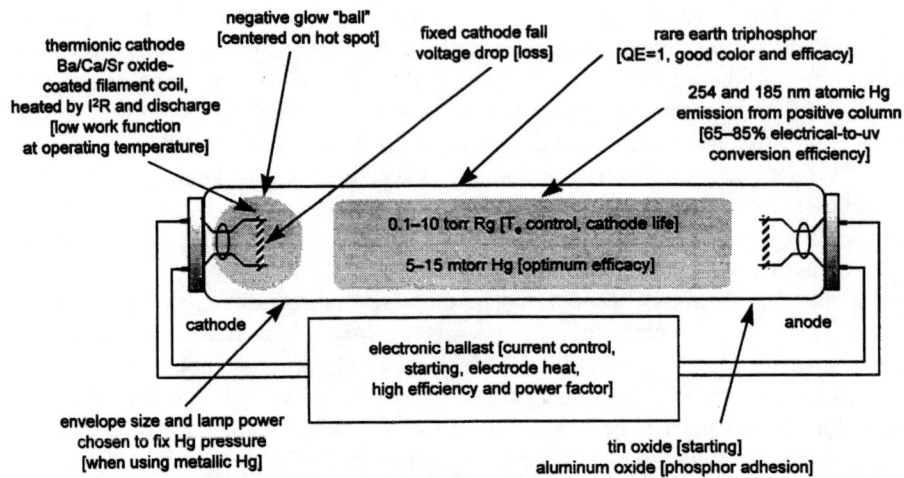

FIGURE 2. Schematic of a conventional fluorescent lamp. The plasma is shown as though the lamp were being operated on direct current or low-frequency alternating current, with the left-hand electrode acting as the cathode.

with a few torr of argon or some suitable rare-gas mixture, and it contains a drop of liquid mercury. During operation the mercury drop finds its way to the coolest location in the tube where it comes to thermal equilibrium with a saturated vapor of mercury in the volume of the tube;[13] the mercury vapor pressure is typically 5–10 mtorr.

Once a discharge is established in the tube, the ballast regulates the average current to a value of typically 200–400 mA. Under these conditions a weakly ionized positive column discharge develops between the two electrodes which converts input electrical power into nearly monochromatic ultraviolet mercury radiation with an efficiency in excess of two-thirds.[14] The electric potential drop along the positive column is on order of 1 V/cm. For these conditions there is negligible electronic excitation of the rare-gas atoms. In lieu of the phosphor the discharge itself is quite dim and appears to the eye as a pale blue characteristic of transitions among the excited levels of mercury.

The electrodes are a bit peripheral to this present topic, but briefly, they are usually tungsten coils coated with an "emission mix" powder and operated so as to emit electrons by thermionic emission. This arrangement is used to reduce the voltage drop between the plasma and the electrode to the minimum practical value, as this voltage

---

[13] The mercury in many newer lamps is introduced as a metal amalgam which is used to further control the mercury vapor pressure and to constrain the location of the mercury source in the tube.

[14] The primary emission occurs on the mercury $6^3P_1 \rightarrow 6^1S_0$ intercombination line at 254 nm, with an additional 10–20 percent emitted on the "true" resonance line $6^1P_1 \rightarrow 6^1S_0$ at 185 nm.

drop consumes power without producing significant ultraviolet or visible emission.[15] The "emission mix" is a mixture of barium, calcium, and strontium oxides. Elemental barium is formed during ordinary operation and provides the low-work-function surface characteristic of a clean surface coated with a monolayer of an electropositive metal.

The phosphor material is usually a refractory oxide doped with a small fraction of metal ions. The phosphor absorbs the ultraviolet radiation generated by the discharge and emits a spectrum of visible light. For example, the phosphor known as LAP is actually $LaPO_4$ doped with both Ce^{3+} and Tb^{3+}; incident ultraviolet photons are absorbed by the Ce^{3+} ions, which then transfer their excitation to the Tb^{3+} ions, which in turn emit strongly in the green according to the energy level and transition rates of the Tb^{3+} ion and the selection rules imposed by the ion site symmetry of the $LaPO_4$ host lattice. Atomic physicists could make significant contributions here; I will return to this subject later.

I said earlier that the discharge converts about two-thirds of the input electrical power into ultraviolet radiation. The phosphor itself then absorbs these ultraviolet photons and converts each incident photon into a visible photon with nearly unit efficiency; that is, the phosphor "quantum efficiency" is nearly 1. Unfortunately, the incident ultraviolet photon carries about 5 eV of energy, while the emitted visible photon carries just over 2 eV on average; hence the phosphor is only about 40 percent efficient in terms of energy. The overall energy conversion efficiency of a fluorescent lamp is therefore around one-quarter, corresponding to 80–100 lumens/watt.

At first glance the phosphor would appear to be the weak link in the energy conversion chain; indeed, there have been significant efforts to develop so-called quantum-splitting phosphors which produce more than one visible photon per incident ultraviolet photon using a controlled photon cascade process. None of these phosphors has been commercially viable. However, it should also be said that, so long as you accept 40 percent phosphor energy conversion, the fluorescent lamp has a very desirable property: the electrical-to-radiation conversion process performed by the discharge is largely independent of the perceived color properties of the emitted visible light. It is therefore possible to separately optimize each, and to develop lamps with different spectral outputs[16] without disturbing the discharge properties. This is in stark contrast with HID lamps, where close coupling between the thermal properties of the lamp, the electrical properties of the discharge, and the spectrum of the emitted visible light necessitates a significant redesign for each lamp wattage and color.

---

[15] It is for this reason that electroded fluorescent lamps tend to be long tubes; you want to make the fixed voltage drop at the electrode (about 15 V) a small portion of the positive column voltage, where the field is typically 1 V/cm, and hence the positive column voltage is about 100 V in a 1 m-long lamp.

[16] There are a myriad of reasons for precise yet subtle control over spectral output. You might wish to match incandescent color for a retail space where the two technologies are used together. Far East markets prefer high color temperatures comparable to sunlight (5000–6500 K). Commercial plant growers need a spectrum tuned for optimum photosynthesis. Even your local grocery store will use red-enhanced fluorescent lamps in the meat case to improve the appearance of red meat.

DEVELOPMENT OF A
NEW LOW-PRESSURE DISCHARGE LAMP

As can be seen in the previous section, fluorescent lamps contain mercury, and mercury is central to the operation of the lamp.[17] There is increasing concern, however, that mercury from spent lamps could leach into groundwater supplies when the lamps are discarded in solid waste landfills, or be emitted into the atmosphere if the trash is incinerated. Hence there has been significant effort and progress to reduce the amount of mercury in fluorescent lamps.[18] It is for these reasons that we have worked recently to develop a replacement for existing fluorescent lamps which is completely free of mercury. In the process we had to revisit nearly every decision and rationale for the features of a general-purpose lamp, and I hope that the story will serve as a good basis for understanding the thought process behind the introduction of new lighting technologies, the continuing improvement of existing light sources, and the role to be played by atomic and molecular data.

The first item to note is that Edison's incandescent lamp contains no mercury, so a practical mercury-free lamp has been around for more than a century. It is probably obvious that there may problems if we were to replace every mercury-bearing fluorescent lamp with a mercury-free incandescent lamp which consumes four or five times more power and has only a fraction of the useful life. What is perhaps more interesting is that, in the global view, replacing fluorescent lamps with mercury-free incandescent lamps would actually result in *more* mercury entering the biosphere (6). Fossil fuels contain mercury, and for the foreseeable future, fossil fuels will provide a major portion of the world's electric power. The first point here is that a global view is sometimes needed. The second point here is that any mercury-free replacement for existing fluorescent lamps should be efficient, ideally comparable to existing fluorescent lamps.

This is an exceptionally stringent requirement, but there yet another difficult requirement imposed by the practical aspects of introducing new lighting technology: compatibility with the installed base. Both lighting users (consumers) and lighting manufacturers have a large investment in existing technology which cannot be easily dislodged. No matter how ideal it may be, any completely new lighting system (by which I mean a new lamp combined with control gear and fixtures which differ from the installed base) will take years to gain widespread use. In particular, a new lamp must have a low selling price to become widely used, but it cannot be made at low cost until made in large volume, and that requires investment in new manufacturing

---

[17] A counterexample is batteries, which until recently contained significant mercury, but where mercury is peripheral to battery operation.

[18] Only about 0.1 mg of mercury vapor is required to give the desired 5–10 mtorr of mercury vapor pressure in a standard "four-foot" fluorescent lamp. However, several milligrams of mercury are needed to ensure sufficient "free" mercury is available to form a suitable vapor, as mercury tends to be chemically bound to the glass, phosphor, and metal surfaces inside the lamp during the course of normal operation. Beyond this, there is the daunting technical problem of reliably dosing small amounts of liquid mercury into lamps on high-speed production machinery. Until recently lamps contained typically 50–100 mg of mercury for the simple reason that this amount constitutes a "drop."

machinery. It is therefore imperative that any new technology be as compatible with the installed base of sockets, control gear, fixtures, and manufacturing equipment as is possible, in order to minimize the time for the new technology to gain widespread acceptance.

Mercury-free fluorescent

The requirements on our new lamp are beginning to form. We have decided that it must not contain mercury, and by logical extension, no other comparable environmental threat. Its efficiency needs to be comparable to existing high-efficiency white light sources, meaning its efficacy must be 80–100 lumens/watt. It needs to be as compatible as possible with the installed base of sockets and manufacturing equipment. Finally, we recognized that although most discharge lamps contain mercury, fluorescent lamps account for the vast majority of mercury in lamps, and therefore should receive our initial attention. So our goal is now clear:

To develop a mercury-free replacement for existing fluorescent lamps.

This is a very tall order. The odds are not in favor of meeting the performance of existing fluorescent lamps alone, and the odds of meeting the additional requirement that the new lamp be electrically compatible with existing sockets has to be close to zero. Nonetheless this outlandish goal is still worth stating, because it reminds us that every step we take toward making the new lamp compatible with existing user installations and manufacturing machinery will make it that much more likely that the new lamp a commercial success rather than a technological oddity.

Technical requirements

We are now looking for some viable path to a mercury-free fluorescent lamp. We need at minimum a mercury-free discharge—that is, a material to replace mercury and its ability to efficiently emit in the mid-ultraviolet. The first basic requirement on this material, which is self-explanatory, is that it needs to have a significant vapor pressure in a device operating near room temperature, say 1 mtorr. We can conceive of methods to manage heat so as to raise the temperature of a 1 m-long tube somewhat above ambient, but not by more than, say, 100°C. The second basic requirement is that the material form an *atomic* vapor. Experience with weakly ionized, low-pressure discharges suggests that, unless the emitting species is atomic, an excessive fraction of the input electrical power will go into excitation of vibrational and rotational modes, as well as molecular dissociation, and none of these processes is promising for emission of useful ultraviolet or visible radiation.

With only these two requirements the choice of materials is rather limited. First there are the rare-gases. Then there are metals which form a reasonable vapor pressure near room temperature: mercury, cadmium, phosphorous, selenium, and so forth, all of which carry similar environmental concerns. There is sodium, a strong visible emitter which is already used in both low-pressure and HID lamps, and the other alkali and alkali earth metals. Finally there are a few interesting longshots like sulfur, which is a known ultraviolet emitter.[19]

The final criterion I will mention is that the material should be chemically compatible with traditional lamp envelope materials such as glass, quartz, and alumina, electrode materials such as tungsten, and electrode mounting materials such as nickel and molybdenum. It is conceivable that a material like sulfur could be used, even though it is exceptionally corrosive to electrode materials, because the discharge could be electrodeless and excited by an antenna external to the bulb. However, while several commercial electrodeless lamps exist, the high cost of the power electronics, particularly the need to prevent electromagnetic interference, has to-date relegated these systems to niche applications.

Once we confine ourselves to the rare-gases, it is the phosphor which leads us to choose xenon. The resonance radiation wavelength of the rare-gases increases with atomic mass, from 74 nm for helium to 147 nm for xenon, corresponding to photon energies of nearly 17 eV down to about 8 eV. I previously described how, even in conventional, mercury-based fluorescent lamps, nearly half the input power is lost when the phosphor converts each 5 eV mercury resonance photon to a visible photon with an average energy of about 2 eV. The power loss is just that much greater if the incident photon carries even more energy; hence the reason we choose the heaviest stable rare-gas xenon and its 8 eV photon over the lighter rare-gases and their more energetic photons.

So now we have settled on xenon, but we already recognize that it may lead to a less-efficient lamp simply because the phosphor energy conversion efficiency will be no better than the ratio of the photon energies: 2 eV to 8 eV, or 25 percent. Recall that our goal is to meet or exceed the efficiency of existing mercury-based fluorescent lamps, about 25 percent. However, there is yet another possibility, which I alluded to before: quantum-splitting phosphors. Existing fluorescent lamp phosphors convert each ultraviolet photon into nearly one visible photon—that is, their quantum efficiency (QE) is close to unity. Many phosphors will emit more than one photon per incident photon through an Auger relaxation process in the host lattice itself (7), but only if the incident photon carries several tens of eV of energy, and hence the process is not useful in fluorescent lamps. However, there are a few examples of phosphors which will "quantum-split" under ultraviolet excitation, and could have quantum efficiencies in lamps in excess of unity. In particular, the phosphor host YF_3 doped with Pr^{3+} was shown in 1974 to emit 1.4 visible photons per incident 185 nm ultraviolet photon (8).

So while it is not certain, it is at least feasible to think of a new, mercury-free fluorescent lamp with overall electrical-to-visible conversion efficiency of 25 percent, comparable to existing mercury-based lamps. The xenon discharge would have to convert electrical power into ultraviolet radiation with an efficiency similar to conventional fluorescent lamps (about two-thirds), and the phosphor would have to convert each incident ultraviolet photon into an average of 1.4 visible photons.

---

[19] However, it is the sulfur dimer S_2 which emits ultraviolet radiation, not atomic S.

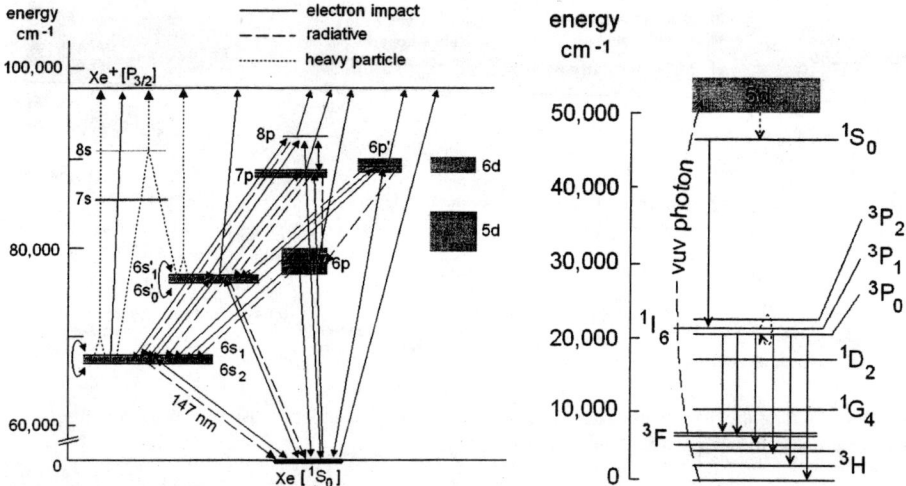

FIGURE 3. (a) Partial energy level diagram of atomic xenon, showing transition processes included in the model. From Ref. (13), used with permission. (b) Partial energy level diagram of trivalent praseodymium, showing a demonstrated quantum-splitting process for a phosphor where Pr^{3+} is doped into a YF_3 host lattice.

[13] Reprinted from *Journal of Physics D,* **29** 729 Timothy J. Sommerer, "Model of Weakly Ionized, Low-Pressure Xenon DC Positive Column Discharge Plasma," pp 772-775, ©1996 with kind permission of IOP Publishing Ltd.

Discharge diagnostics and modeling

Efforts to characterize, model, and optimize a weakly ionized, low-pressure discharge for lighting inevitably center around an accurate and absolute estimate of the emitted flux of ultraviolet radiation. The schematic energy level diagram for atomic xenon in Fig. 3(a) shows the important electronic energy level labeled $6s_1$; it is the dipole-allowed decay of this level to the ground state which results in the emission of the desired 147 nm radiation. It is this radiation which is absorbed by the phosphor, is shown in Fig. 3(b) and converted into visible light. The fact that xenon resonance radiation is deep in the vacuum ultraviolet adds an enormous complication to the experimental diagnostic work. Both the experimental diagnostics and computational modeling rely on the availability of atomic data: resonance line structure, level-to-level radiative transition rates, electron-impact excitation cross sections, and so forth.

Xenon atom data for the experimental measurements

Integrating sphere photometry is the traditional diagnostic of visible radiation from lamps, and it can be used at other wavelengths into the ultraviolet and infrared. However, it is not practical for use at 147 nm, both because the sphere would need to be purged of oxygen to prevent absorption of the short wavelength radiation, and also because the lamp itself would need to be made from a material with near unit transmission at 147 nm. We instead relied on two independent approaches to characterize 147 nm emission: emission spectroscopy, and absorption spectroscopy, which I will describe briefly here; the details can be found in Ref. (9).

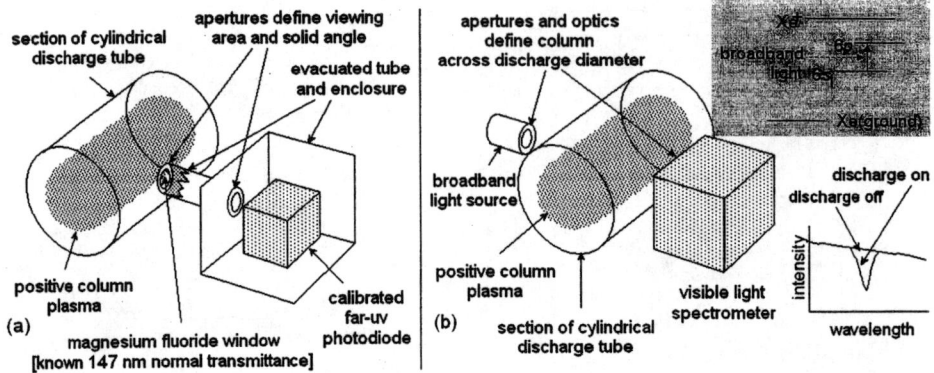

FIGURE 4. Schematic of (a) emission and (b) absorption diagnostics used to measure absolute ultraviolet emission from a positive column discharge.

A calibrated photodiode[20] was used for the emission diagnostic. It was connected as shown in Fig. 4(a) so as to view across a diameter of the positive column through a magnesium fluoride window set flush to the interior of the discharge tube. The volume between the window and the photodiode could be evacuated, eliminating the oxygen from the optical path, and the window transmission could be accurately characterized for normally incident 147 nm radiation.

However, in order to determine the total power emitted per unit area of discharge surface, one still needs to characterize the angular distribution of emitted radiation, as it is well-known that discharge emission is distinctly non-Lambertian (non-cosine). Fortunately, the angular distribution can be computed with high confidence through the use of a Monte Carlo simulation combined with the availability of crucial atomic data. and on The requisite data includes all the information needed to compute the detailed lineshape: isotopic splitting, hyperfine structure, and plasma-induced broadening mechanisms. Also needed are values for the $6s_1$ vacuum lifetime and the xenon isotopic abundances. The plasma-induced broadening mechanisms deserve special note, as they can dominate the emission lineshape. They include resonance broadening during collisions with xenon ground-state atoms (a well-characterized process), broadening caused by collisions with other atoms in the discharge[21] (which must be computed or measured for each collision partner species), and numerous broadening processes attributable to the presence of charged particles. Data for many of these plasma-based broadening mechanisms are not available; we were fortunate in that the poorly characterized processes could be neglected for the most part, based on reasonable guesses for the true values and some worst-case analyses.

---

[20] The photodiode was maufactured by Ball Aerospace and calibrated by the NIST Far Ultraviolet Physics Group.
[21] Nonresonant collisional broadening is not relevant in a pure xenon discharge, but it can be important in discharges in gas mixtures such as the rare-gas/mercury mixture found in conventional fluorescent lamps.

The second experimental approach, absorption spectroscopy, measures the line-averaged density of Xe($6s_1$) atoms across a diameter of the discharge, and combines this measured line density with a computational estimate of the effective lifetime of the $6s_1$ level to estimate the total decay rate of Xe($6s_1$) atoms, and hence the production rate of 147 nm radiation. In the experiment, illustrated in Fig. 4(b), Xe($6s_1$) atoms are excited to a selected higher-lying $6p$ level by a broadband light source external to the discharge. The absorption lineshape occurs near 828 nm, so it can be characterized using rather conventional spectroscopic techniques, modified by the necessity for sufficiently high resolving power. A curve-of-growth analysis is then used to convert the measured equivalent width of the absorption feature into a line-averaged Xe($6s_1$) atom density; this analysis requires the same detailed lineshape information as was needed to compute the angular distribution of resonance radiation: isotopic and hyperfine contributions, and collisional broadening mechanisms of all types.

The effective lifetime of the $6s_1$ level can be computed with high confidence both because of the sophistication of the theory, which includes incomplete spectral redistribution (10), and because the estimates can be compared with detailed measurements available in xenon for similar conditions. However, because the absorption diagnostic combines the measured density with the effective lifetime of the radiating level, the ultimate accuracy of this approach is directly dependent upon the accuracy of the vacuum lifetime of the Xe($6s_1$) level. For the purpose of evaluating a new discharge light source such as this, it is desirable for the uncertainty of the estimated efficiency or output to be less than, say, ±10 percent.

In the final analysis both techniques gave similar results for the 147 nm radiance of xenon positive column discharges, and both were in reasonable agreement with visible light photometry of complete lamps (that is, discharge tubes with suitably characterized phosphors), the predictions of the discharge model, and an overall analysis of the discharge power balance (11).

Xenon atom data for the computational model

For the discharge conditions in the positive column of a fluorescent lamp the dominant mechanisms for populating and depopulating a given atomic excited level are electron-impact collisions and radiative emission and absorption. Because of relatively low collision rates the electrons are distinctly out of equilibrium with the atoms and ions, and in fact are not even thermal. Significant computational effort is therefore expended to compute the distribution of electron energies in the discharge, and this distribution function is then combined with the electron-impact collision cross sections to determine transition rates between the various atomic levels, both discrete and continuum. The populations of most important excited atom levels are out of equilibrium with any temperature in the discharge, and hence the detailed kinetics must be modeled. The actual transport of electrons, ions, and atoms in the bulk plasma (that is, apart from the boundary layers at the tube wall and near the electrodes) can be adequately represented using fluid equations. Proper description of the radiative transitions is generally straightforward (12) so long as one accounts for the large optical depth of the plasma near resonance transition wavelengths,[22] and for moderate optical depth on some transitions which terminate on excited levels.

I want to walk through the sources of information we used in assembling a data set for the computational discharge model. The full details of the model, including both the numerical technique and sources for the data set, can be found in Ref. (13). Here I will try to give some sense for the sources ultimately used and not get bogged down in the specific references. I also want to mention that this data set is already a few years old, and some recent publications have added significantly to the available data.

Somewhat to our surprise, we found there to be little information even on the most basic process, the electron-impact excitation of ground state xenon. Several references report optical excitation cross sections (OECSs), but this information is not directly useful here in lieu of information about cascade processes. We therefore chose to begin with cascade-corrected cross sections for the analogous process in neon, and to scale them for use with xenon based on the OECS data. So in effect we assumed that the cascade correction for neon was applicable to xenon.

Electron-impact excitation and ionization of $Xe(6s_1)$ atoms are the dominant nonradiative loss mechanisms for this level; however, we could find no suitable xenon data for these processes, either. In the case of electron-impact excitation of $Xe(6s_1)$ we noted that the electronic configuration of this excited atom, with one $6s$ electron promoted to $6p$, is very similar to ground state cesium, and the literature does contain some information on electron-impact excitation of ground-state cesium. We used the cesium excitation directly for the xenon re-excitation process. In the case of electron-impact ionization of the $Xe(6s)$ and other excited levels we relied on semiempirical expressions which are also widely used to describe mercury ionization in fluorescent lamps. I note here that electron-impact ionization of the $Xe(6s)$ levels is sometimes the dominant ionization process in these discharges, and hence this process directly affects not only the generation of ultraviolet light, but also the electrical properties of the discharge.

The final electron-impact process I will mention is collisional "mixing" of important closely spaced xenon energy levels: $6s_1$, $6s_2$, $6s'_1$, and $6s'_0$. The high density of low-energy electrons, combined with the large cross sections for these nearly resonant transitions and the large optical depth of the resonance lines, allows these electron-impact collision rates to compete with other collisional and radiative rates. For instance, these mixing processes can "convert" an otherwise useless metastable $6s_2$ atom into a useful radiating $6s_1$ atom. For some discharge conditions we concluded that these levels were out of statistical equilibrium with each other, and hence a detailed kinetic treatment was required. However, we found only minimal data in the literature for these types of collision process. We therefore used temperature-dependent rate coefficient data and some cross section data for neon and argon, and then used some intuition to derive crude cross sections for xenon.

---

[22] The effective lifetime of the mercury resonance level 6^3P_1 in a fluorescent lamp is 50–100 times the vacuum lifetime. The effective lifetime in xenon is typically 1000 times the vacuum lifetime.

Note that to this point I have not mentioned *any* genuine xenon collisional data. There were only two collision processes for which multiple reliable data sources were found for xenon: electron-impact ionization of ground state xenon, and momentum transfer during elastic collisions of electrons with ground state xenon atoms.

Level-to-level radiative transitions are also important for determining the Xe($6s_1$) population, particularly the infrared emission generated during the radiative decay of Xe($6p$) atoms to the Xe($6s$) levels. Fortunately, sufficient lifetime and branching ratio data is available to properly compute these decay rates. Note that detailed level-to-level transition rates are used to compute the transition rate even when we choose to lump all the Xe($6p$) levels into a single composite level for computational expediency.

There are a number of additional process included in the model, but for which the data is either known or not critical. Good data is available to describe ion and neutral atom transport (14). Collisions between pairs of excited xenon atoms probably contribute to the overall ionization rate, but the contribution seems to be small, and we contented ourselves with setting the xenon value to the known rate coefficients for helium, neon, and mercury.

Tests of the xenon data set

Now that I have shown that the so-called xenon data set actually contains very little xenon data, it seems appropriate to show that the data set does in fact lead to realistic predictions. Two means were used to verify the data set: electron transport properties from the cross section set and actual positive column discharge properties. The comparisons are shown in Fig. 5.

Electron swarm experiments are designed to measure electron transport properties through specific gases and gas mixtures. Typical compiled data (15) are transport properties—electron drift speed, diffusion coefficient, ground state ionization coefficient, and so forth—as a function of the electric field E reduced by the gas density N. One can therefore use available codes (16) to compute transport coefficients from the assembled data set and compare the predicted transport coefficients with the experimental compilations.

The second method used here was to use the data set, in concert with the discharge model developed for this project, to predict positive column discharge characteristics, and compare with published values. Typical points of comparison are the axial electric field and the electron density and "temperature."

Even the combination of these two tests is not entirely satisfactory. Conditions in swarm experiments are deliberately chosen so that excited atom populations are negligible, electrons interact only with ground state atoms, and space charge can be neglected. The transport properties are therefore strongly dependent upon the electron-atom momentum transfer cross section, are weakly dependent on processes such as electron impact excitation, and are independent of collisions between electrons and excited atoms. The second test utilizing discharge properties has different strengths and weaknesses. On the plus side, a good match between predicted and measured discharge properties gives one good confidence in the overall data set and computational technique. On the minus side, it is difficult to account for any

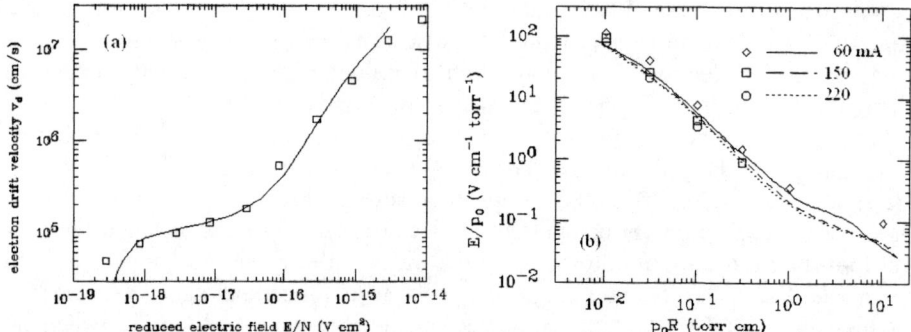

FIGURE 5. Validation of xenon data set against (a) electron swarm transport values (here the ion drift speed versus reduced electric field) and (b) measured positive column data (here the axial electric field versus the product of the discharge radius and the gas pressure). The numerous experimental data points were fitted with the lines shown, and the isolated data points shown here correspond to model predictions. From Ref. (13), used with permission.

Reprinted from *Journal of Physics D,* **29** 729 Timothy J. Sommerer, "Model of Weakly Ionized, Low-Pressure Xenon DC Positive Column Discharge Plasma," pp 772-775, ©1996 with kind permission of IOP Publishing Ltd.

mismatch in such comparisons, because the problem might be inaccuracies or gaps in the data set, or it might arise from approximations in the discharge model itself. A particular difficulty is that reliable discharge measurements for the conditions of interest are unlikely to be available; for even a seemingly mundane rare-gas discharge like xenon we relied on unpublished conference proceedings (17) where the current density was significantly above the level of present interest.

Phosphor development

Phosphor development is ordinarily thought of as the province of solid state chemists. However, there is at least one area where atomic data contributes significantly, and that is in the arena of atomic ion energy level data. Most modern lamp phosphors, along with phosphors used in certain other devices, are based on rare-earth ions doped into refractory metal oxides (18). The ion site symmetry in the host lattice determines the transition selection rules, and can also modify the energy level structure of the ion. However, the initial development of a phosphor, and much of the later understanding, is based on thoughtful consideration of the energy level diagrams of the atomic ions.[23]

Phosphor chemists are interested both in the energy level locations as well as their associated terms. The energy levels will determine where a given ion can absorb incident radiation; if the ion cannot absorb such radiation, energy level information will suggest other ions which can be used to absorb the initial photon, then transfer the excitation to the emitting ion in a near-resonant process. It is then the terms of the initial and final levels of the desired phosphor emission which lead to the selection of

---

[23] This is obvious once it is realized that the energy levels of "trivalent ion" of a solid state chemist are identical to those of the "triply ionized ion" of an atomic physicist.

a lattice with the proper ion site symmetry which will allow the desired emission and which will simultaneously inhibit undesirable competing decay processes.

Figure 3(b) illustrates the operation of a particular so-called quantum-splitting phosphor based on the Pr^{3+} ion (8). Ultraviolet light from the discharge excites ground state Pr^{3+} into the $Pr^{3+}(4f5d)$ band. The excitation decays nonradiatively (by phonons) into the $Pr^{3+}(^1S_0)$ level, which in turn radiates to the $Pr^{3+}(^1I_6)$ level, emitting a photon near 405 nm. A second nonradiative relaxation then couples the $Pr^{3+}(^1I_6)$ excitation with the $Pr^{3+}(^3P)$ levels, which subsequently decay to the $Pr^{3+}(^3F)$ and $Pr^{3+}(^3H)$ levels near the ground state, emitting blue and red photons. Note that the entire process occurs within the Pr^{3+} ion (as opposed to the lattice); hence the reason that knowledge of the ion energy levels by themselves can be so valuable.

The host lattice (here YF_3) determines the extent of the crystal-field-split $Pr^{3+}(4f5d)$ band. We know that the $^1S_0$ level must be lower than the lower edge of the $4f5d$ band to avoid direct luminescence of the $4f5d$ band to the low-lying triplet levels. We can predict that $^1S_0$ will couple strongly to $^1I_6$ because we know the selection rules set by the lattice ion site properties. We know that $^1I_6$ will likely couple to the ^3P levels because we know the energy level spacing and the phonon spectrum of the host lattice. We then once again draw on our knowledge of the selection rules to predict that the ^3P levels will decay to ^3F and ^3H.

Further energy level knowledge may help us to refine the phosphor. It may be that the phosphor does not absorb the discharge radiation wavelength. In that case we might peruse the energy level data for other rare-earth ions, looking for ions with energy levels which might both absorb the discharge wavelength and also resonantly transfer the excitation to the $Pr^{3+}(4f5d)$ or $Pr^{3+}(^1S_0)$ levels. Knowledge of the other rare-earth ion energy levels might also help us to improve the visible spectrum of this phosphor by suggesting another ion which could absorb the $Pr^{3+}(^1S_0 \rightarrow {}^1I_6)$ emission, which is nearly ultraviolet, and in turn emit a more useful green photon.

One would really need to consult an active phosphor chemist to determine which ions are of greatest interest at the present time. However, the general statement can be made that the trivalent (triply ionized) rare-earth ions are of great interest because of success in the recent years of turning these ions into commercially viable phosphors. The energy levels and terms are generally known for these ions from the ground state up to 40,000 or 50,000 cm^{-1}, but little is known for higher energies. The energy range of interest extends from these lower energies up to the bandgap of practical solids, say 75,000 cm^{-1}. It is knowledge of these higher-lying energy levels which may lead to the development of the first commercially successful quantum-splitting lamp phosphor

CLOSING REMARKS

I want to close with some comments about the usefulness of various data sources for modeling weakly ionized, low-pressure gas discharges such as lamps, as well as discharges used in the semiconductor processing industry and reactors used for all manner of thin film coating and processing. This is not a comment on the general validity of different data sources, only how some seemingly minor aspects can greatly affect the utility of the data for use in diagnosing and modeling weakly ionized gas discharge plasmas.

A proper discharge model generally requires data in sufficient detail to determine the atomic level-to-level transition rates, both collisional and radiative. Electron-impact collisions on neutral atoms tend to dominate discharge behavior, and electron-impact cross sections are needed for each important level-to-level transition. The energy range of interest is 0–20 eV for many important applications, although higher energy processes can be relevant. For electron-impact excitation processes it is important to have cascade-corrected cross sections, as opposed to optical emission cross sections. Electron-impact collisions with excited atoms can be of overwhelming importance in practical discharges, but cross sections for these processes exist for only a few select atoms. Electron-impact data which are not so useful are total electron-impact cross sections, which are typically useful only as a check, and detailed differential ionization cross sections with secondary energy or scattering angle. The discharge collisionality tends to erase any feature which might arise from detailed inclusion of this latter information.

Ion transport in such discharges can often be treated using ion mobility data, although some applications require a more detailed treatment of ion collision processes, notably elastic collisions and charge-exchanging collisions. However, it is generally not useful to have information such as state-selective charge transfer cross sections, for the same reason that the discharge collisionality tends to wash out many of the effects these processes might otherwise have.

Sufficient radiative data is necessary to compute level-to-level transition rates. For important processes such as resonance radiation decay it may be necessary to have detailed knowledge of the emission lineshape, sufficient to compute an accurate curve-of-growth. The required data include both atomic properties such as isotope and hyperfine splittings, as well as plasma effects: collisional line broadening by atoms, ions, and electrons, plus contributions from plasma collective effects such as electric fields.

We data users are generally not expert in the methodologies used by data producers, and would ideally want to rely on data producers for information to judge the validity and accuracy of published data. We are most grateful to occasionally find that a reliable third party has already performed the task of finding, critically evaluating, and publishing compiled data collections. In lieu of that, some general guidance and commentary on the reliability of certain data, methods, and analysis procedures would be quite valuable. Of particular note and difficulty is the vast amount of data which can be found in the untranslated (mainly Russian) literature.

We data users can effectively assist in guiding the generation of data by identifying important processes deserving careful investigation. We can also be quite adept at assembling a data set for a given physical problem, balancing the need for an exhaustive data set against shortages of available data and the need to keep the model sufficiently compact to aid understanding or reduce computational effort.

ACKNOWLEDGMENTS

Numerous people contributed to the mercury-free fluorescent lamp work; here I would like to acknowledge contributions directly related to the xenon atom data set. D. A. Doughty and J. Katine developed the initial data set. H. J. Oskam pointed out the unpublished rare-gas positive column discharge data, and C. W. Leiby kindly agreed to its use. C. C. Lin and Y. -K. Kim contributed unpublished cross section data.

REFERENCES

1. J. R. Coaton and A. M. Marsden, *Lamps and Lighting* (Wiley, New York, 1997). A large, applications-oriented reference is M. S. Rae, ed., *IES Lighting Handbook* (Illuminating Engineering Society of North America, New York, 1994).
2. John M. Anderson and John S. Saby, *Physics Today*, October 1979.
3. Laurence Bigio and Peggy Y. Chang, J. Electrochem. Soc. **141**, 2024 (1994).
4. Two recent introductory articles describe the operation of discharge lamps: J. T. Dakin, IEEE Trans. Plasma Sci. **19**, 991 (1991); and J. F. Waymouth, IEEE Trans. Plasma Sci. **19**, 1003 (1991). These references describe low-pressure and HID lamps, respectively.
5. G. Wyszecki and W. S. Stiles, *Color Science* (Wiley, New York, 1982); also http://cvision.ucsd.edu.
6. A quantitative analysis of this tradeoff can be found in R. Clear and S. Berman, J. Illum. Engr. Soc. **23**, 138 (1994).
7. J. K. Berkowitz and J. A. Olsen, J. Luminescence **50**, 111 (1991).
8. W. W. Piper, J. A. DeLuca, and F. S. Ham, J. Luminescence **8**, 344 (1974).
9. D. A. Doughty and D. F. Fobare, Rev. Sci. Instrum. **66**, 4834 (1995).
10. H. M. Anderson, S. D. Bergeson, D. A. Doughty, and J. E. Lawler, Phys. Rev. A **51**, 211 (1995).
11 Timothy J. Sommerer and Douglas A. Doughty, unpublished.
12. T. Holstein, Phys. Rev. **72**, 1212 (1947); **83** 1159 (1951).
13. Timothy J. Sommerer, J. Phys D **29**, 769 (1996).
14. The sources of information here were compilations of ion transport data by Ellis et al., Atomic Data Nuclear Data Tables **17**, 177 (1976); **22**, 179 (1978); and **31**, 113 (1984); and Lennard-Jones parameters compiled in a NASA Technical Report: R. A. Svehla, NASA Lewis Research Center Tech. Report R-132 (NTIS N63-22862), unpublished.
15. A particularly useful compilation appeared in J. Dutton, J. Phys. Chem. Ref. Data **4**, 577 (1975).
16. A particular code is ELENDIF: W. L. Morgan and B. M. Penetrante, Comput. Phys. Commun. **75**, 60 (1990).
17. C. C. Leiby and C. W. Rogers, Bull. Am. Phys. Soc. **13**, 211 (1968).
18. Relevant background information on phosphors can be found in K. Butler, *Fluorescent Lamp Phosphors* (Penn State, University Park, 1980) and G. Blasse and B. C. Grabmaier, *Luminescent Materials* (Springer, Berlin, 1994). However, these references include limited information on rare-earth-based lamp phosphors.

Understanding the High Intensity Discharge Lamp: The Need for More Data

Helmar G. Adler

OSRAM SYLVANIA
71 Cherry Hill Drive, Beverly MA 01915

Abstract. High intensity discharge (HID) lamps, in particular the metal halide (MH) lamps, comprise a group of light sources with high efficacy and good color rendering. The basic principle behind this lamp type is well known; however, improvements in light output and life of these lamps can only be achieved by studying detailed processes involving advanced diagnostics and modeling, and materials research. After a brief description of the behavior of a metal halide lamp, examples of research areas and diagnostic methods are shown with the emphasis on data needs.

INTRODUCTION

High intensity discharges (HID), especially metal halide lamps, play an ever increasing role in lighting applications. Their efficiencies and thus their economic value with regard to energy consumption can be very high - however, they still have not reached their theoretical limit (1). A better understanding of the basic physical processes in these lamps will lead to improved performances with respect to light output and lifetime.

The light output of a lamp depends critically on the fill chemistry, since radiating atomic and molecular species at the elevated temperatures of the discharge plasma are responsible for the radiation characteristics. The real art of lamp chemistry design is to introduce species with vapor pressures high enough to make significant contributions to the light output without introducing life-shortening chemical reactions with envelope or electrode materials.

Understanding of the chemical processes and transport mechanisms in these discharge lamps is essential for developing new lamps and fills. The interactions between the plasma and the electrodes, the plasma/gas and the wall materials, and the molten salts with the walls are complicated and need further research. All these processes can not be treated independently if a real lamp should be modeled and understood.

The modeling of a HID-lamp depends on a large number of different material functions and parameters. As a particular example, the emission of photons in the hot arc core, absorption and reemission on the way out strongly influence the energy-balance. Thus, broadening parameters of the spectral lines, especially resonance- and

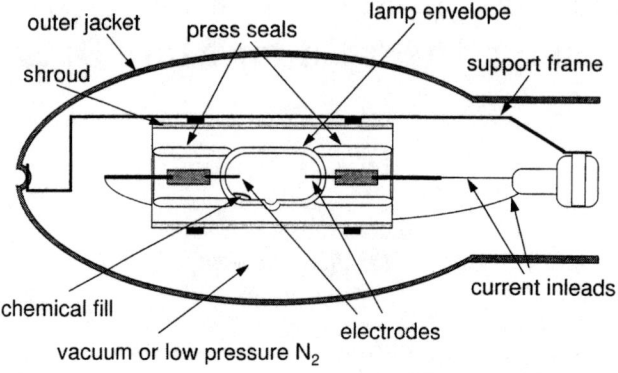

FIGURE 1: The basic construction of a high pressure Hg-lamp.

van der Waals-broadening parameters, have to be known to model the radiation transport correctly.

Probing the discharge plasma using spectroscopic methods is necessary not only for characterization of the lamp performance, but also to determine species distributions and temperatures as a function of location in the discharge and time. These measurements are essential for a quantitative understanding of the discharge mechanisms and for providing a test bed for simulation efforts. The advances in both computer power and sophisticated diagnostic techniques provide an opportunity for a better understanding of these lamps. This contribution will show examples of general problems and data needs in HID-research and focus on some diagnostic techniques.

Historical Development of the Metal Halide Lamp

The high pressure mercury lamp, developed in the 1960s, has been used for a fair number of years in the United States, in particular for street lighting applications. Figure 1 displays a simplified version of a Hg-lamp, displaying the most important parts of a working lamp. The actual light source consists of a more or less cylindrical discharge vessel with two pinch-sealed electrodes; in between an arc discharge is created by applying electrical power through the outer leads. In the simplest form, the vessel is filled with small amounts of mercury and an inert gas like argon at a pressure of 40-100 torr. For visible light the vessel is usually made of transparent vitreous silica also known as quartz, or more recently, of translucent ceramic material. The electrodes have to provide electrons by thermionic emission and thus consist of material like tungsten to withstand the necessary high temperatures. To lower the work function, doped tungsten or emissive coatings are used. The discharge vessel can be surrounded by a so-called shroud which acts as a protector against possible but unlikely explosions and - maybe more importantly - helps in the redistribution of heat. In addition, by using appropriate transmissive materials, it can have the role of a filter against the strong UV-radiation from Hg-lamps. The discharge vessel is usually mounted in an outer jacket which provides mechanical means for mounting, electrical feedthroughs and for example a screw-in socket. The volume between jacket and vessel can be held under vacuum or filled with nitrogen to circumvent or change convective cooling mechanisms which influence the wall temperature distribution of the vessel.

During the starting phase a low pressure glow discharge is formed, the electrodes are heated up. The - at room temperature condensed - Hg starts to vaporize, and the discharge develops from a glow to a high pressure Hg arc (of the order of a few bar). After a few minutes, when the arc is fully developed, typical arc core temperatures reach about 6000 K, the vessel wall temperatures are around 1000 K and electrode tip temperatures reach values of 2000 - 3000 K.

It is evident that at these elevated temperatures the material strains can be severe, in particular electrode erosion and chemical reactions at the discharge walls play life-time (typically about 10000 hours) limiting roles.

Despite the great success of this type of lamp in street lighting, however, the Hg-lamp shows a severe lack of spectral balance: There is no red emission in the visible spectrum (400-800 nm), the radiation is concentrated in only a few spectral lines below 579 nm. Figure 2 displays a typical midplane side-on, line-of-sight spectrum of a high pressure Hg-lamp at 5 bar. Objects illuminated by this light source seem to be strongly color shifted and take on a "ghostly" appearance. A good and bright "white" light source has to show good spectral balance and has to provide a large portion of emitted light around 555 nm - the position of maximum eye sensitivity.

To improve on the color rendering capabilities, measured as Color Rendering Index (CRI) [1], several ideas were developed. Attempts were made to use phosphors similar to fluorescent lamps. However, the overall conversion efficiencies for phosphors in these lamps turn out to be fairly low.

Metals usually have a large number of transitions in the visible region which, in principle, could fill up the regions in between the few visible Hg-lines. Thus it seemed to be almost natural to introduce metals as emitters into the discharges. The vapor pressures of the ingredients of a discharge fill are determined by the "cold-spot" - i.e. the coolest part of the inside of a discharge vessel. It turned out, that only a limited number of metals exist with vapor pressures high enough to make a substantial contribution to emission at typical cold-spot temperatures of 1000 K which are unreactive with silica. As examples one could think of introducing zinc or cadmium which provide stronger red emission. However, they radiate also heavily in the blue which leads to an overall low luminous efficacy (LPW) $\eta_v = \Phi_v / P$. Here Φ_v is the photopically weighted integrated spectral output power and P is the electrical power input. Another example, sodium, shows very severe reactions with silica. Nevertheless, changing the envelope material to polycrystalline alumina (PCA), which is Na-reaction resistant, led to the development of the high-pressure Na-lamp.

---

[1] CRI: measure of the degree of color shift objects undergo when illuminated by a light source as compared with the color of the same objects when illuminated by a reference source of comparable color temperature

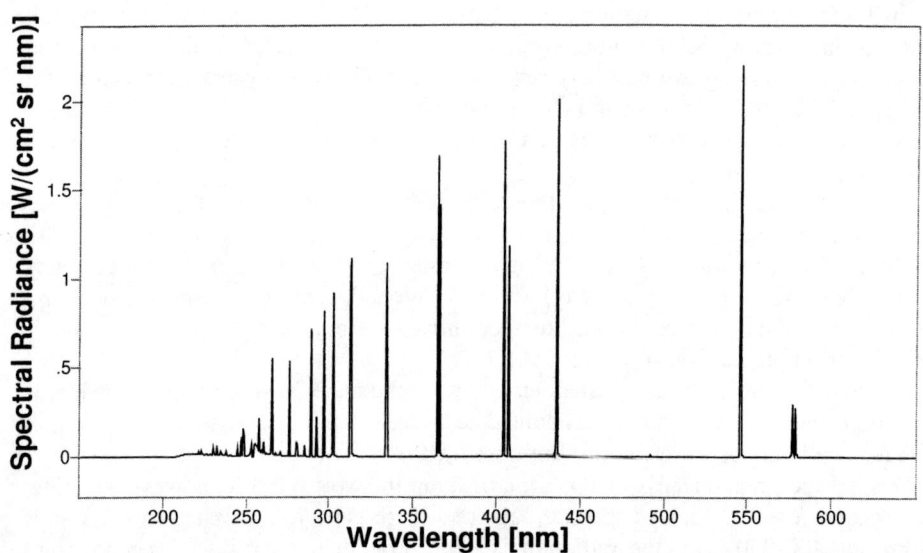

FIGURE 2: A line-of-sight spectrum of a high pressure (about 5 atm.) Hg-lamp.

The real breakthrough was achieved when metal halides, generally in the form of bromides or iodides, were used to introduce the metals into the hot arc core region. The idea was here that most metal halides are not very reactive with hot silica (except the fluorides). The metal halides vaporize at the walls, diffuse into the discharge plasma and dissociate into the atomic components, thus introducing a high concentration of metals into the arc.

The Metal Halide Lamp

To determine lamp industry typical values of CRI and LPW, measurements are usually performed by mounting the light sources in integrating spheres and measuring spectral power distributions (SPDs) with low resolution spectrometers. CRI and LPW are then calculated by defined calculation algorithms. However, to get an understanding of the physical and chemical processes that are occurring in the arcs, higher resolution measurements are needed, ideally with high spatial and time resolution. Figure 3 displays an experimental set-up which provides line-of-sight data acquisition capabilities. By replacing single detectors, very often photo-multiplier tubes at the exit slit of a scanning monochromator, by 1D- or even 2D-detector arrays (here we use a 2D intensified CCD-camera) and making use of spectroscopic imaging systems, we can gain tremendous improvements in data acquisition capabilities. In the following a series of lamp spectra will be shown.

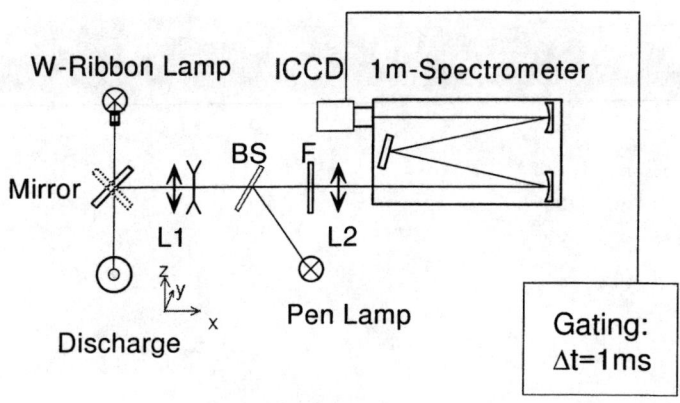

FIGURE 3: An experimental set-up to measure line-of-sight spectra with spatial and time resolution.

Atomic Radiation

The basic idea of the function of a metal halide lamp is well displayed in Figure 4, which shows the temporal development of a part of the visible spectrum during the first 150 s after starting the lamp, here measured with a low resolution spectrometer. It is apparent that in the spectral range displayed during the first 30 s a pure Hg spectrum develops with strong increasing radiation from the Hg 546 nm and 577/579 nm lines. After about 50 s the chemical fill salts are heated up enough to melt and vaporize. They diffuse into the arc core where they dissociate into, for example, Sc and Na. The spectral features of Sc and Na are growing, the intensities of Hg lines are decreasing, indicating a lowering in arc temperature (see the 577/579 nm line intensities which act as a temperature monitor), until after about 300 s the spectrum has reached a stable distribution.

Whereas only few Hg lines exist in the visible, the spectral gaps in between those are filled up by the large number of scandium lines in this lamp. The spectrum from 400 to 600 nm is shown in Figure 5, where the measured spectrum of a Na:Sc:Hg lamp (top) is compared with intensity/wavelength data for scandium, taken from the NIST data tables (2). The good resemblance between both spectra might not be very surprising, since the data of the NIST tables were determined from a scandium arc spectrum. However, it is obvious that good data sets can provide good information in regard to predictions for useful lamp chemistries. Modern lamp fills contain more and more rare earth chemistry. For those elements, however, good data sets are not as easily available. Only in the recent past, rare-earth elements like thulium and dysprosium have been addressed, by combining laser spectroscopic life-time measurements with branching ratio analysis from Fourier-Transform-Spectroscopy measurements (3).

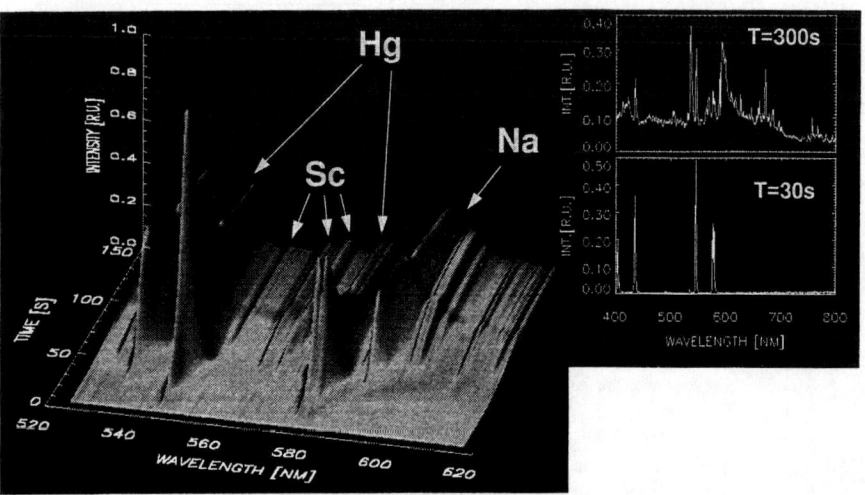

FIGURE 4: The temporal development of a metal halide lamp spectrum during the starting phase. The inserts on the right display the spectrum after 30s over the whole visible range, and the fully developed metal halide spectrum after 300s.

To show the vast amount of additional spectral lines a rare-earth fill can provide, Figure 6 shows the spectral region from 380 to 560 nm of a modern RE-MH-lamp. The few Hg-lines are blended in to show the strong filling-up of the gaps in between the Hg-lines. This translates to a very much improved color rendering index, reaching values of CRI=95 or above.

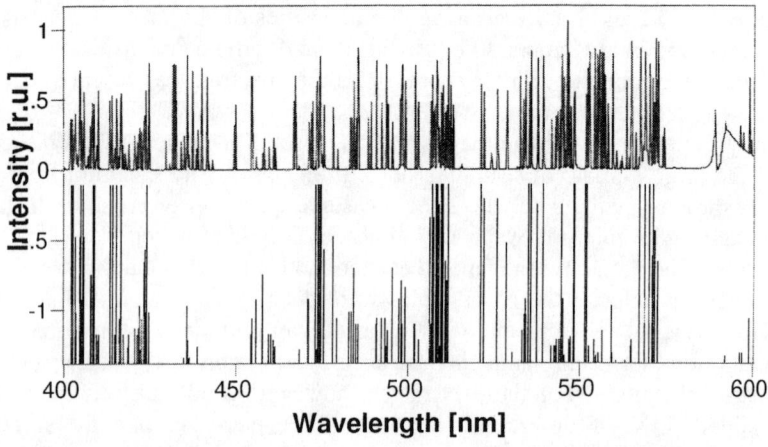

FIGURE 5: Comparison between a measured HID-spectrum (top) and Scandium data (bottom).

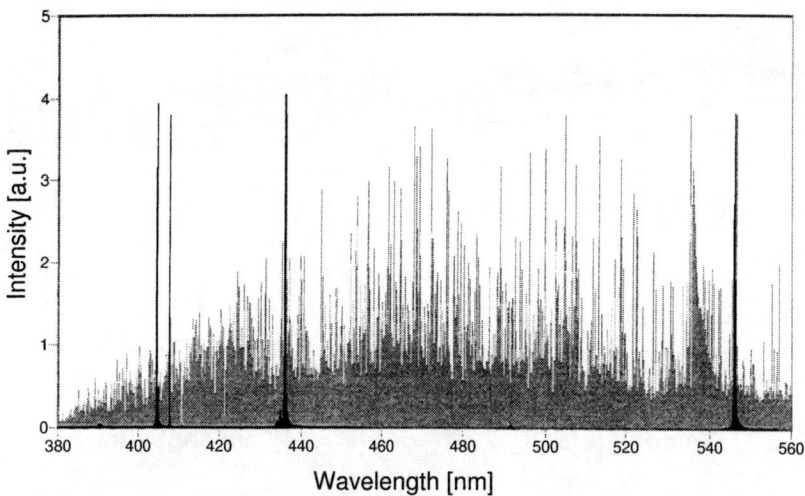

FIGURE 6: A modern rare-earth MH-lamp spectrum (gray) in comparison with a Hg-spectrum (black).

Molecular Radiation

Up to now we have only considered atomic line spectra. However, even a simple mixture like Hg-I-Ar gives rise to molecular spectral features. Thermodynamic calculations for this simple mixture result in densities for the different species present, shown in Figure 7 as a function of temperature (4). Radial gas temperature distributions have typical maxima of about 5000 K in the discharge core, and temperatures of the order of 1000 K at the walls. Thus there must exist temperature regions with molecules of sufficiently high densities to contribute to the emission from the discharge. An example is shown in Figure 8, displaying the HgI radiation around 440 nm. This radiation originates from HgI($B^2\Sigma^+$-Σ^2S^+) which is a fragment of the parent molecule HgI_2.

A modern metal halide lamp fill often contains a larger number of chemical elements, for example Hg, Na, Tl, I, Dy, Ho and Tm. If one starts to write down possible molecules from pure stochiometric arguments one can easily see how the number of species in the lamp is growing fast. To describe the lamp chemistry and the chemical transport inside the lamp completely might become an impossible task, since too many species with too many unknown parameters (e.g. vapor pressures and diffusion coefficients) are formed. Table 1 gives an impression of the number of simple molecules which might form from the chemical elements given above. The list could easily be extended, and the possibility of complex-formation (5) has not even been touched at this point.

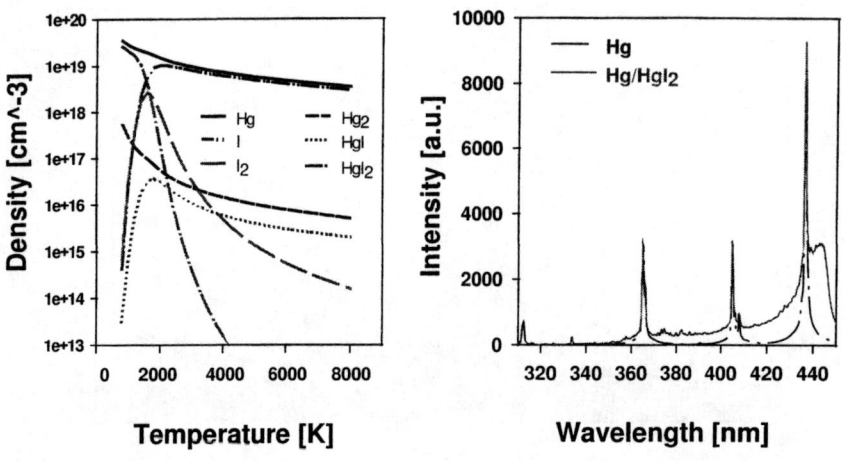

FIGURE 7 (left), **FIGURE 8** (right): Atomic and molecular species of Hg and I; on the left densities as a function of temperature, on the right spectra of Hg and of an Hg/HgI_2 mixture.

Electrodes

Electrodes play a very important part in the behavior of metal halide lamps beside the fact that they enable the energy input into the discharge lamp. During the starting phase, electrode material, usually tungsten, can be sputtered from the electrode surfaces and deposited on the walls. This "wall-blackening" can result in reduction of the light output and change of the thermal balance of the lamp by changing the temperature of the cold spot, thus changing vapor pressures and introducing color changes.

During steady state operation of the lamps, the electrodes work as thermionic emitters, thus producing the current for maintaining the discharge and heating the Hg and the salt melt. The heat and energy balance of electrodes and their intricate

Table 1: Possible stochiometric atoms and molecules in a lamp containing Hg, Na, Tl, I, Dy, Ho, Tm

Hg	Na	Tl	Dy	Ho	Tm
Hg_2	Na_2	Tl_2	DyI	HoI	TmI
HgI	NaI	TlI	DyI_2	HoI_2	TmI_2
HgI_2	Na_2I_2	TlI_2	DyI_3	HoI_3	TmI_3
Hg_2I_2	Na_3I_3	TlI_3	Dy_2I_6	Ho_2I_6	Tm_2I_6
I					
I_2					

interaction with the arc and the vessel have been an ongoing field of research. The electrode-plasma interaction is in many cases not well understood, and various theoretical and experimental studies have been reported. Electrode material parameters like work function and thermal or electrical conductivity are still of high interest, especially at elevated temperatures (typical electrode tip temperatures range from 1500-3000 K) and high pressures.

Figure 9 shows images of electrodes working in a vertically burning HID lamp. The top electrode (9a) is here in the anode phase, the bottom electrode (9b) in the cathode phase. The cathode shows a contracted discharge with high intensity and plasma temperature. This mode is usually referred to as the spot mode of operation. By increasing the current into the electrode(s) the spot mode usually goes

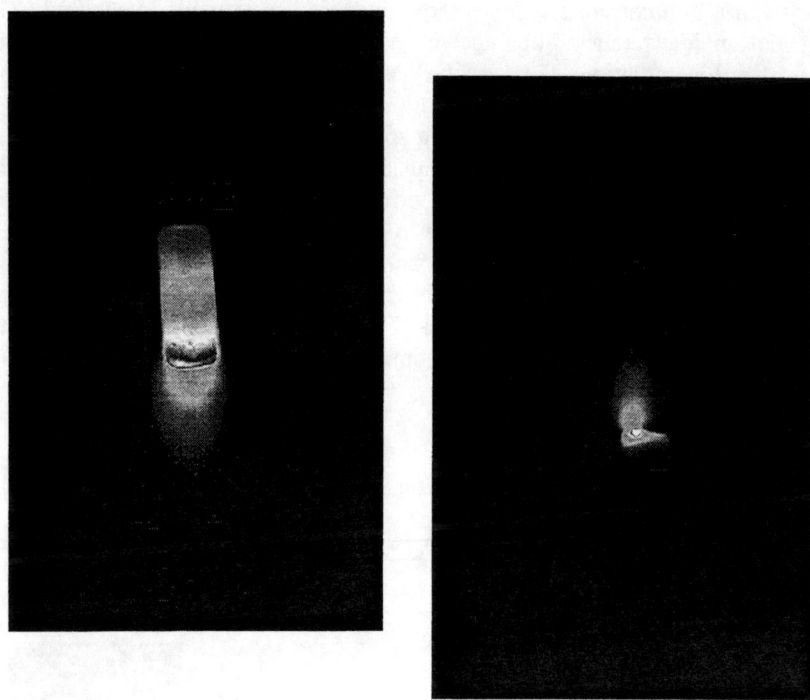

FIGURE 9. Images of electrodes in an HID-lamp: a) left: top electrode in anode phase, b) right: bottom electrode in cathode phase and "spot mode".

over into a diffuse mode of operation where the plasma seems to attach to the electrode over the whole electrode diameter and eventually, by further increasing the current, even engulfe the electrode tip. These different operational modes are still heavily discussed (6).

Diagnostics

Since we are talking about light sources, spectroscopy seems to be a natural way to examine the HID lamps. But we can do much more than determining LPW or CRI values which are the parameters most commonly used to describe and compare light source performance. These parameters are easy to obtain - each light source company has setups for standard measurements of those values, and they have become daily routine. However, if we want to understand the physical and chemical processes determining the performance of a lamp we need to look in a much more detailed way literally into the light source.

Modern HID spectroscopy diagnostics are able to supply us with spatial and time-resolved information about species temperature and density distributions. A combination of a 2-dimensional array detector system (for example intensified CCD-cameras) and an interference filter enables us to obtain time-resolved information from optically thin radiation. This can be used, at least in principle - after Abel-inversion - to determine temperature distributions of the emitting part of the discharge.

We can use a combination of an imaging spectrometer and 2D-detectors to obtain time-resolved spectral information as a function of one spatial dimension, either radially or axially. Figure 10 shows the principal setups. Figure 11 displays an example of the spectral region around 530 to 630 nm of a vertically burning metal halide lamp with Na and Hg components, operated with ac at 60Hz. The Na-doublet around 589 nm has been used extensively in the past for diagnostics of the Na-high-pressure lamp for Na density determinations (7). The top part of Figure 11 shows the intensity image at the exit of a spectrometer during the cathode phase of the bottom

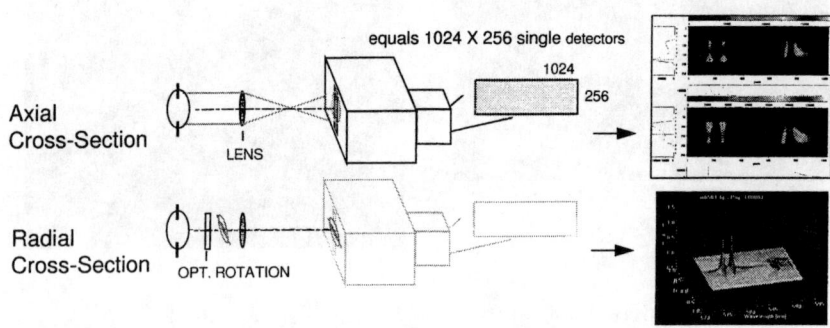

FIGURE 10: Spectroscopic setups with 2-dimensional detector arrays supplying spectral and spatial resolution; top: axial cross-section, bottom: radial cross-section measurements of lamp discharges.

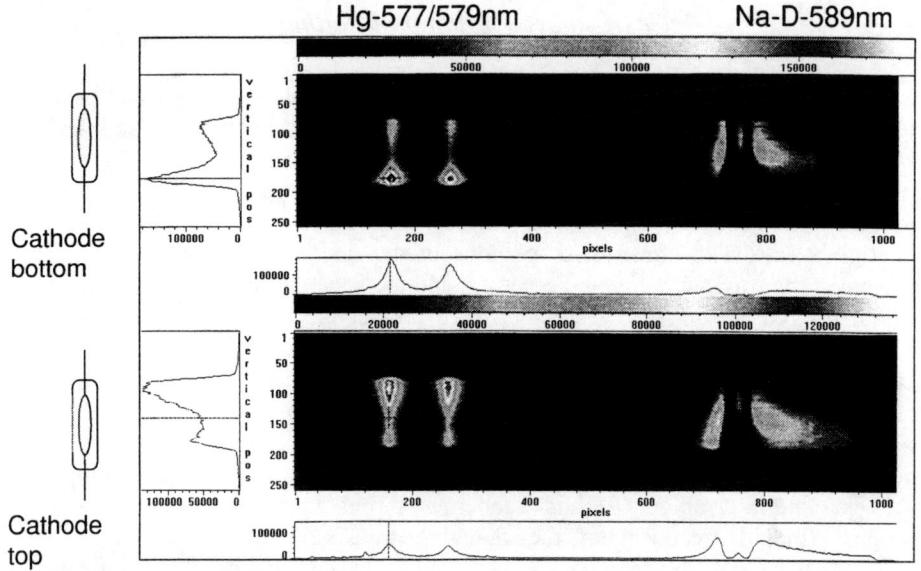

FIGURE 11: The output images of the axial cross-section set-up (Figure 10) with spectral features of Hg-577/579 nm on the left, and Na-D-598 nm on the right. The vertically burning lamp shows on the top a bottom cathode, on the bottom a top cathode. For further explanations see text.

electrode. The Hg lines at 577 and 579 nm are usually optically thin and can be used as a direct temperature indicator. The high plasma temperature at the contracted plasma hot spot in front of the cathode is easily seen. The Na-doublet shows – typical for this discharge type - strong self-absorption with merged self-reversed features of both spectral lines; the location of the peak intensity of the blue wing is an indicator of the Na-density. This peak is strongly shifted to the blue side in the lower part of the discharge. If the upper electrode is in the cathode mode the plasma hot spot shifts to this electrode, too. The features of the Na-doublet do not change too much, particularly the location of the peak intensity of the blue wing does not show a strong variation. This is an indication that the Na density is higher in the lower part of the discharge during the whole phase. This phenomenon is called segregation and can be explained by the interaction of convection, ionization of Na in the hot core and radial ambipolar diffusion (8).

It is apparent from the images shown that the analysis of measured spectral line shapes could be a very powerful tool to determine and examine plasma processes in the arc discharges. Since many of the spectral lines show self-absorption, radiation transport becomes an important, if not determining part of the energy-balance in a high pressure lamp.

In the following a case study will be described which utilizes optically thick spectral lines to determine temperature distributions in a lamp. At the same time it will be shown that there is a need for an improved understanding of radiation transport in these discharges and the line-broadening mechanisms which lead to the heavily broadened and self-absorbed spectral lines in high-pressure lamps.

Case Study: Hg-lamp at 8 atm

As mentioned above the objective of this study was twofold: a) Measure radial temperature distributions and b) Determine line-broadening parameters for visible self-reversed Hg-lines. At typical pressures and temperatures of HID lamps the broadening of spectral lines in the positive column is mostly governed by resonance and van-der-Waals broadening, Doppler- and Stark-broadening play only a minor, if not negligible role. The emission of photons in the hot arc core, absorption and reemission on the way out define the spectrum and the spectral line shapes which can be measured from the outside of the lamps. To predict the line shapes, models for radiation transport have been developed with various degrees of sophistication. In general the radiation transport has to be treated in a 3-dimensional form. However, computing time increases dramatically with the dimensionality involved, especially if more complex chemistries are to be modeled with more complicated spectral features and numerous frequency points. We used an approach which was formally developed by, for example Stormberg (9) and Liebermann and Lowke (10), using a 1-dimensional (radial) treatment of the radiation transport utilizing net-emission coefficients. This approach is assumed to work well in Hg-discharges which generally in the positive column (away from the electrodes) show very similar radial temperature distributions as a function of axial position.

To simplify the experiment-model comparison even more we filled a lamp (Figure 12a) with pure mercury only and ran the discharges in vertical position with square-wave power supplies so that we were able to model the region of interest in the arc as if in dc-mode. The amount of mercury was determined to give a reasonable voltage drop across the arc length of about 100 V. Voltage and current traces are shown in Figure 12b. The operating Hg- pressure was determined from the half-width of the Hg 491.6 nm line which depends, following an analysis by Asselman (11), mainly on the Hg pressure, amounting to about 8 atm.

Already in the 1950s, Bartels (12) had shown that for cylindrically symmetric arcs the maximum temperature along a line-of-sight perpendicular to the axis is directly related to the peak intensity of self-reversed spectral lines. His methods were in the recent past extended by Karabourniotis (13). We determined the radial temperature distributions from the line shapes of the Hg 546 nm line by using Karabourniotis analysis.

The vertically burning discharge was imaged onto the entrance slit of a 1m spectrograph; an intensified CCD-camera at the exit slit served as the gateable 2D-detector. By turning the image with the help of two mirrors by 90^0 and choosing appropriate magnification we could image the full diameter of the lamp. For comparison of spectral line shapes with the model calculations we started with images of the area in the middle of the discharge. The vertically burning discharge was imaged onto the entrance slit of a 1m spectrograph and an intensified CCD-camera at the exit slit served as the gateable 2D-detector. By turning the image with the help of

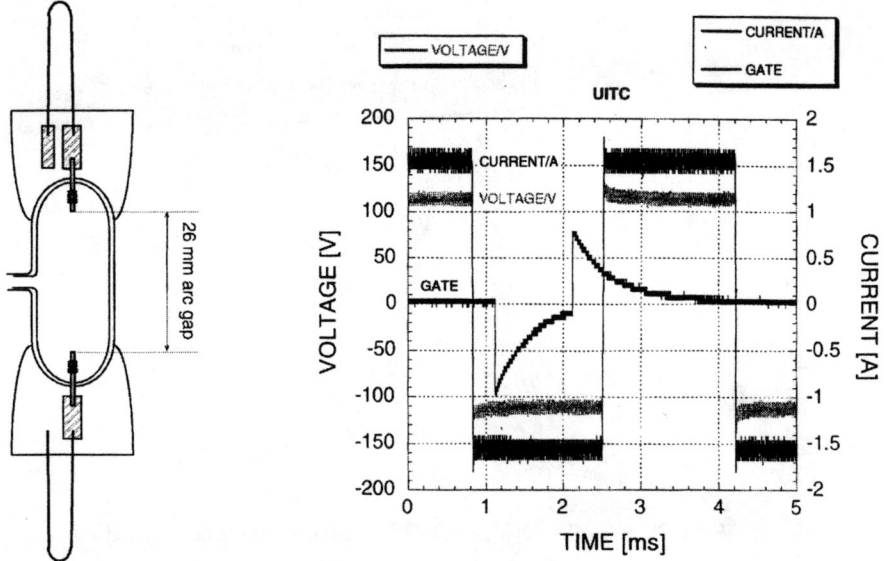

FIGURE 12: The geometry of an Hg-discharge lamp for radiation transport studies (right). Current and voltage traces are shown on the right; in addition a gate monitor signal of the ICCD-camera is displayed to show the time window for acquisition of spectra.

two mirrors by $90°$ and choosing appropriate magnification we could image the full diameter of the lamp. For comparison of spectral line shapes with the model calculations we started with images of the area in the middle of the discharge.

Since the analysis of the maximum temperature can be used for all view-lines which show self-reversed line features, we can determine maximum temperatures at all radial positions where the Hg 546 nm line appears to be self-reversed. Thus by using the 2-dimensional absolutely calibrated images we are able to determine very efficiently radial temperature distributions.

The radial temperature distributions cover about half of the discharge radius and were extended to the wall by extrapolating to the measured wall temperatures of about 1000 K. These temperature profiles were then used as inputs for the model calculations.

The measured line profiles of several Hg lines taken as line-of-sight measurements from the center of the discharge were compared with calculated profiles. We varied parameters for resonance and van-der-Waals broadening until we reached best fits.

Figure 13 shows an example for such a comparison. Here the influence of the resonance broadening mechanisms is shown. A best fit was achieved when we used a resonance broadening parameter which was doubled compared to earlier studies (9).

Figure 14 displays a comparison of a modeled with an experimental image of a radial "slice" as it was used for the temperature determinations. For the calculations we used the broadening parameters found as described above. The similarity of both images can be described as good. However, at higher pressures we expect stronger deviations particularly in the line wings since the assumption of a simple $1/R^6$ potential for the van-der-Waals interactions might break down.

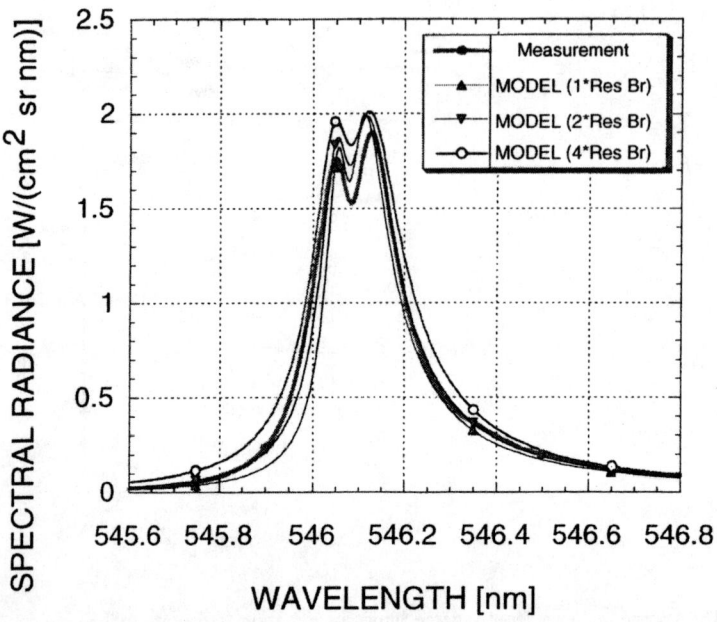

FIGURE 13: Measured and calculated Hg-546 nm line shapes. For the calculations the resonance broadening parameter was varied.

To give an example of the usefulness of the temperature diagnostics described above we moved the lamp along the axis by increments of 1mm to cover the whole arc length of 26 mm. Assuming local thermodynamic equilibrium due to the high pressure (which might not be valid in the regions very close to the electrodes) we determined radial temperature distributions for each axial position. These distributions can be combined to give a complete radial and axial temperature map of the arc core region, as shown in Figure 15. The hotter areas close to the electrodes are clearly visible. In addition, the upper part of the discharge (20-25mm) seems to have a widened temperature distribution compared to the lower part (0-5mm). This behavior is expected due to convection effects in these lamps.

Temperature and density maps of this kind are very useful for comparison with modeling efforts of lamp discharges. However, since "real" lamp chemistry includes many more components than just mercury, it has still to be shown that the methods used for pure mercury will be applicable to more complicated fills. The broadening of spectral lines in the blue wings is mainly determined by collisions with atoms of the same kind, i.e. Hg-Hg (resonance broadening). The broadening of the red wings is, in addition, influenced heavily by collisions with other partners, for example Hg-Na. The

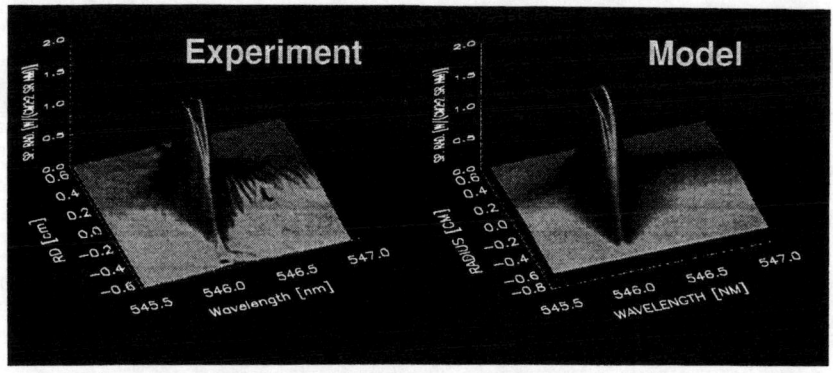

FIGURE 14: A comparison of the radial cross-section intensity distribution of the optically thick Hg-546 nm: the experiment on the left, the model simulation on the right.

transfer of the methods shown to different elements and the inclusion of appropriate collision physics is still an open field of research. Most of the molecular potentials (the collisions leading to very extreme line-wing broadening can be viewed as molecular processes) are not known, but determine the shape of the heavily broadened spectral lines, thus the radiation transport. Radiation transport - for a long time the domain of astrophysics research - has not found sufficient entry into the research of

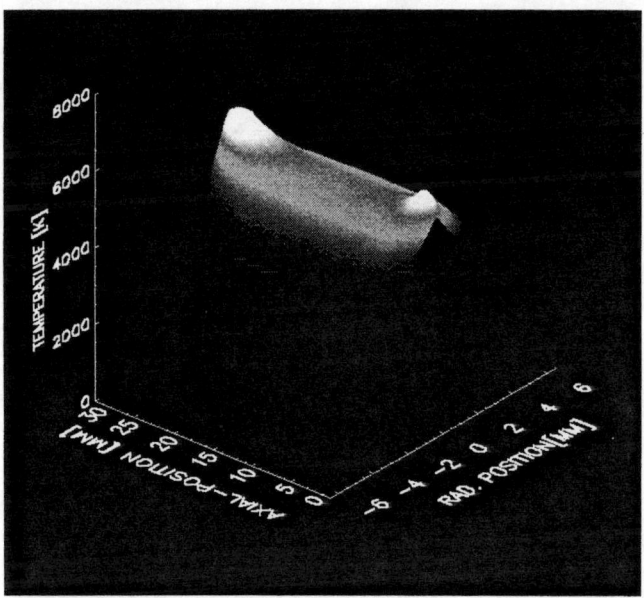

FIGURE 15: The temperature map of the core region of a Hg-lamp (Figure 10), derived from analysis of optically thick Hg-546 nm radiation analysis.

light sources, but is believed to play a determining factor in the understanding of the energy balance of HID-lamps.

SUMMARY

The aim of this paper was to give a brief overview of a few current research interests in the domain of HID light sources emphasizing the need for more data. Since books have already been written about controlling mechanisms of these discharge lamps (14), only a short description of the main ideas concerning the behavior of Hg- and metal halide lamps was given.

TABLE 2: Areas of Research and Data Needs

Arc	Wavelengths, transition probabilities Broadening parameters Ionization cross-sections Molecular formation Complex formation Transport coefficients (diffusion etc.)
Electrodes	Work-functions under lamp conditions (high pressure) Thermal and electrical conductivities Emissivities
Melt	Composition Thermodynamics of molten salts Vapor pressures of compounds
Tube	Material properties

To be able to understand in particular the metal halide lamps several areas need to be understood. Table 2 gives an admittedly limited overview of research areas and the data needed. For example, light emission from the arc discharge is governed by atomic and molecular radiation. Thus, wavelength and transition probability data are needed for the species of the lamp fills. Appropriate broadening parameters are needed for a description of spectral line shapes and radiation transport.

For an improved understanding of plasma and gas transport phenomena in the lamps, better coefficients for diffusion and ionization cross-sections are needed. Electrodes play an important role in the operation and maintenance of high pressure lamps. Since they are working as thermionic emitters, work-function data under lamp conditions (high pressures, high temperatures) are needed. The thermal balance of electrodes depends on material functions such as thermal and electrical conductivities and spectral emissivities. The chemicals of lamp fills are usually introduced as salts which will melt at the high wall temperatures. The composition of complicated salt melts are not well understood; thermodynamics of molten salts and vapor pressures of the compounds is a broad and not well understood area of research. Last, but not least, the material properties of the lamp vessels are extremely important in regard to resistivity to chemical reactions with the fill chemistries.

In conclusion it is hoped that this contribution was able to show that despite the fact that the basic principles behind HID light sources seem to be understood, there remains a broad area of unknowns and data needs which have to be explored in order to produce and improve these efficient light sources to their full potential.

ACKNOWLEDGMENTS

The assistance of Ernest Davey Jr. in making and filling lamps for our experiments is gratefully acknowledged. I would like to thank Graeme Lister for cooperation in regard to radiation transport modeling and Lothar Hitzschke for Figure 8. The efforts of Walter Lapatovich in reading the manuscript have made this paper possible. I would like to thank the management of OSRAM SYLVANIA in allowing and supporting our contributions to more fundamental lamp research.

REFERENCES

1. Waymouth, J. F., "Where We Are in Light Source Performance; Where We Would Like to Get", presented at APS Conference of the 1996 Division of Atomic, Molecular, and Optical Physics Annual Meeting, Ann Arbor, Michigan, May 15-18, 1996
2. Wiese, W. L. and Fuhr, J. R., J. Phys. Chem. Ref. Data **4**, 263 (1975)
3. Den Hartoog, E. A., Curry, J. J., Anderson, H. M., Wickliffe, M. E. and Lawler, J. E., "Radiative Lifetimes and Atomic Transition Probabilities of Rare-Earth Elements", submitted abstract to this conference
4. Hitzschke, L. private communication
5. Shea, M. J., Keeffe, W. M. and Struck, C. W., " Measurement of Enhanced Thulium Abundance in Metal Halide Discharge Lamps", in *Proceedings of the Third International Symposium on High Temperature Lamp Chemistry*, New Orleans, October 10-15, 1993, *pp. 129*
6. Waymouth, J. F. , J. Light & Vis. Env. Vol. 6, No. 2, 1982, pp. 5
7. de Groot, J., and van Vliet, J., "The High-Pressure Sodium Lamp", Philips Technical Library, Kluwer Technische Boeken B. V., Deventer - Antwerpen, 1986, pp. 84
8. Rogoff, G. L., Feuersanger, A. E., Drummey, J. P., and Rothwell, H. L., "Determination of Two-Dimensional Temperature and Additive Density Distributions in a High-Intensity-Discharge Arc", J. Appl. Phys. **62** (10), 15 November 1987

9. Stormberg, H. P. and Schaefer, R., J. Appl. Phys. **54**, 1983, pp. 4338
10. Liebermann, R. W. and Lowke, J. J., J. Quant. Spectr. Radiat. Transf. **17**, 1976, pp. 253
11. Asselman, A., Aubes, M., Damelincourt, J. J., Salon, J., J. Appl. Phys. **71**, 1992, pp. 4739
12. Bartels, H., Z. Phys. **127**, 1950, pp.243; **128**, 1950, pp. 546
13. Karabourniotis, D., "Self-Reversed Emission Lines in Inhomogeneous Plasmas", in *Radiative Processes in Discharge Plasmas*, edited by Proud, J. M. and Luessen, L. H., Plenum Press, New York and London, Nato ASI Series, Series B: Physics, Vol. 149, 1986, pp.171
14. Waymouth, J. F., "Electric Discharge Lamps", MIT Press, Cambridge, Massachusetts and London, England, 1971; also Reference 7

Industrial Applications of Low Temperature Plasmas

J. Norman Bardsley

*Lawrence Livermore National Laboratory
and the
United States Display Consortium*

Abstract. The use of low temperature plasmas in industry is illustrated by the discussion of four applications, to lighting, displays, semiconductor manufacturing and pollution control. The type of plasma required for each application is described and typical materials are identified. The need to understand radical formation, ionization and metastable excitation within the discharge and the importance of surface reactions are stressed.

INTRODUCTION

Gaseous discharges have been used in industry for over a century, beginning well before the discovery of the electron and the consequent understanding of the mechanisms by which charge is separated and transported in gases. The important early empirical progress included the discovery of arc lamps by Davy around 1801, with application to lighting in mines, and of the dielectric barrier discharge in 1857 by Siemens (1), which led to the efficient production of ozone.

Until the discovery of the transistor and the development of integrated circuits, the use of ionized gases was essential to the control of electric currents within the electrical power and communications industries, creating the interdisciplinary field known as gaseous electronics. Although most of the gaseous (often inappropriately called "vacuum") components within electronic systems disappeared with the development of semiconductor technology, the manufacture of semiconductor devices is dependent upon the use of gas discharges through plasma processing techniques. Meanwhile, understanding of ionized gases is still essential to the lighting and display industries, despite the advent of light emitting diodes, solid-state lasers and liquid crystal displays.

This paper will not attempt a comprehensive review of all industrial applications of low temperature plasmas. The focus will be on four areas, lighting, electronic displays, semiconductor manufacturing and pollution control. These four industries are at very different stages of evolution and thus have varying ability to support research.

The lighting industry is relatively mature. For mass-produced lamps, approximately 75% of the global market is held by three multinational companies, the pace of technological evolution is relatively slow and profit margins are thin. More companies are involved in producing specialty lighting products, and the opportunities for innovation and double-digit profit margins are greater.

The display industry is at a very exciting stage, in that the dominance of the cathode ray tube is being challenged by flat panel display (FPD) technologies. There is a great need for research, but most of the manufacturing activity is concentrated in Asia. Furthermore, although the global FPD industry has invested over $20B in fabrication facilities during the last ten years, this investment has yet to produce any sustained profits.

The semiconductor industry is now dependent on plasma processes for the production of devices with sub-micron features. The manufacture of plasma reactors has become a large international industry, with annual revenues in excess of $10B. The business is still extremely volatile. As this article was being written, during the early weeks of 1998, several leading U. S. suppliers of semiconductor manufacturing equipment curtailed operations in response to the financial problems in Asia, but by the time that this volume reaches the reader, the expansion may have resumed.

The use of plasma techniques in pollution control has received much attention in recent years. Although ozonizers have been long used for water treatment and electrostatic precipitators for dust control, most of the newer methods have not yet been accepted commercially. The two most important remaining challenges are to reduce the cost of manufacture and operation of the equipment and to be able to guarantee that no harmful by-products are created as the target pollutants are destroyed.

Even within the restricted scope of this review, much more good work has been done than can be reported in the space available. The selection of references is somewhat arbitrary, and apologies are extended to those authors whose work is not acknowledged in this brief report.

LIGHTING

This section will provide brief descriptions of the operation of several of the most common lamps. In the first two examples, electrons and ions play minor roles, but the interaction between a heated gas and the various surface materials is critical to the performance of the lamp.

There are three major characteristics that are used in comparing the performance of different lamps, efficiency, color characteristic and lifetime. Other essential factors that constrain lamp design are size, shape, safety in both fabrication and use, and environmental impact. By far the most important consideration is cost. For almost all application cost constraints rule out many technological options and cost minimization within traditional designs is critical in maintaining market leadership.

Lamp efficiency is a measure of the apparent brightness of the light (in lumens) produced for each watt of electrical power. The lumen unit allows for the wavelength dependent response of the human eye. The efficiency of the nominal eye in the blue (450 nm), green (555 nm) and red (610 nm) is 51.5, 683 and 343 lumens per radiant watt, respectively. The theoretical efficiency of a three-wavelength emission source with acceptable color rendering is approximately 300 lumen per watt (lpw).

The efficiency of commonly used lamps rose substantially between 1940 and 1980, as shown in Figure 1, but has risen little since that time. The most efficient commercial sources with acceptable color mix are the fluorescent and metal halide lamps, which have an efficiency just over 100 lpw, a figure which has changed little over the past twenty years. The sodium lamp produces about 20% more useful light, but with poor color balance.

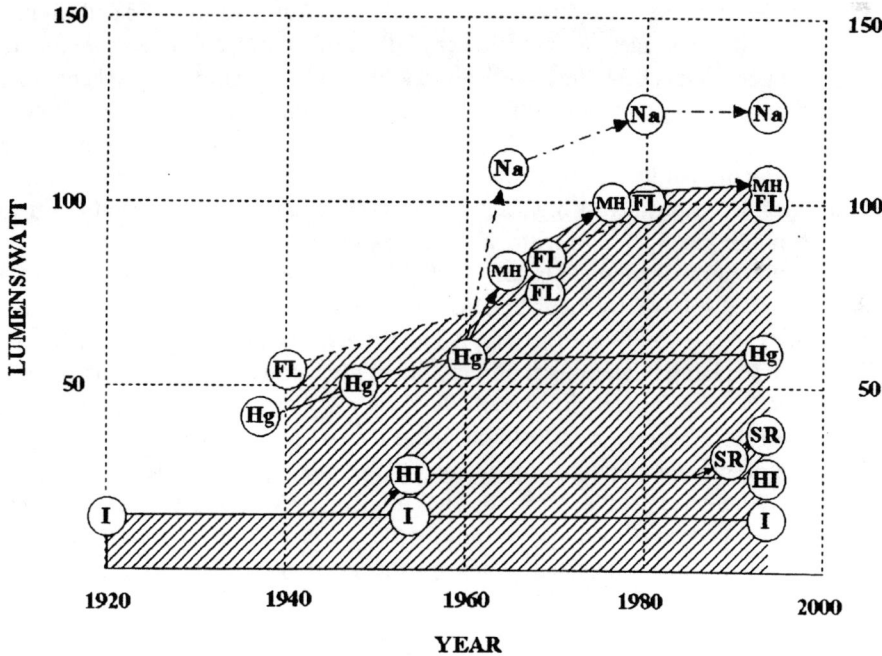

FIGURE 1. The evolution the efficiency of lamp technology since 1920: I, incandescent; HI halogen -cycle incandescent; SR, same with IR reflector; Hg, high-pressure mercury lamp for street-lighting; FL, fluorescent, MH, metal-halide lamp; Na, high pressure sodium arc lamp (Courtesy: J. F. Waymouth).

Color balance can be achieved using a continuous spectrum, such as black-body radiation at a temperature around 5500K. However, such high temperatures are difficult to sustain in compact lamps without going to high pressures. Alternatively, one can use a mix of discrete sources distributed across the visible spectrum. The desired emission frequencies can be produced either through direct excitation of visible transitions, or through the creation and down-conversion of UV light.

Lamp lifetime is determined mainly by the stability of the various surfaces exposed to the radiation and fill gases. Particular care has to be taken to avoid the growth of hot spots in which damage grows exponentially. For example, localized evaporation from incandescent filaments can increase the resistance, leading to more heating and enhanced evaporation. Deleterious heating can also be caused by deposition of opaque material on the glass envelope of the lamp.

Incandescent Lamps

In incandescent lamps, light is produced by radiation from a heated filament. Although the first commercial incandescent lamps of Edison and others used carbon filaments, tungsten has been the material of choice for the last forty years, because of a combination of three properties. With a melting temperature close to 3650K, tungsten has the lowest evaporation rate of all natural metals. Its emission spectrum enhances the production of visible light by about 30% over that of a black-body source. Finally, tungsten retains its rigidity at high temperature, allowing extended operation at over 90% of its melting temperature without undue sag or distortion.

In conventional incandescent bulbs, the light comes from the glowing filament and gas is introduced solely to reduce the rate of evaporation of W atoms from the filament and thereby increase the life of the lamp. The presence of the fill gas also reduces the efficiency of the lamp by cooling the filament. With a straight wire, the cooling effect would dominate, but for coiled filaments, the heat loss is reduced significantly and the combined effect is beneficial. Most household lamps are filled with Ar gas at around 600 Torr at room temperature, producing a pressure close to one atmosphere in operation. Another positive effect of the fill gas is to moderate the increase in temperature of the glass envelope, permitting the use of inexpensive soft (lime) glass. Typically, deposition of tungsten atoms on the glass reduces the transmittance by about 10%.

Halogen Lamps

The performance of incandescent lamps can be be improved by increasing the fill pressure, which increases the possibility of explosive failure, or using the heavier, but more expensive, krypton or xenon. The chance of explosions and the extra cost of heavier gases can be ameliorated through the use of more compact bulbs made from glasses, such as fused silica or quartz, that sustain higher temperatures.

With compact lamps, deposition of W atoms on the glass wall causes major problems. Not only does this reduce the amount of light escaping from the bulb, but the resulting heating of the glass envelope can lead to softening and failure, even for quartz. The solution is to add a component to the gas fill to remove the W atoms from the glass. This can be achieved using halogen atoms, which react with the W atoms to form tungsten halide molecules. At normal operating temperature, the tungsten halide vapor pressure at the wall is above the condensation point, so that the build-up of tungsten on the wall is avoided. In the hotter regions near the filament, the tungsten halide molecules dissociate, and many of the W atoms return to the filament, whereas the halogen atoms are free to diffuse back to the walls.

The active ingredient in early tungsten halogen lamps was iodine. However lamp manufacture is then more difficult, since iodine is a solid at room temperature and oxygen needs to be added to produce a stable tungsten-halogen cycle. Further, iodine absorbs blue light and reduces the perceived color temperature of the lamp. The element preferred in current halogen lamps is Br, which can be introduced in several forms, such as HBr, CH_3Br or CCH_2Br_2. Both fluorine and chlorine have been found to be too corrosive for use in commercial products.

The efficiency of both incandescent and tungsten-halogen lamps can be increased through the use of filters which reflect infrared light but transmit visible light. These often are constructed as multilayers, for example with a mixture of Ta_2O_5 and TiO_2 for the high refractive index material and SiO_2 for the low index layers.

Fluorescent Lamps

The introduction of active plasmas into lamp technology around 1940 led to a three-fold increase in efficiency. The first key advance was the use of mercury vapor as a source of radiation. Some of the positive characteristics of Hg as an emitter are the wide spacing and low excitation energy of the key radiative levels and the presence of neighboring metastable states, which feed excitation into the desired levels. The hyperfine line structure of natural mercury reduces the imprisonment of the radiation, and it is relatively easy to maintain Hg vapor at mTorr pressures. The fact that the primary radiation from mercury is in the UV at 254 nm is not a major disadvantage, since efficient, stable phosphors exist to convert this radiation into visible light.

The major disadvantage of the use of mercury is the environmental impact. Although relatively small amounts are used in each lamp and the low vapor pressure at room temperature reduces the risk of air-borne dispersion, the very large numbers of lights in use and the high toxicity of mercury causes concern.

In fluorescent lamps, the radiating vapor atoms are excited by electron impact, either directly from the ground state or indirectly through metastable or ionic states.

Over 80% of the applied electrical power is delivered to the positive column of the discharge rather than to the sheaths, where the energy is mostly lost to electrode heating. Over two thirds of this fraction is converted to UV radiation and the efficiency of the phosphors in converting to visible light is around 40%. Allowing for a small amount of direct visible light from the Hg atoms, the efficiency of transfer from electrical power to visible radiant power is close to 30%. The spectrum can be tuned to give around 350 lumens per radiant Watt.

In low pressure sodium lamps, the efficiency with which energy is transferred to visible light is also around 30%, but the higher response of the eye to yellow light results in greater overall efficiency. However, the color spectrum is not acceptable for most applications.

The partial pressure of mercury in fluorescent lamps is typically around 5-10 mTorr. The remainder of the 1-10 Torr is made up with a rare gas. Krypton provides highest efficiency for low power lamps, argon is better at intermediate power levels and neon is best at high power.

The major limitation of traditional fluorescent lights is their size and shape, which is necessary to provide an effective positive column at low pressures.. In recent years, considerable effort has been expended in the design of more compact lamps suitable as replacements for less efficient incandescent bulbs. Another innovation has been the use of electrodeless RF discharges for longer lifetimes.

Analysis of the data needs for the low pressure discharge lamps outlined above is given in the accompanying article by Sommerer.

Electrodeless Lamps

In discharges with electrodes, the strongest fields are found in the cathode fall region. In glow discharges, potential drops of the order of 100V are needed to ensure sufficient flow of positive ions to the cathode to maintain the discharge through secondary emission. In arc plasmas the potential drop is smaller, but the ion flux must be able to heat the cathodes to temperatures at which field-enhanced thermal emission of electrons occurs. Acceleration of ions across the cathode fall and consequent electrode heating leads both to energy loss from the discharge and to cathode erosion through sputtering and enhanced evaporation. Loss of material from the electrodes not only limits the lifetime of the lamp, but also the deposition of the cathode material on the glass envelope reduces its transparency. Operating lifetimes in excess of 50,000 hours can be obtained in lamps without electrodes, allowing street lights to function for around 15 years without replacement.

Electrodeless discharges were discovered over 100 years ago (see e.g. Hittorf (2) and Thomson (3)) and crude lamps were constructed by Tesla (4). These discharges have been used for many years in spectroscopy, but study of their properties has been intensified in recent years because of their suitability for plasma processing. When driven by induction coils in the Radio-Frequency (RF) regime, there are several modes of operation. In the absence of external magnetic field, the "E-mode" is observed at relatively low power levels and is dominated by radial or axial electric fields. At higher power levels, the "H-mode" takes over, in which azimuthal electric fields and magnetic fields are more important. The system acts like a transformer, with the induction coil acting as the primary and the circulating plasma current as the secondary. At the higher drive frequencies of the microwave regime, these two modes merge. Microwave discharges can be excited either using standing or traveling waves.

One of the advantages of electrodeless lamps is that they can be turned on and off quickly. The absence of electrodes in the lamp also allows one to use more corrosive gases, such as fluorine or chlorine, giving one greater flexibility in tuning the gas mix to give higher efficiency or better color balance. The length and shape restrictions of traditional fluorescent lamps can be overcome and many forms of compact discharges have been demonstrated. The major remaining challenges (5) appear to be in the design of electronic controls that are compact, inexpensive and do not create unacceptable electromagnetic radiation (EMI). The space and EMI problems can be solved more easily if the discharge and electronic controls are in separate units. The popularity of microwave ovens has led to significant price reductions in microwave generators, but still most consumers are prepared to pay more for an oven than for a lamp.

There are niche applications, such as the curing of inks or glues, for which electrodeless lamps are well established, and efficient lamps with long lifetimes for household and office lighting are now available as replacements for incandescent bulbs.

High Pressure Arc Lamps

Fluorescent lamps use non-thermal low-pressure plasmas in which electrons are accelerated to energies much higher than that of the neutral atoms or molecules. Although ions do not pick up much energy in the positive column, they are accelerated across the sheaths and contribute significantly to electrode erosion. The velocity with which ions strike the electrodes can be reduced through the use of higher pressures. In arc lamps, the neutral gas is also heated. The high temperature leads to reduced neutral densities in the arc channel, thus facilitating electron transport. The pressure in these arc lamps range from 1 to 100 atmospheres and peak temperatures are between 4,000K and 10,000K. The envelopes must be more rugged and are often made from fused quartz, polycrystalline alumina or translucent ceramics. These lamps, often called High Intensity Discharge (HID) lamps, and offer high brightness, high efficiency and long operating life, typically 10,000 to 30,000 hours.

The simplest HID lamp uses mercury vapor and tungsten electrodes and is contained by a fused quartz arc tube, with Mo foil seals to allow current to pass through the quartz envelope to the electrodes. The central temperature is around 6000K, whereas the temperature of the quartz tube does not exceed 1000K, since thermal diffusion is moderated by the pressure of several atmospheres. Despite the significant temperature gradients, the discharge is close to Local Thermal Equilibrium (LTE). Typically about 15% of the input electrical energy is converted to visible light.

The efficiency of high pressure Hg lamps can be improved through the addition of sodium, with xenon added as a buffer gas to facilitate ignition. The color dominance of the Na resonance lines is reduced by self-absorption and the presence of Hg and Xe leads to intense excimer emission in the red. By appropriate selection of gas mixture, pressure and surface materials, high pressure sodium lamps can be tuned to give luminous efficiency over 150 lpw, color balance similar to that of incandescent lights, or lifetimes in excess of 20,000 hours. Unfortunately, simultaneous achievement of these characteristics has not yet been achieved.

High pressure Hg lamps are often modified through the addition of metal halide salts, such as NaI and ScI_3. Although most of the salts remain as a condensate on the arc tube wall, the wall temperature is sufficiently high to support a partial vapor pressure of order of 0.01 atmosphere. Because the excitation and ionization energies of Na and Sc are much lower than those of the denser Hg atoms, the minor components can dominate both the discharge formation and the radiation at the center of the metal-halide discharge. This can lead to a doubling of the fraction of energy transferred to visible light. The dominance of Hg atoms in the outer, cooler regions is important in setting the arc impedance and the thermal conductivity of the discharge.

Iodine is the most common halogen in metal halide (MH) lamps, although bromine and chlorine have also been used. A combination of metals is usually included to provide the desired color balance. Na and Tl are commonly used to provide emission in the red and green portions of the spectrum, while the blue can be supplied by In, Sc, Dy, Ho or Tm.

The chemistry and transport in HID lamps are extremely complex and fully warrant the special attention that is provided by Adler in the accompanying paper. Earlier reviews were provided by Work (6) and by Dakin et al. (7).

Mercury-Free Arc Lamps

The use of mercury in light sources has led to considerable concern about its environmental impact, and there has been growing interest in the development of mercury-free lamps. However, in analyzing the health impact of alternative lamps, one must also consider the indirect effects of changes in efficiency. Burning fossil fuel in power plants releases Hg to the atmosphere in a form that is more dangerous to life than the Hg in lamps that are discarded in land fills. So, for example, the substitution of incandescent lamps for Hg-containing fluorescent lights leads to an increase in the absorption of Hg by biological systems, as well as to additional emission of global warming gases.

A prerequisite of mercury-free lights is therefore that they are at least as efficient as the lamps they replace. One possible substitute is sulfur, but low pressure S lamps are inefficient and high-pressure S lights produce a yellow-green color. Xenon discharges can be designed to be efficient producers of UV light at 148nm or 172 nm, but most of the energy will then be lost in conversion to visible light, unless one can find phosphors which give more than one visible photon for each UV photon absorbed. Another candidate is barium, which can be vaporized at modest temperatures and emits directly in the visible. Unfortunately, the primary emission is a single line in the green, with a little emission in the red and hardly any in the blue. Nevertheless, investigation of the efficiency of light production in barium discharges is underway.

The growth of the flat panel display industry has provided further incentive for the development of mercury free lamps. Multi-colored images are often produced by the use of filters that separate out the red, green and blue components of the light. Strong emission lines that fall near the borders of these three ranges are either lost or cause difficulties in color control. Both mercury and sodium lamps often have strong yellow components that can be troublesome in this respect. Lamps with just three strong lines, one near the center of each portion of the spectrum, could increase the efficiency of the color modulation system. If the three lines come from different elements, the amount of each component can be tailored to achieve a wide range of color temperature.

The alkaline earth elements, barium, calcium, magnesium and strontium have relatively simple outer shell structure and possess suitable strong lines in the green (Ba) and blue (Ca, Mg and Sr). However, these elements do not function well in arc lamps. Shaffner (8) has suggested the use of indium for blue, thallium for green, and lithium for red, but the practicality of this recipe has yet to be tested. Obtaining a strong source of blue light for flat panel displays seems to be the most difficult challenge, in metal-halide lamps as well in light emitting diodes and electroluminescent sources.

Dielectric Barrier Discharge Lamps

Dielectric barrier discharges (9) are created by covering metallic electrodes with a thin layer of insulating material. When driven by AC circuits over a wide range of frequencies, these discharges consist of a series of many narrow streamers with radii of order 100μm and lifetimes of a few ns. Non-thermal plasmas can thus be produced even at pressures above 1 atmosphere and electrons with energy of 1-10 eV can be generated efficiently. Electron densities of over 10^{14} cm^{-3} are achieved, giving currents of about 0.1A in each micro-discharge. Conversion efficiencies of up to 60% have been reported in the conversion of energy deposited into Xe discharges into UV light at 148 nm and 172 nm. There are as yet no phosphors to convert this UV radiation to visible light with high efficiency, but these discharges provide excellent sources of UV light for material modification (10) or pollution control applications (11).

Shorter wavelengths can be obtained by using pure Ar (126 nm) or Kr (146 nm) excimers, and longer wavelengths can be produced efficiently by mixing F or Cl with the rare gas.

ELECTRONIC DISPLAYS

For the past sixty years the electronic display industry has been dominated by cathode ray tubes. Technological progress has been steady, but relatively slow; the major concern is to minimize costs as the performance levels increase. The technological challenges relate to electron beam formation and transport, phosphor efficiency and stability, and general electronics issues. Access to improved atomic and molecular data is not a high priority in this segment of the industry.

The search for alternative technologies has been driven by a desire to reduce the size and weight of the display, while improving the quality of the image. Although laboratory studies have been underway for many years, the availability and performance of commercial products were very limited until the 1980's, when Japanese companies courageously began to invest billions of dollars in the development and fabrication of liquid crystal displays, first as viewfinders in cameras and camcorders and then in portable computers. This foresight led to almost complete domination of the FPD industry, with over 95% of the production coming from Japan. This dominance is now being challenged, primarily by Korean and Taiwanese companies, but the manufacturing experience, industry infrastructure and broad patent portfolios of the pioneering companies give them great advantages in maintaining business leadership.

Liquid Crystal Displays

The major scientific bases underlying liquid crystal displays (LCD) are in condensed matter and optical physics. For conventional LCDs the impact of atomic and molecular data is mainly in the design of the light source. In the large area displays used for notebook computers, compact fluorescent tubes are used as backlights or on the edge of the screen, with deflector systems to direct the light through the pixels. Both hot and cold cathodes are used in these miniature tubes. The cold cathodes require higher starting voltage and are less efficient, but they reduce problems from overheating and give longer lifetimes.

Small liquid crystal panels are now being used to create the images within projection systems. The demands for long lifetime and high efficiency leads to the use of high pressure lamps with very small arc gaps, filled with mercury, metal-halide mixtures or xenon. There is currently a particular need for the development of efficient lamps with lifetimes in excess of 10,000 hours that can be focused onto light modulators of diagonal size less than 1".

Since the transmission efficiency of liquid crystal modulation systems is low, usually 3-5%, much current research is aimed at the development of reflective light valves, which make use of ambient light, or emissive displays, which provide their own light. Emissive displays are discussed below.

The transmission of polarized light through liquid crystal cells depends on their orientation, which can be modified through the application of electric fields. In passive matrix devices, the potential difference across each pixel is controlled directly from the voltage applied to the relevant row and column drivers. However, this simple approach provides limited contrast and response speed, which can be improved by the introduction of non-linear elements. In active matrix LCDs this is usually achieved by fabricating thin-film transistors at each pixel. Although millions of TFTs can now be made on a single panel, the yield of good panels falls rapidly with increasing substrate size. Gas discharge physics provides an intriguing alternative.

Plasma Addressed Liquid Crystal Displays

In active matrix LCDs, each pixel is controlled by a separate thin film transistor (TFT). In PALC displays (12), the thousands of TFTs in each row are replaced by a single plasma channel. The role of these channels is not to create light, but solely to act as a switch. This replacement avoids many of the complications of AMLCD manufacture and provides a simple way to build large area LCD panels.

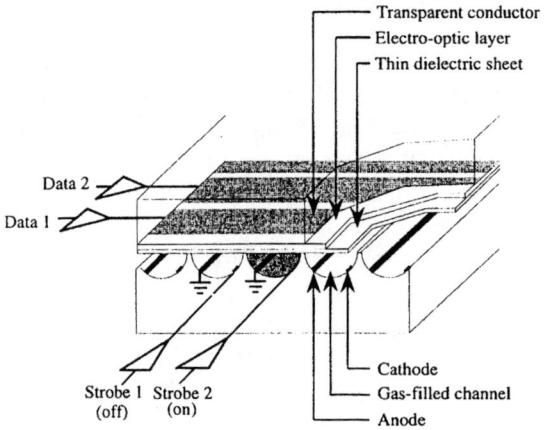

FIGURE 2. Structure of the Plasma-Addressed Liquid Crystal Panel (Courtesy: K. J. Ilcisin)

As shown in Figure 2, the liquid crystal material is contained by the color plate on one side, with embedded column electrodes, and a dielectric microsheet on the other. The plasma channels are formed between this microsheet and the second glass substrate, with barrier ribs separating the rows. Parallel cathode and anode electrodes are formed along each channel, either on the surface of the second substrate or on the barrier ribs. When a voltage of about 350 volts is applied between the anode and cathode, a plasma is formed, creating a virtual electrode on the surface of the dielectric layer at a potential close to that of the anode. The data voltage of about 70V applied between this virtual electrode and a column electrode is sufficient to align the liquid crystal molecules.

The two major concerns in the selection of gases are the panel lifetime and the switching speed. Since the electrodes are exposed to the plasma, the use of highly reactive gases would reduce the lifetime. The first choice is thus one of the rare gases, or perhaps a mixture, and VGA panels have been constructed using both He and Xe. Gas breakdown is relatively rapid and so the switching speed is determined primarily by the rate of decay. In single-component rare gases, this decay is slowed significantly by the formation of metastable states. Nevertheless, the 30μs cycle times required for VGA panels can be achieved. For high-resolution monitors or HDTV, cycle times of 15μs are required. One way to quench the metastables is through a Penning mix, but this may just lead to the replacement of one metastable by another. Another solution is to add a molecule that can dissociate to absorb the extra energy. Ilcisin et al (13) have shown that decay times less than 5ms can be achieved with a He-H_2 mix, allowing the addressing of 3000 lines within a 60 Hz frame rate. The use of H_2 provides challenges in attaining a long lifetime, but the preliminary results are extremely encouraging.

Emissive Displays

Multicolor flat panel displays use solid-state phosphors to create the light. In electroluminescent (EL) and field-emission displays (FED), the phosphors are excited by electron impact. Field emission devices are leading to a renaissance in the art of vacuum tube technology, but require little new atomic or molecular data. In some senses, they are like cathode ray tubes in that electrons are accelerated at low pressures onto a phosphor screen. In CRTs the electrons are drawn from a single cathode and directed onto the appropriate spot on the screen by electric and magnetic fields. The distance between cathode and screen must be of the same order as the screen diagonal and large screen CRTs are necessarily bulky. In FEDs, the gap is very small and each pixel has its own source of electrons. Cold cathodes are used, made from low work function materials often shaped into narrow cones to enhance the electric field strength near the surface.

EL displays and light emitting diodes (LED) employ solid state technology and are even further removed from atomic and molecular physics. This leaves only one emissive FPD technology, the plasma display, for our attention.

Plasma Displays

The early plasma display panels produced monochromatic light and were essentially arrays of neon lamps. Today's color PDPs (13) are more analogous to fluorescent tubes, in that the gas discharge produces UV radiation which is converted to visible light by phosphors. Since the vapor pressure of mercury is too low at room temperature, xenon is used as the active gas.

Plasma panels are better suited to larger displays, indeed it is very difficult to manufacture plasma displays with pixel sizes less than 100 microns. Diagonal sizes around 40" are now standard for TV applications. Semiconductor manufacturing techniques are much less important than in other types of display.

Panel Architecture and Operation

Most plasma displays consist of an array of parallel discharge channels of height and width around 200μ, separated by barrier ribs, between two glass substrates. The channel surfaces are coated, on three sides if possible, by phosphor material, with red, green and blue phosphors in separate channels. The fourth side is transparent to let out the light. The channels are filled with rare gas mixtures at a pressure of about 400 torr. The active xenon gas is usually a minor constituent in a buffer gas of helium, neon or a He-Ne mixture.

Plasma displays are driven by applying high frequency voltages, either in AC or DC mode. In the more common AC devices, the electrodes are embedded in a dielectric film, so that current flow leads to the formation of a wall charge. Each half-cycle, the charge that was deposited on the dielectric during the previous half-cycle magnifies the applied field and contributes to discharge growth. This reduces the voltage needed to sustain the light production and provides a memory effect. If the voltage of the sustain pulse is just insufficient to cause breakdown, light is created only in those pixels with wall charge. The dielectric film is protected with a thin layer of MgO, which has a high coefficient of secondary electron emission to maintain the discharge during each pulse. Despite many searches, a better emitter of secondary electrons than MgO has not yet been found

In AC panels, at least three different pulse shapes are needed, one to turn individual pixels on, by creating a wall charge, another to turn pixels off, and the third to sustain the discharges in all pixels that are "on". The addressing of individual pixels is accomplished by orienting the electrodes on the two substrates in perpendicular directions.

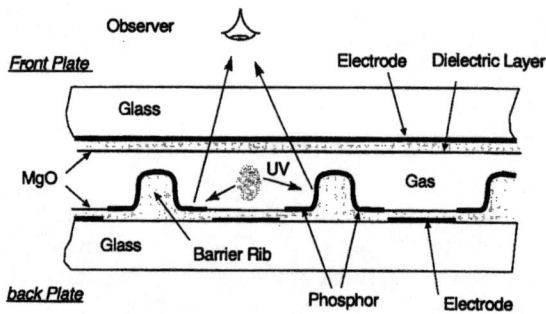

FIGURE 3. Opposed electrode structure of a plasma display panel.
(Courtesy: J. P. Bouef).

There are two forms of AC display. The simpler structure, shown in Figure 3, has two sets of electrodes, one on each substrate, that are used both to supply current and to switch the discharge on and off. This form is referred to as the dual-substrate, opposed electrode or matrix structure. The electrodes run in perpendicular directions, with those on the lower electrode running down the channels defined by the barrier ribs. The presence of these electrodes reduces the amount of surface area that can be covered with phosphors.

The second structure, as shown in Figure 4, has dual (parallel) electrodes on the upper substrate that create a surface discharge between each pair. This form has been called the single-substrate, surface discharge or coplanar structure. The barrier ribs are again built upon the lower substrate, running in a direction perpendicular to that of these electrode pairs, to define the discharge channels. A single electrode runs along each channel on the lower surface and is used solely for addressing. The addressing electrodes are covered with a dielectric layer, but are not subjected to bombardment by energetic ions and so can be covered with phosphor material rather than MgO.

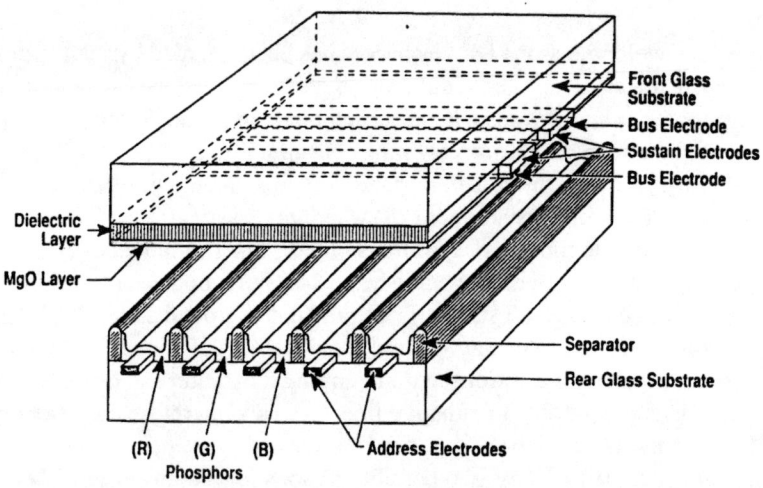

FIGURE 4. Structure of the surface-discharge plasma display panel. (Courtesy: Fujitsu)

The dielectric layer in AC displays also limits the current growth during each half-cycle. In DC discharges, the electrodes are in direct electrical contact with the plasma and a capacitor must be added to prevent runaway. Memory is usually achieved through auxiliary discharges which act as triggers for subsequent pulses. In some DC devices, barriers are constructed in both directions, so that each pixel is separated from its neighbors, to reduce "cross talk".

Gas Discharge Physics

The major challenges in discharge physics are to choose the gas mix, wall materials, pixel geometry and electrical pulse shapes for optimal performance.

The efficiency of transfer of electrical energy to emitted visible light is typically around 1%, which is almost two orders of magnitude less than that of a fluorescent light and leads to a relatively high power consumption for plasma displays. Problems arising from variations in the breakdown voltages at different points along each channel, and over time as the display ages, can be reduced by increasing the "voltage margin", that is the difference between the minimum and maximum voltages that will sustain the light in "on" pixels without causing breakdown in "off" pixels. Panel lifetime is another major concern; sputter damage to phosphors and MgO should be minimized and the gas mixture should remain constant. Operating lifetimes of 30,000 hours are now claimed.

Color control is obtained not by changing the intensity of the light but by varying the length of time that each pixel is on. Up to 256 brightness levels for each color have been obtained, but more would be desirable for the elimination of artifacts in portraying moving objects. This would be facilitated by faster discharge formation and decay.

Experimental development of PDP discharges has been retarded by the lack of detailed diagnostic data. Computer simulations have been developed in recent years by several groups. 1-D simulations can be carried out using fluid, kinetic and hybrid methods, but the kinetic methods are very time consuming in 2-D or 3-D. 2-D fluid models (15-17) are valuable, but it is important to understand the limitations of these models and to develop hybrid techniques (18) where necessary.

Most of the relevant atomic physics is already present in 1-D models. Veerasingham et al (19) have recently published a 1-D fluid model for a He-Xe panel including 3 states of neutral He, 7 states of neutral Xe, the ground states of He^+ and Xe^+ and Xe_2 excimers. No molecular ions are considered. Collisions of excited atoms are very important, both with electrons and in Penning ionization and dimer formation. Collisional broadening controls the radiative line width and therefore the extent of trapping of the atomic UV radiation.

Veerasingham et al (19) suggest two possible reasons for the low efficiency of plasma displays in comparison with fluorescent lamps. The first is the small size. The efficiency of miniature fluorescent lamps is also less than that of the larger versions, but progress is being made through their use as backlights for LCDs. The second factor is the need for rapid switching of PDP discharges. Both of these factors mean that the more of the discharge energy is deposited in the cathode fall region, rather than in the positive glow region. Perhaps a more useful comparison could be obtained with dielectric barrier excimer lamps, which attain high efficiency of UV production with relatively small electrode gaps and short-lived microdischarges. However, Falkenstein and Coogan (20) have shown that the efficiency of UV production from XeBr in dielectric barrier discharges decreases from 7% to 4% when the gap size is reduced form 5.5 mm to 2,5 mm. This is still well above the 200 μm gap typical of plasma displays. A third factor reducing the overall efficiency comes from the greater energy loss in conversion from UV to visible light using single-photon phosphors.

The leading commercial producers of PDPs currently use Ne-Xe mixtures, for which models have been developed by the Toulouse group (15,21,22). These models do include molecular ions, but their densities are small compared to that of Xe^+. For the particular parameter set assumed in reference (15), the models confirm that over half of the energy deposited into the plasma goes into ion acceleration, with only about 15% going into xenon excitation.

Discharge Diagnostics

To address issues concerned with display uniformity and the interactions between neighboring pixels, electrical and optical data on individual pixels is valuable. Measurements of the UV output are particularly challenging, since the materials used in PDP construction absorb the UV radiation. Special diagnostic panels must therefore be built, using transparent materials, such as MgF or allowing for observation along the channel. Unfortunately, this latter approach does not provide information on individual pixels. Pioneering work in this field has been performed by Tachibana et al (23), but much more study is needed.

SEMICONDUCTOR MANUFACTURING

Plasma processing involves the use of ionized gases in the production of solid or liquid materials with special properties. Since low temperature plasmas interact only with the surface of the condensed matter, they are most useful in the formation of thin films and so are particularly valuable in the semiconductor manufacturing industries. Almost all of the scientific effort in the U. S. concerned with plasma processing is focused upon semiconductor applications. The effort is more broadly based in the rest of the world, but the global economic impact is also dominated by the application to the microelectronics industry. An excellent introduction to the basic physics involved in plasma processing can be found in the book by Lieberman and Lichtenberg (24).

Plasma processing can be divided into three categories, depending on whether the goal is to add, remove or modify surface material. Once again, this article is not designed to be completely comprehensive; it is focused upon the most important subtractive and additive processes.

In surface modification, the plasma is used to change the chemical composition of the outer layers of the material or to change the physical or chemical properties of the surface. Usually the goal is to improve the resistance of the surface to external damage, whether by chemical attack, abrasion or impact, or to passivate the surface so that it becomes mechanically or chemically inert. For example, plasma nitriding (25, 26) is often used to improve the strength of metal surfaces. Sometimes, however the purpose of the plasma treatment is to make the surface more active. For example, plasma processing can increase the absorption of paint or inks by plastics.

One topic that has received much attention in the research community is plasma-immersed ion implantation (PIII) (27) . Most commercial ion implantation is done with accelerators, creating ions at low pressure and accelerating them onto the work-piece at energies at energies between 10^4 and 10^7 eV, depending on the desired doping depth. This is well suited to the treatment of planar surfaces. In PIII the work-piece is immersed in a plasma and suddenly given a high negative bias. Ions are then drawn into the surface at high velocity. PIII can be used not only for planar surfaces, but also for small components that need to be implanted from all directions. However, despite many years of intense effort, the adoption of PIII by industry has been slow.

The simplest type of plasma reactor has two parallel electrodes that are driven by an RF potential, usually at 13.56 MHz. Such reactors are often called reactive ion etchers (RIE) and are most efficient at pressures around 100 mTorr. The energy is deposited through capacitive coupling. One major disadvantage of simple diode reactors is that the plasma density and ion energy cannot be varied independently. The addition of magnetic fields, in magnetically enhanced reactive ion etchers (MERIE), provides an extra variable parameter and allows operation at lower pressures, as does the introduction of a third electrode in triode reactors.

Much attention has been given in recent years to low pressure high density reactors, in which the electron density can exceed 10^{12} cm^{-3} at pressures around 1 mTorr. The type used most commonly by U. S. industry is the inductively coupled plasma (ICP) reactor (28, 29, 137-140). The plasma is created through inductive coupling from RF current flowing in a coil which is separated from the plasma by a dielectric window. The substrate holding the work-piece is driven by a second RF circuit and develops a DC bias, which draws the plasma ions onto the wafer. Variation of the relative power levels in the two RF circuits provides separate control over plasma density and ion energy. Inductive reactors seem to scale well in size. The transition in semiconductor processing from 200 mm wafers to 300 mm wafers should cause few problems, since ICP reactors have been used successfully for uniform etching of flat panel display substrates measuring 600 mm by 720 mm.

In electron cyclotron resonance (ECR) reactors (30), the plasma is created by the resonance absorption of microwave radiation, usually at 2.45 GHz, in an external magnetic field. These reactors can be operated at even lower pressures, below 1 mTorr, and are popular in Japan. However, they have not been widely adopted in the U. S., probably due to their extra complexity. ICP plasmas can also be enhanced through the use of magnetic fields. One such device is the helicon (31), in which the electron density is increased by the propagation of very low-frequency waves. Although helicon reactors have led to considerable academic interest, they have achieved only limited commercial success.

Physical Vapor Deposition

Physical vapor deposition, or sputtering, is used to create layers of metals, alloys or insulators in a wide variety of industries, including semiconductor manufacturing. In its simplest form, physical sputtering, a plasma is created to remove atoms from a target by ion impact and transport the neutral atoms over to the work-piece. Argon plasmas provide efficient sputtering with minimal reactions of the sputtered atoms in transit. Pressures of a few mTorr are used to reduce losses due to elastic collisions. DC glow discharges cannot easily be maintained at such pressures and magnetron discharges are favored. In magnetron reactors, permanent magnetic fields are used to trap electrons and increase the probability of the electron impact ionization that is needed to maintain the discharge. Physical sputtering is used for a broad variety of metals and alloys.

Reactive Sputtering

A greater range of deposition layers can be obtained through the use of reactive sputtering (32-34). Molecular gases are introduced and the plasma parameters optimized so that the chemical reactions lead to the desired stoichiometry in the deposited material. The properties of the layer can often be improved by controlling the temperature of the substrate and the energy and flux of the ions that strike the substrate.

Reactive sputtering is widely used to deposit dielectrics, such as oxides and nitrides, as well as carbides and silicides. Common reactive gases are O_2 and H_2O for O atoms, N_2 and NH_3 for N atoms, CH_4 and C_2H_2 for C atoms and SiH_4 for Si atoms. Reactive sputtering is often necessary even when the deposited and target materials are the same. For example, depositing insulating layers of SiO_2 from an SiO_2 target directly by physical sputtering leads to silicon-rich films. The 1:2 Si/O stoichiometry can be restored by the addition of O_2 to the Ar feed gas. Alternatively, SiO_2 layers can be produced from pure silicon targets, by using gases containing oxygen (33).

Ionized Physical Vapor Deposition

In many deposition tasks, the goal is to deposit a layer of uniform thickness over the whole workpiece. However, the development of semiconductor devices with multiple metal layers and the continued reduction in feature dimensions require the deposition of metals to the bottom of narrow trenches and holes. Since neutral atoms approach the surface from all directions, they will stick on the sides of the feature and prevent access to the bottom. The solution is to create a high-density plasma in which the majority of the sputtered atoms are ionized and accelerated by the substrate bias voltage at right angles to the work-piece. This approach (35,36) is now called ionized physical vapor deposition (I-PVD).

Both plasma and surface processes are important in reactive sputtering, to assure the correct stoichiometry and the required physical properties of the deposited films. Surface reactions on the reactor walls, as well as on the sputter target and workpiece, can affect the balance of ions and radicals in the plasma. Although there have been recent empirical studies of the correlations between film quality and the plasma parameters, the development of comprehensive models has not received as much attention as for chemical vapor deposition (CVD) and etching.

Chemical Vapor Deposition

The deposition of elements that can be extracted from gaseous compounds was traditionally achieved by chemical vapor deposition. The desired atoms are freed thermally, so that the major adjustable parameters are temperature, pressure and the flow rates of each component in the gas mixture. Ionizing the gas provides much more control over the process, producing more radicals as well as ions and avoiding the need for high substrate temperatures, which may lead to damage. This process is usually called plasma-enhanced chemical vapor deposition (PECVD).

The deposition of hydrogenated amorphous silicon (a-Si:H) plays a very important role in the manufacture of active matrix liquid crystal displays and in photovoltaic devices, such as solar cells. The common approach is to use an SiH_4/H_2 discharge in an RF reactor. 2D models, involving electrical, thermal and chemical modules, have been recently developed for these discharges and have been compared against laser-induced fluorescence measurements of radical densities (37,38).

Three types of ions are included in this model, light positive ions, such as H_2^+, along with the high mass +ve and -ve "mean" ions, SiH_m^+ and SiH_m^-. The electrons are described by three moments, density, momentum and energy, in the standard fluid approximation. However, the chemistry is dominated by the neutral radicals. The source of these radicals is assumed to be electron impact dissociation of the molecules SiH_4, Si_2H_6 and Si_3H_8. The radicals that are followed are H, SiH_n (n-0-3) and Si_2H_n (n=2-5). The Palaiseau group has provided excellent analyses of the collision processes in the plasma (39) and on the substrate (40).

The growth of particulates in plasma reactors for semiconductor manufacturing can be very deleterious and there has been considerable study of the growth and transport of dust particles. Much of this work has been focused upon silane plasmas, but there appears still to be disagreement over the relative roles of SiH_2 radicals, SiH_3 radicals and negative ions in the nucleation and growth of particulates (41-43).

Silane mixtures are also used in the production of dielectric materials by PECVD. For silicon nitride (Si_3N_4), the added gas may be N_2 (44,45) or NH_3 (45,46), whereas NO_2 can be added to give the oxynitrides SiN_xO_y. The most commonly used dielectric, SiO_2, can be produced from mixtures of SiH_4 with an oxygen donor such as O_2 (47), NO, N_2O or H_2O_2, usually diluted in Ar (48). A typical chemistry model for SiO_2 deposition can be found in the paper by Meeks et al. (47). Since silane is explosive at room temperature, a popular alternative Si source is tetraethoxysilane ($Si(O\ CH_2\ CH_3)_4$ or TEOS), which is a relatively inert liquid (49). A fascinating study of the ions and neutral fragments produced in a TEOS/Ar discharge has recently been presented by Basner et al (50).

The need to reduce the wiring capacitance in integrated circuits as metal spacimgs are reduced to below 0.5 micron has led to the search for intermetal dielectric films with lower dielectric constant (k) than that of SiO_2, which is 3.9. One possibility is to use organic polymers. Another approach is to dope SiO_2 with boron nitride, boron oxide or fluorine. Fluorinated SiO_2 films can be deposited by PECVD by modification of the gas mix. Han and Aydil (51) have produced films with k = 3.2 using a simple SF_4/O_2 mix, whereas Yoshimaru et al (52) use a more complex gas containing TEOS, He, O_2 and C_2F_6.

Dry Etching

The need for dry (plasma) etching, rather than wet (chemical) etching, has increased as the dimensions of semiconductors have shrunk. The creation of narrow trenches or holes with vertical sidewalls demands anisotropic etching, and the need to cut through one layer without damage to very thin underlying layers of different materials requires excellent selectivity in the etching process. Etching rates inside holes or trenches depend on the aspect ratio of the feature as well as on the density of nearby features that are being etched simultaneously. The gas mix is chosen to minimize these effects and to maintain control over feature shape, avoiding notches and protrusions. Plasma processing is ideal for etching in that the directionality of the incident ions provides a capability to fabricate vertical walls. Plasmas can also be turned off rapidly compared with etching times scale (tens of seconds) so that when one layer is completely etched, the process can be stopped with minimal damage to the underlying layer. A change in the material being etched often leads to changes in the plasma behavior that can be detected, either from the optical emission or the response to applied RF fields. This provides a signal that can be used to terminate etching.

In semiconductor equipment manufacturing companies, etch processes are often sub-divided into three categories, silicon, metal and dielectric.

Silicon Etching

Although the etching of amorphous silicon is important in the AMLCD and photo-voltaic industries, the demands for narrow feature size are not so demanding and so most attention has been paid to the etching of polysilicon and silicides for the IC industry. As an example of the challenges that will soon be faced, Tennant et al. (53) envisage a transistor with a 70 nm channel length. At its heart is a gate stack with 80 nm of tungsten silicide (WSi_x) on top of 100 nm of polycrystalline silicon. Underneath the stack is dielectric layer of SiO_2 of less than 2 nm in thickness. Etching through the stack and removing all the unwanted polysilicon, without breaking through the dielectric, remains the benchmark task.

Although fluorine etches Si rapidly, it is in fact too aggressive, since the etching is isotropic and the selectivity is poor. Chlorine and HBr are most commonly used as the active etchants, with possible admixtures of Ar and O_2. There is always interest in looking for alternative etchants, such as HI (54) and NF_3.

Great progress has been made in recent years in the analysis of the etching of polysilicon in chlorine plasmas. Etching proceeds through the formation of chlorinated surface layers, through the absorption of Cl, Cl_2, Cl^+ or Cl_2^+, followed by ion-induced sputtering of $SiCl_x$ molecules. Although both Cl atom and Cl_2 molecules stick readily to pristine Si surfaces, as the Cl coverage increases, the absorption of Cl_2 molecules drops and the presence of Cl atoms is essential to maintain a high etch rate. Fourier-transform infrared (FTIR) absorption measurements (55) show that the major reaction product leaving the surface is $SiCl_4$, although other $SiClx$ molecules are observed in the surface layers.

The etch rate depends not only on the energy and flux of neutrals and ions striking the surface, but also upon the flux of reaction products returning to the surface and of molecular fragments from resist erosion. Recent experiments by Choe et al (56) have shown that the levels of ionization and dissociation are both sensitive to the condition of the reactor walls and change slowly as more etching is accomplished. This is attributed to reduction of the rate of chlorine atom association on the walls due to the deposition of etching products. Molecular dynamics simulations of chlorine atom association on several of the surfaces found on wafers and reactor walls have been reported by Kota et al (57). More experimental checks of these calculations would be valuable.

This progress has been achieved through the work of many groups, using a wide variety of techniques. Beam studies of the etching reactions (58-62) have shown how the etch rate varies with the variation of single parameters, such as the energy or angle of approach of the incident ions or the degree of pre-absorption of chlorine into the surface layers. However, great care must be taken in the application of such data to etching in a real discharge, since there are many variable parameters. There appears to be an effective threshold around 20 eV for rapid ion-induced etching of Si by Cl. The corresponding threshold for SiO_2 removal is over 50 eV, and high selectivity can be obtained by arranging the impact energy of the incident ions to lie mainly between the two thresholds. The chemical state of the surface during etching has been studied by in-situ infra-red spectroscopy (55), by laser-induced fluorescence studies of laser desorbed molecules (56) and by X-ray photoelectron spectroscopy (63). Optical emission actinometry (64), two-photon laser-induced fluorescence (65) and mass spectrometry (141,142) have been used to measure the degree of dissociation of the chlorine plasma. Plasma models (66-68) have been extended to include surface reaction chains as well as collision processes within the discharge. The models attempt to predict the evolution of feature profiles as well as average etch rates and so provide more opportunities for checks against experiment.

Many forms of silicon are used in the semiconductor industry. In addition to variations in the crystal structure, silicon is often doped with electron donors or acceptors. Usually the etching chemistry must be adapted for each particular variety. Sometimes one wishes to etch continuously through two or more silicon layers in a single step and thus must make a compromise in the choice of chemistry, but the greatest challenge is to be able to etch one variety of silicon selectively with respect to another.

Metal Etch

The metal that is most commonly used in semiconductor chips is aluminum, often in alloy form with small amounts of another metal, such as copper or neodymium, added to reduce electromigration and prevent the formation of surface hillocks during deposition. The aluminum may be sandwiched between metallic thin films, such as TiN, to improve electrical contact. Aluminum can be etched anisotropically using chlorine based gas mixtures. When Al is exposed to air, a layer of aluminum oxide is formed rapidly. This layer is usually called native oxide. BCl_3 is usually added to the etching gas to break through this oxide layer. Other gases, such as Ar, O_2, or N_2 are often added to optimize the process. For example, the addition of O_2 can lessen the dependence of the etch rate on aspect ratio as well as increasing the rate (69).

Kazumi et al (70) have recently published a valuable analysis of the dissociation and ionization pathways in BCl_3 plasmas and the relationship between the plasma chemistry and etching parameters in BCl_3/Cl_2 mixtures, including a discussion of the etching of TiN and resist material. Another comprehensive plasma chemistry model has been developed by Meeks et al. (71). The effects of the negative ions in such plasmas have been studied carefully by the Sandia (Albuquerque) group (72,73).

Although the degree for selectivity required in etching metals and dielectrics is not as high as for polysilicon, minimization of the etching of the walls of narrow features is critical. It has been suggested that deposition of photoresist fragments on the walls reduces the etching (74), which could explain the observed dependence on the ratio of resist to metal.

There has been much discussion about the replacement of aluminum by copper. Although its higher conductivity makes it attractive, the difficulty of etching copper has delayed its introduction for several years. An alternative is to deposit the metal into pre-existing channels and remove the excess by chemical-mechanical polishing (CMP).

Dielectric Etch

The controlled, anisotropic etching of dielectrics such as SiO_2 selectively with respect to Si is a formidable challenge. High ion energies are required for rapid etching and energetic ion often sputter indiscriminately. The key is to deposit protective polymer films on surfaces that should not be etched, such as the walls of dielectric trenches and the underlying silicon layers. This is achieved through the use of fluorocarbon gases, the simplest of which is CF_4.

The feedstock gas must provide an appropriate mix of F atoms to do the etching and radicals to build the fluorocarbon polymer layer. The simple CF_4 gas usually provides too many F atoms. The balance can be improved by using a compound with a greater C:F ratio (75), such as C_2F_4, C_2F_6, C_3F_6, C_3F_8 or C_4F_8, or by introducing hydrogen, which extracts F atoms to form HF. The latter can be achieved by adding H_2 (76), CH_4 (77), or through the use of incompletely fluorinated methane constituents, CH_xF_y (78). Although hydrogen is very effective at abstracting F atoms from the discharge and the polymer layers, Bjorkman et al (79) have suggested that its presence will cause problems in the etching of features with very high aspect ratios.

Although polymer deposition enables selective etching of dielectrics with respect to polysilicon, the simultaneous deposition and etching adds considerable complications to models of profile evolution in oxide and nitride etching. The polymer films may also be deposited on the reactor walls. Since the plasma chemistry is sensitive to the condition of the walls, significant changes can occur through continued use of the reactor. The etching parameters drift with time and wall cleaning is a major concern in the design of etching systems for dielectrics.

Standaert et al (80) have shown that during etching by such mixtures, the protective fluorocarbon layers on polysilicon surfaces are typically 2-7 nm thick and conclude that the Si etch rate is determined by the rate of F atom diffusion through the layers. They suggest that the incorporation of H into the fluorocarbon layer slows this diffusion through the formation of HF molecules and thus suppresses Si etching even further.

The mechanism by which the protective layers are built up is as yet uncertain. Stoffels et al (81) have used electron attachment mass spectrometry to study polymer formation within the discharge. They find fully saturated polymers $F(C_nF_{2n+2})$ in CF_4 plasmas. As the amount of fluorine in the parent gas decreases, the polymers become less saturated. They are then more reactive and may more easily attach to receptive surfaces. This work supports the thesis that the polymers are formed in the discharge and then transferred to the silicon. Others suggest that it is small radicals that stick. Inayoshi et al. (82) have injected the long-lived CF_2 radicals into Ar and H_2/Ar plasmas and studied their adsorption on Si surfaces. They find that the presence of the plasma enhances the sticking rate, more in the case of Ar than in the H_2/Ar mix. The effect is so large that they believe that the ion bombardment from the plasma makes the surface more receptive to CF_2, but that H adsorption passivates the surface layer. They do not discuss the possibility that the plasma is dissociating the CF_2 to form more reactive radicals, F and CF. Clearly more studies of this kind are needed.

The interaction of fluorocarbon and hydrocarbon ions and radicals with reactor walls also deserves attention. Sugai et al (83) have studied collisions of CF_4^+ and CH_4^+ ions on aluminum. As soon as the ion beams are turned on, the target surface becomes coated with thin films of fluorocarbon or hydrocarbon and so the measurements are not on pristine surfaces. This might not meet with the approval of a Ph. D. thesis examiner, but probably provides more useful information than an experiment performed on pure Al. Sugai et al find that at low energies most of the ions are neutralized upon impact. They study the probability of dissociation among the reflected ions, but unfortunately do not detect the reflected neutrals, which are more numerous. Once again, the difficulty of observing neutral products is retarding progress, as it has with respect to electron impact reactions.

Plasma Cleaning

Control over semiconductor processing requires frequent cleaning, for the reactor itself between wafers as well as for the wafers before or after each processing step. Two problems of special interest are the removal of deposits from reactor walls and the removal of the resists that are used in lithography from the semiconductor wafers.

In deposition systems some of the feedstock gas can be deposited on the reactor walls instead of the work-piece, while in etching the wall deposits can either be etch products or the polymer layers that are used for passivation on the substrate. Photoresist mask materials are primarily long-chain organic polymers consisting mostly of hydrogen and carbon or, in the so-called hard masks, are composed of Si_3N_4 or SiO_2.

Plasma cleaning differs from plasma etching in that the process can be isotropic, but selectivity is still important to avoid damage to the work-piece or reactor. The process is also known as ashing or stripping.

Although the details of cleaning techniques vary with the primary process, the type of reactor, the gas mixture and the wall material, the basic idea is to deliver low energy etching radicals, such as F or O atoms to the surface. Typical sources of these radicals are O_2, NF_3, SF_6, C_xF_y, CF_4/O_2 and CF_4/H_2O, often mixed with rare gases or N_2 to ease plasma formation.

The radicals have traditionally been created in situ by creating a relatively quiet plasma that will do minimal damage to the system. There is growing interest in the use of downstream systems, in which the radicals are created in a separate discharge and are transported to the reactor. Clearly it is important to minimize radical recombination, either in the gas phase or on surfaces, as the gas moves from the source to the processing chamber. A data set for downstream etching in an NF_3/O_2 mix has been given by Meeks et al. (84)

As feature sizes shrink, there is growing concern about the deleterious effects of ions on semiconductor devices. Not only can energetic ions cause collateral damage, but the charging of dielectric surfaces can perturb the trajectories of electrons and ions that arrive later or can cause shorting. Thus all opportunities to perform dry chemistry using radical rather than ions are being sought, with the radicals being formed in a remote reactor (85,86).

Data Needs

A survey of data needs for the semiconductor manufacturing industry was carried out recently under the auspices of the U. S. National Research Council and their report was published in 1996 (87). This report covers structural, radiative and collisional processes in both the gas phase and on surfaces. With respect to the assembly of data on electron collisions, two promising developments since that time have been the series of articles by Christophorou, Olthoff and Rao (88) on CF_4, CHF_3, CCl_2F_2 and C_2F_6, and the internet site maintained by Kinema Research (89). The review by Oehrlein (90) provides a valuable supplement to the NRC report in regard to surface reactions.

Although the nature of the required data varies from one application to the next, a common need is to understand all the processes that govern the creation, transformation and destruction of radicals and ions. Most of the molecules that make up the plasma are created by reactions, either in the discharge or on surfaces, and the effects of reaction products on charge and energy balance and on particle transport must be considered. Excitations leading to metastable states are almost always significant, and accurate data on those specific excitations that lead to diagnostic radiation are especially valuable.

POLLUTION CONTROL

Plasma techniques for treating hazardous gases are usually divided into two classes, thermal and non-thermal methods. In the thermal approach, the applied electrical power is mostly used to heat the gas to a temperature at which the chemical reactions lead to the destruction of the unwanted species. These methods draw less on physical data and are not discussed here.

Non-thermal plasma processing operates by producing a plasma in which the majority of the electrical energy goes into the production of energetic electrons. These plasmas are characterized by electrons with kinetic energies much higher than those of the ions or molecules. Even though the electrons are short-lived under atmospheric conditions and collide relatively rarely with the pollutant molecules, they undergo many collisions with the dominant bulk-gas molecules. Electron-impact dissociation and ionization of the background gas molecules create a mix of reactive species, in the form of radicals, ions and secondary electrons, that permits unique and diverse chemical reactions to be possible even at relatively low temperatures. The potential of the approach to gas cleanup arises from the fact that these species react selectively with the pollutant molecules, which are often present in very small concentrations.

Non-thermal plasmas can be formed using electron beams or electrical discharges. In the electron beam method, electrons are accelerated by high voltage in a vacuum region before being injected through a thin foil window that serves as a vacuum seal The high-energy electrons going through the thin window can then be used to produce a large volume of plasma as they collide with the gas molecules in an atmospheric-pressure processing chamber.

In the electrical discharge method, the high voltage electrodes are immersed in the atmospheric-pressure gas, instead of a vacuum. The electrons collide with and transfer energy to the gas molecules as they drift along the high voltage region. The electrical discharge method therefore results in average electron energies less than those obtained from the electron beam method. The numbers of secondary electrons, ions and reactive free radicals are strongly influenced by the electron energy distribution. A comparison of the relative efficiencies of the two approaches for four volatile organic compounds (ethylene, o-xylene, toluene and benzene) has been given by Penetrante et al. (91).

The two major issues facing the commercial implementation of non-thermal plasmas to pollution control are the process efficiency and the identity of the byproducts (92). For example, systems to treat the exhausts from engines or power plants will usually be of little interest if they consume more than 10% of the generated power or produce molecules that are more hazardous than those being removed. There are two kinds of efficiencies of concern: (a) electrical conversion efficiency, and (b) chemical processing efficiency. The electrical conversion efficiency refers to the efficiency for converting wall plug electrical power into power deposited by the electrons into the plasma. The chemical processing efficiency refers to the amount of pollutant removed or decomposed for a given amount of energy deposited into the plasma. The chemical processing efficiency is often expressed in terms of the specific energy consumption in units such as eV per pollutant molecule, or grams of pollutant per kW-hr. The electrical conversion efficiency is highly dependent on the plasma reactor configuration and power supply used, and so will not be discussed here. On the other hand, the chemical processing efficiency is a more basic quantity, making it possible (under the same gas conditions) to compare the radical, ion or electron production in different reactors regardless of the reactor configuration or power supply.

It is often difficult to assess and compare the performance of various kinds of plasma reactors. The data presented in the literature using different kinds of reactors often were measured under different gas conditions. In many cases, the data are presented in a way that makes it impossible for the reader to determine the energy consumption of the reactor.

In this section we will examine typical processes by which non-thermal plasmas can treat various gas phase pollutants. The kinetic analysis of the deposition of energy into contaminated air will be reviewed. The collisions of electrons with the air molecules result in the formation of ions, secondary electrons and reactive free radicals. The role of these plasma species in the decomposition of various pollutant molecules will be illustrated by studies of the removal of nitrogen oxides (NO_x), methylene chloride, carbon tetrachloride, methanol and trichloroethylene. A more detailed review of the kinetics has been presented by Penetrante et al (93).

Energy Deposition

The intent in using a non-thermal plasma is to selectively transfer the input electrical energy to the electrons. In the kinetic analysis of non-thermal plasma methods, the first step is to understand the deposition of energy into contaminated air. This is controlled primarily by the major components, N_2 and O_2. Because of the resonance in electron scattering by N_2 near 2 eV, low energy electrons lose considerable energy through vibrational excitation (94, 95) which does little to enhance the desired reactions. Thus raising the average electron energy well above 2 eV is necessary for efficient treatment.

The most useful energy deposition into N_2 and O_2 is usually associated with the production of N and O atoms through electron impact dissociation (96,97) and ionization (94,95), producing either atomic or molecular ions. One must also account for the creation of metastables, especially in the dissociated atoms O and N. For example, the reaction rates for the metastable $O(^1D)$ are almost always larger than those for the corresponding reactions with ground state $O(^3P)$ (98-100). This can enhance both desired and undesired reactions. Detailed analyses of electron transport and electron excitation processes in N_2-O_2 discharges have been given recently by Guerra et al. (101) and by Slavik and Colonna (102).

The electron energy distribution in a plasma reactor is important because it determines the types of radicals produced in the plasma and the input electrical energy required to produce those radicals. In discharge processing, the rate coefficients for electron-impact dissociation reactions strongly depend on the electron mean energy in the discharge plasma. In coronal discharges, the non-thermal plasma is produced through the formation of statistically distributed microdischarges known as streamers (103-107). The electrons dissociate and ionize the background gas molecules within nanoseconds in the narrow channel formed by each microdischarge. The electron energy distribution in the plasma is complicated because the electric field is strongly non-uniform (e.g. because of strong space-charge field effects) and time dependent. During the microdischarge formation phase, the electron number rises drastically. Due to field strength enhancement in the ionization wave, the highest electron energies occur during this phase. The mean electron energy reaches values of more than 10 eV - suitable for considerable dissociation and ionization of the gas. However, since this is a highly transient phase, and since the ionization wave covers only small parts of the gap at any one time, this phase may be less important in producing most of the active radicals.

The Livermore group (93) believes that most of the species responsible for the chemical processing are generated during the main current flow in established microdischarge channels. Other authors (106,107) believe that most of the active species are produced in the strong field regions near the tip. This is an important issue, not just a debating point between theorists. If the number of ions and radicals depends critically on the maximum strength of the electric field in the streamer tip, there should be scope for innovative devices to optimize the pulsed power system that drives the streamers. On the other hand, if the average field strength is more important, there is little to be gained by complex electrode geometries or pulse shapes. The Livermore group believes that the electrical engineers should concentrate more on the wall-plug efficiency of the power system and the lifetime and cost of the device than on tailoring the system to increase the electric field in the streamer tip.

The dissipation of input electrical power as a function of the average kinetic energy of the electrons in a dry air discharge is shown in Figure 5. The electron mean energy in most electrical discharge reactors operating at atmospheric pressure is typically between 3 and 6 eV. In this range, a large fraction of the input power is wasted in vibrational excitation of N_2 and a significant fraction goes into dissociation of O_2. The electron mean energy in electrical discharge reactors is optimal for the electron-impact dissociation of O_2, which is important for the production of O radicals. These oxidizing radicals play a key role in the generation of ozone (108-110). However, for reaction chains that are initiated by dissociation of N_2 or by the creation of ion pairs, the higher energies obtained in the electron beam method usually lead to higher chemical efficiencies. More detailed comparisons of energy deposition mechanisms in electron beam and coronal discharge plasmas can be found elsewhere (92,93,111).

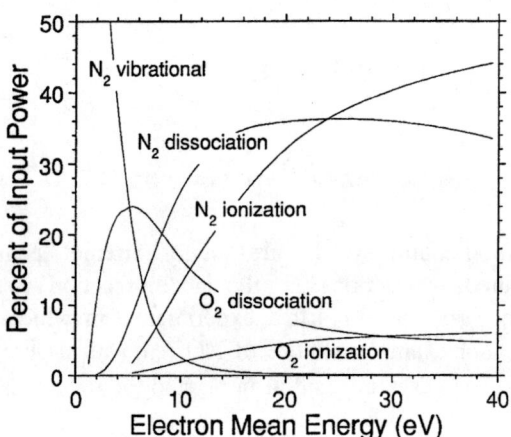

FIGURE 5. Electrical power deposition in a dry air discharge showing the percentage of input power consumed in various electron-impact processes (Courtesy: B. M. Penetrante).

In humid air mixtures (112,113), OH radicals can be produced in a variety of ways. In discharge reactors for which the electron mean energy is low, the OH radicals are produced via three types of reactions:
electron attachment:
$$e + H_2O \rightarrow H^- + OH \tag{1}$$
direct dissociation by electron impact:
$$e + H_2O \rightarrow e + H + OH \tag{2}$$
dissociation by $O(^1D)$:
$$O(^1D) + H_2O \rightarrow 2\ OH \tag{3}$$
In electron beam reactors, the OH radicals come mainly from the positive ions reacting with H_2O. The sequence of fast steps are as follows:
electron-impact ionization:
$$e + O_2 \rightarrow 2e + O_2^+ \tag{4}$$
and similar ionization processes to produce molecular ions N_2^+, H_2O^+, CO_2^+
electron-impact dissociative ionization:
$$e + O_2 \rightarrow 2e + O + O^+ \tag{5}$$
and similar dissociative ionization processes to produce N^+, H^+

charge transfer reactions to form additional O_2^+ ions, such as:

$$N_2^+ + O_2 \rightarrow N_2 + O_2^+ \tag{6}$$

formation of water cluster ions:

$$O_2^+ + H_2O + M \rightarrow O_2^+(H_2O) + M \tag{7}$$

dissociative reactions of water cluster ions to form OH:

$$O_2^+(H_2O) + H_2O \rightarrow H_3O^+ + O_2 + OH \tag{8}$$

$$O_2^+(H_2O) + H_2O \rightarrow H_3O^+(OH) + O_2 \tag{9}$$

followed by

$$H_3O^+(OH) + H_2 \rightarrow H_3O^+ + H_2O + OH \tag{10}$$

Benchmark Experiments

The kinetics of pollution control systems are usually extremely complex, involving tens of species and hundreds of reactions, and the data on reaction rates is incomplete. Thus it is important to carry out simplified experiments, in which the number of components is limited. For example, studies of NO removal in pure N_2 (114) have helped to check the data on energy deposition in N_2 and the reduction of NO through the reaction

$$N + NO \rightarrow N_2 + O \tag{11}$$

It must be stressed that simplified experiments of this kind are valuable only for checking reaction rate data and not as direct indicators of what happens in real systems. To complete the analysis of NO removal from exhaust gases, it is necessary to account for the effects of other gas-phase constituents, such as O_2, CO_2 and H_2O, the gas temperature, and surface reactions on particulates as well as chamber walls.

For very dilute concentrations of NO in N_2, the input energy required for NO reduction is determined by the energy required for dissociation of N_2. The input electrical energy is consumed in electron-impact reactions with N_2 and the removal of NO proceeds mainly via reduction by the N atom. By doing experiments using this mixture it is therefore possible to examine the dependence of the dissociation rate of N_2 on the different types of plasma reactors. These experiments (114) also provide a validation of the calculated N_2 dissociation rates.

Penetrante et al (115) presented a systematic comparison of data on plasma-induced NO reduction obtained from three different laboratories using independently constructed discharge reactors and different chemical diagnostic methods. By using identical gas mixtures, the plasma processing performance of various reactors could be compared. The important control parameter is the energy density input, which is the power input into the gas divided by the total gas flow rate. The concentration of NO (in units of parts per million) as a function of the input energy density (in units of Joules per standard liter) is shown in Figure 6. The six coronal reactors used a wide variety of electrode materials, geometries and power supplies, but showed very similar chemical efficiencies. The specific energy cost for NO reduction using the various electrical discharge reactors ranged from 0.8 to 1.25 ppm-liters/Joule.

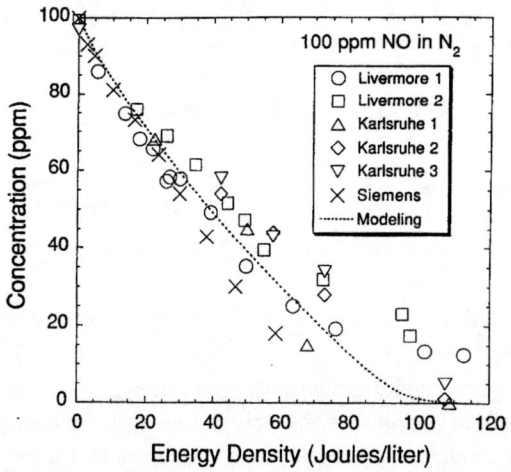

FIGURE 6. The destruction of NO molecules as a function of deposited energy in pulsed corona and dielectric-barrier discharges, beginning with 100 ppm NO in N_2 at atmospheric pressure.

The dotted line in Figure 6 shows the results of models using the cross sections for electron impact dissociation of N_2 as measured by Cosby (97). The agreement is much poorer when the cross sections of Winters (96) are used. It would be very useful if a third experimental measurement could be made to determine which measurement is correct.

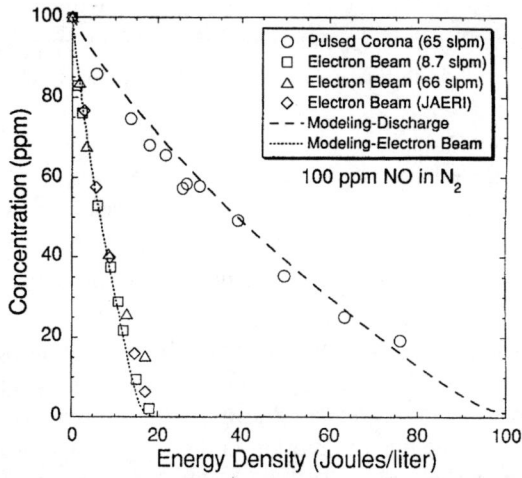

FIGURE 7. Electron beam and pulsed corona processing of 100 ppm NO in N_2.

363

In figure 7, these results obtained with pulsed corona are compared with those obtained in three electron beam experiments and with models using the same cross section set. Full details can be found in the paper by Penetrante et al (93). The greater efficiency of NO destruction in the electron beam experiments is due to the larger dissociation rates that arise from the higher average electron energy.

Reaction Mechanisms

When small amounts of NO are destroyed in N_2 discharges, the process is reductive, leading to N_2 and O_2, driven by reaction (11). However, in the presence of O_2, the removal process is primarily oxidative (116), driven by the reaction

$$O + NO + M \rightarrow NO_2 + M. \tag{12}$$

In the presence of water, acids are usually formed through reactions like

$$OH + NO_2 + M \rightarrow HNO_3 + M. \tag{13}$$

The presence of water droplets can introduce surface reactions, which must be taken into account (117). In stationary applications, such as power plants burning fossil fuel (118,119) and automobile tunnels, these acids can be neutralized, for example through the addition of ammonia (120) to form phosphates. These can be sold as fertilizer, if they are not contaminated with toxic by-products. Since the easiest way to remove NO is to turn it into other oxides of nitrogen, it is important not to base analyses purely on the disappearance of NO.

Much of the research into non-thermal treatment of gaseous pollutants has been focused upon volatile organic compounds. An excellent review of this work has been given by Vercammen et al. (121). A major concern is with chlorinated carbon and hydrocarbon compounds, which are widely used as industrial solvents and are often found in hazardous concentrations at government and industrial cleanup sites. For example, methylene chloride (CH_2Cl_2) is used for removing paint from aircraft. An examination of the rate coefficients for methylene chloride decomposition in a non-thermal plasma (122) show that the most likely mechanism is the reaction with nitrogen atoms

$$N + CH_2Cl_2 \rightarrow \text{products} \tag{14}$$

If the decomposition of methylene chloride is dominated by reaction (14), we would expect that, for the same initial concentrations, the energy consumption for processing methylene chloride in N_2 should be the same as the energy consumption for processing NO in N_2. The rate controlling step is the production of N atoms. This has been verified experimentally (122), as shown in Figure 8. Deviations between methylene chloride and NO are observed only after the concentration has dropped down to around 15 ppm. This deviation may be caused by the consumption of N atoms in reactions with the intermediate products during the decomposition of methylene chloride. The destruction of methylene chloride in discharge reactors becomes much more efficient if the temperature is raised to 300C (122,123).

In dry air, methanol is primarily destroyed (124) by ions, through the dissociative charge transfer reaction

$$N_2^+ + CH_3OH \rightarrow CH_3^+ + OH + N_2 \tag{15}$$

Additional reactions may result from the formation of OH radicals from the initial decomposition reaction (15). Comparison of the decomposition rates in pure nitrogen and dry air suggests that O radicals are not particularly efficient at destroying methanol.

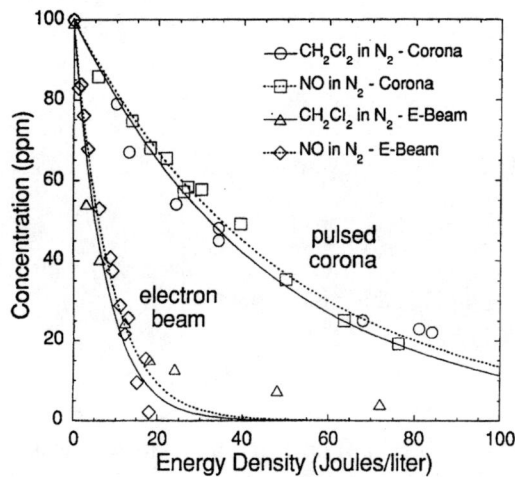

FIGURE 8. Comparison of the destruction rates of NO and methylene chloride in electron beam and pulsed corona reactors.

Examination of the rate coefficients for carbon tetrachloride decomposition in a non-thermal plasma (125-128) show that the rate limiting step is the dissociative attachment of carbon tetrachloride to secondary electrons

$$e + CCl_4 \rightarrow Cl^- + Cl_3 \tag{16}$$

Since an electron-ion pair is produced during an ionization event, the energy consumption for producing electrons should be the same as that for producing ions. In the case of carbon tetrachloride, the electrons do the decomposition. In the case of methanol, the positive ions are the prime agent. Experiments (124,125) show that the energy required to consumption to destroy carbon tetrachloride in dry air is the same as that for decomposing methanol. This suggests that the rate limiting step is electron-ion pair formation in methanol as well as in CCl_4.

There has been significant recent progress (129-131) in understanding the complex reaction chain responsible for the destruction of trichloroethylene (TCE or C_2HCl_3). At least in the electron-beam treatment, the initial reaction appears to be electron impact dissociative attachment

$$e + C_2HCl_3 \rightarrow Cl^- + C_2HCl_2 \tag{17}$$

In discharge processing, the electron density is lower and TCE can also be broken up by attacks from O, OH or O_3 upon the C=C double bond (92,129,130). The observation that the rate of destruction increases as the removal proceeds has led to the conclusion that there must be an autocatalytic process, involving a chain of reactions initiated by Cl radicals, that not only destroys one TCE molecule, but also creates several new Cl radicals.

Analysis of the by-products is extremely important in pollution treatment. Two products that may arise from the destruction of chlorinated hydrocarbons are phosgene ($COCl_2$) and dichloroacetyl chloride (DCAC or $CHCl_2COCl$). Although phosgene is extremely toxic, it can be easily removed by bubbling through water.

Scavenging Reactions

Not all reactions of radicals or ions are beneficial. For example, the three-body reaction

$$Cl + CCl_3 + M \rightarrow CCl_4 + M \tag{18}$$

destroys two radicals and recreates the unwanted molecule. Fortunately, the presence of O_2 scavenges the CCl_3 through the fast reaction

$$CCl_3 + O_2 \rightarrow CCl_3O_2 \tag{19}$$

Scavenging reactions can also be detrimental. For example, electrons can be removed through attachment to O_2 or H_2O rather than to CCl_4. After the concentration of CCl_4 has decreased to a few tens of ppm, the three-body attachment of secondary electrons to oxygen molecules

$$e + O_2 + O_2 \rightarrow O_2^- + O_2 \tag{20}$$

$$e + O_2 + N_2 \rightarrow O_2^- + N_2 \tag{21}$$

becomes a significant electron loss pathway compared to reaction (16). The attachment frequency of secondary electrons to O_2 in dry air at atmospheric pressure is

$$v_{O2} = k_{(20)} [O_2]^2 + k_{(21)} [N_2][O_2] \approx 0.8 \times 10^8 \text{ s}^{-1}. \tag{22}$$

For 100 ppm CCl_4, the attachment frequency to CCl_4 is

$$v_{CCl4} = k_{(17)} [CCl_4] \approx 10^9 \text{ s}^{-1}. \tag{23}$$

When the concentration of CCl_4 is down to around 10 ppm, the electrons will attach to oxygen molecules as frequently as to CCl_4 molecules.

Humidity enhances the attachment of electrons to O_2 via

$$e + O_2 + H_2O \rightarrow O_2^- + H_2O \tag{24}$$

In humid air, the attachment frequency of secondary electrons to O_2 is

$$v_{O2} = k_{(20)} [O_2]^2 + k_{(21)} [N_2][O_2] + k_{(24)} [H_2O][O_2] \approx 1.5 \times 10^8 \text{ s}^{-1}. \tag{25}$$

Figure 9 compares the results of experiments on electron beam processing of 100 ppm of CCl_4 in dry air and humid air. Note that humidity is deleterious to the decomposition of CCl4. Humidity enhances the attachment of electrons to O_2, thus effectively decreasing the efficiency for decomposition of CCl4.

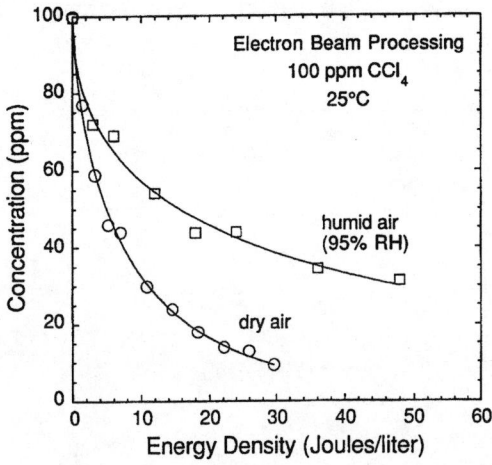

FIGURE 9. Electron beam processing of 100 ppm CCl_4 in dry air and humid air at atmospheric pressure.

Effect of Excited Species

The amount of NO reduction is directly proportional to the number of N_2 dissociations that can be achieved in the plasma. Unfortunately, not all of the N atoms resulting from the dissociation of N_2 lead to the reduction of NO. Dissociative excitation of N_2 contributes a large fraction to the total N_2 dissociation (131). A significant species produced by dissociative excitation of N_2 is the long-lived metastable, $N(^2D)$. For electron beam reactors, over half of the total N radicals produced are in the excited metastable states. The rate constants characterizing the interaction of the metastable species $N(^2D)$ with various gases are large In the treatment of NO, there are two competing reactions involving the $N(^2D)$ metastable species:

$$N(^2D) + NO \rightarrow N_2 + O \qquad (26)$$
$$N(^2D) + O_2 \rightarrow NO + O \qquad (27)$$

With 1000 ppm NO and 10% O_2, the $N(^2D)$ species is ten times more likely to react with O_2 than with NO. This means that $N(^2D)$ is consumed in the production of NO rather than in the reduction of NO. Whereas the reaction of ground state N atoms,

N(^{4}S), with O_2 can proceed only at very high temperatures, the reaction of excited N atoms, N(^{2}D), with O_2 can proceed even at room temperature. Since almost half of the total N atoms produced in the plasma are in this excited state, the reduction of NO by the ground state N atoms is almost completely counterbalanced by the production of NO by the excited N atoms. What is left in terms of NO reactions is the oxidation reaction (12)

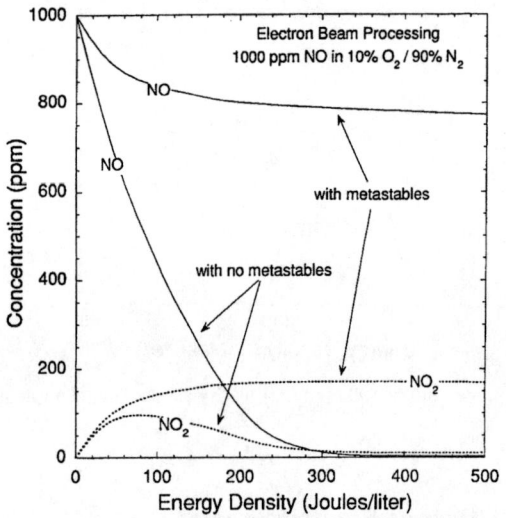

FIGURE 10. Calculations showing the effect of N(^{2}D) metastable atoms on electron beam processing of 1000 ppm NO in a 90% N_2/ 10% O_2 mixture at atmospheric pressure.

Calculations showing the effect of the metastable species N(^{2}D) on electron beam processing of 1000 ppm NO in a 90% N_2 /10% O_2 mixture is shown in Figure 10. The agreement between these results and experiments (143) at the Japan Atomic Energy Research Institute (JAERI) on electron beam processing of 500 ppm NO in 97% N_2/3% O_2 provides support for the thesis that N(^{2}D) is responsible for the deleterious effect of O_2 in electron beam processing.

Catalytic Surface Reactions

Despite the progress that has been made over the past decade, the economics of pollution control using coronal discharges seems marginal at best and their is little incentive to chemical engineers to adopt a new technology. The prime approach to cost-effective chemical processing is to use catalytic reactions (132) and so there is growing interest in the combination of plasma techniques and catalysis. Removing

NO_x from diesel engines and lean-burn gasoline engines is a particularly attractive target for such work, since the catalysts that work so well for automobile engines are quickly poisoned due to the added oxygen and water content of the diesel exhaust.

One of the first approaches to combine plasma and surface reactions was through the use of a packed bed reactor (133,134), in which pellets are used both to enhance the electric fields and to provide surfaces for heterogeneous reactions. Unfortunately, this work has become so sensitive that unsuccessful attempts are published (135), whereas more successful ones are not. Billions of dollars are at stake! Once again, the paucity of data concerning the interactions of plasmas with real (far from pristine) surfaces is seriously hampering research.

Further Data Needs

By understanding what plasma species is responsible for the decomposition of a pollutant molecule, it is possible to establish the electrical power requirements of the plasma reactor and help identify the initial reactions that lead to the subsequent process chemistry. There is much work still to be done in understanding the full chemical kinetics for many pollutant molecules. Nevertheless, the kinetics with respect to electron-impact reactions can been studied thoroughly enough so that it is possible to identify which plasma component (electron, positive ion, nitrogen atom or oxygen radical) is mainly responsible for the initial decomposition of various pollutant molecules. The effort to assemble and assess data on the neutral reactions in the atmosphere (99,100,136) is also very useful in understanding the susequent chemistry. It would be helpful if the data sets on atmospheric ion-molecule reactions could also be updated.

Although the execution of benchmark experiments and the development of complex reaction schemes have led to considerable progress in understanding the kinetics for relatively simple gas-phase systems, much more work is needed on the effects of minor constituents in the exhaust, surface reactions and temperature variations. Determination of the energy efficiency, by-products and scalability of the treatment process are critical to the commercial success of this technology, and detailed understanding of the underlying science will help in each respect.

CONCLUSIONS

Almost all industrial plasmas are unconfined. This means that for neutral atoms and molecules or ions, collisions with surfaces are at least as important, and often more important, than gas phase collisions. For electrons, collisions within the plasma are more significant, since the approach of electrons to the walls is usually retarded by the plasma potential. Thus electron impact dissociation of molecules is usually balanced by association on surfaces. On the other hand, electron impact ionization is often balanced mainly by gas-phase recombination, either through dissociative recombination or ion-ion reactions.

The major need for data currently is in support of computer simulations of industrial processes. These are usually used for off-line analysis and in support of empirical development of new processes and new equipment. When databases are more complete and simulations more accurate, it will become possible to use plasma diagnostics for real-time control of plasma processing and for computer-aided design in process development (137). This could lead to significant savings in time and expense, but has not yet been achieved in most industrial applications of plasmas.

The list of species and processes in this review is by no means complete, but shows the need for much work by the atomic, molecular and optical physics community. One characteristic is perhaps worthy of mention. The atoms and molecules that are listed here are not the easiest for academic study, either experimentally or theoretically. The industrial interest often arises because the gases are reactive and therefore nasty to handle, in one way or other. Thus there is often a need to know the details of a particular process that has been studied by researchers for other, more benign, species. It is then very difficult for academics to obtain funds to make the appropriate measurement or calculation. It is essential that government, academia and industry continue to work together to identify those processes that are of paramount importance and to provide support for the generation, assembly and dissemination of the necessary data.

ACKNOWLEDGEMENTS

This work was performed partly at Lawrence Livermore National Laboratory under the auspices of the U.S. Department of Energy under Contract Number W-7405-ENG-48 and partly at the U. S. Display Consortium with support from the Defense Advanced Projects Agency. The author is grateful for the assistance of many of his colleagues in the preparation of this manuscript, particularly to Bernie Penetrante, Jong Shon and Peter Vitello.

REFERENCES

1. Siemens, W., *Ann. Phys. Chem.* **102**, 66-122 (1857).
2. Hittorf, W., *Ann. Physik* **21**, 90-139 (1884).
3. Thomson, J. J., *Phil. Mag.* **32**, 321-36, 445-64 (1891).
4. "Tesla's experiments with alternating current at high frequency", *Electrical Engineer*, **7**, 549-50 (1891).
5. Wharmby, D. O., "Electrodeless lamps for lighting", *IEE Proceedings A* **140**, 465-73 (1993).

6. Work, D.E., *Light. Res. Technol.* **13**, 143 (1981).
7. Dakin, J. T., Rautenburg, T. H., and Goldfield, E. M., *J. Appl. Phys.* **66**, 4074-88 (1989).
8. Shaffner, R. O., "Prime Color Light Source Development for Helmet Mounted Displays" in *SPIE Proceedings* **2735**, 136-40 (1996).
9. Kogelschatz, U., Eliasson, B., and Egli, W., *J. Physique IV* **7**, Colloque C4, 47-66 (1997).
10. Kogelschatz, U., *Appl. Surf. Sci.* **54**, 410-23 (1992).
11. Rosocha, L. A., "Processing of hazardous chemicals using silent-discharge plasmas", in *Plasma Science and the Environment*, (eds. Mannheimer, W., Sugiyama, L. E. ,Stix, T. H., AIP, New York, 1997) pp. 261-98.
12. Ilcisin, K. J., Buzak, T. S., and Parker, G. J., *J. Physique IV* **7**, Colloque C4, 225-34 (1997).
13. Ilcisin, K., Buzak, T., Hinchliffe, R., Martin P., Roberson, M., Kakizaki, T., Tanamachi, S., Hayashi, M. amd Morita T., "Breakthrough gas mixtures for HDTV performance in plasma addressed displays", in *Eurodisplay 96* (Society of Information Display, 1996) pp 595-99.
14. Boeuf, J.-P., Punset, C., Hirech, A. and Doyeux, H., *J. Physique IV* **7**, Colloque C4, 3-14 (1997).
15. Punset, C., Boeuf, J.-P., and Pitchford, L. C., *J. Appl. Phys.* **83**, 1884-97 (1998).
16. Veerasingham, R., Campbell, R. B., and McGrath, R. T., *IEEE Trans. Plasma Sci.* **24**, 1411 (1996).
17. Choi, K. C., and Whang, K. W., *IEEEE Trans. Plasma Sci.* **23**, 399-404 (1995).
18. Fiala, A., Pitchford, L. C., and Boeuf, J.-P., *Phys. Rev E* **49**, 5607-22 (1994).
19. Veerasingham, R., Campbell, R. B., and McGrath, R. T., *Plasma Sources Sci. Technol.* **6**, 157-169 (1997).
20. Falkenstein, Z., and Coogan, J. J., *J. Phys. D* **30**, 2704-10 (1997).
21. Meunier, J, Belanguer, P., and Boeuf, J. -P., *J. Appl. Phys.* **78** 731-45 (1995).
22. Boeuf, J.-P., and Pitchford, L. C., *IEEE Trans. Plasma Sci.* **24**, 95 (1996).
23. Tachibana, K. Kosugi, N., Sakai, T., *Appl. Phys. Lett.* **65**, 935-7 (1994).
24. Lieberman, M. A. and Lichtenberg, A. J., *"Principles of Plasma Discharges and Materials Processing"*, New York: John Wiley, 1994.
25. Ricard, A., *J. Phys. D.* **30**, 2261-9 (1997).
26. Baldwin, M. J., Collins, G. A., Fewell, M. P., Haydon, S. C., Kumar, S., Short, K. T., and Tendys, J., *Jpn. J. Appl. Phys.* **36**, 4941-8 (1997).
27. Jones, E. C., Shao, J.,Denholm, A. S., and Cheung, N. W., *Jpn. J. Phys.* **36**, 4935-40 (1997).
28. Keller, J. H., Forster, J. C., and Barnes, M. S., *J. Vac. Sci. Technol. A.* **11**, 2487 (1993); Barnes, M. S., Forster, J. C., and Keller, J. H., *Appl. Phys Lett.* **62**, 2622 (1993).
29. Carter, J. B., Holland, J. P., Peltzer, E., Richardson, B., Bogle, E., Nguyen, H. T., Melaku, Y., Gates, D., and Ben-Dor, M., ., *J. Vac. Sci. Technol. B.* **11**, 1301 (1993).

30. Asmussen, J., Grotjohn, T. A., and Perrin, M. A., *IEEE Trans. Plasma Sci.* **25**, 1196-221 (1997).
31. Chen, F. F. and Boswell, R. W., *IEEE Trans. Plasma Sci.* **25**, 1245-57 (1997).
32. Yoshimura, K., Miki, T., and Tanemura, S., *J. Vac. Sci. Technol. A.* **15**, 2673-6 (1997).
33. Nakashima, H., Furukawa, K. Liu, Y. C., Gao, D. W. Kashiwakazi, Y., Muraoka, K., Shibata, K., and Tsurushima, T., *J. Vac. Sci. Technol. A.* **15**, 1951-4 (1997).
34. McCormick, C. S., Weber, C. E., Abelson, J. R., Davis, G. A., Weiss, R. E., and Aebi, V., *J. Vac. Sci. Technol. A.* **15**, 2770-6 (1997).
35. Hopwood, J. and Qian, F., *J. Appl. Phys.* **78**, 758-65 (1995).
36. Dickson, M., and Hopwood, J., *J. Vac. Sci. Technol. A.* **15**, 2307-12 (1997).
37. Leroy, O., Jolly, J., and Perrin, J., "2D Electrical, thermal and chemical modeling and diagnostics of SiH_4-H_2 RF discharges", *Proceedings of the 3rd Int. Conf. on Reactive Plasmas*, Nara, January 1997, pp. 13-4.
38. Leroy, O., Dorval, N., Hertl, M., Jolly, J., and Pealat, M., "Theoretical and experimental study of SiH radical density and thin film growth in PECVD H_2/SiH_4 radio frequency discharges", in *Proceedings of the 23rd Int. Conf. on Phenomena in Ionized Gases, Vol IV,* Toulouse, July 1997, pp. 216-7.
39. Perrin, J., Leroy, O., and Bordage, M. C., *Contrib. Plasma Phys.* **36**, 3-49 (1996).
40. Perrin, J., Shiritani, M., Kae-Nune, P., Videlot, H., Jolly, J., and Guillon, J., *J. Vac. Sci. Technol. A.* **16**, 278-89 (1998).
41. Kawasaki, H., Ohkura, H., Fukuzawa, T., Shiritani, M., Watanabe, Y., Yamamoto, Y., Suganuma, S., Hori, M., and Goto, T., *Jpn. J. Appl. Phys.* **36**, 4985-8 (1997).
42. Kim, D.-J., and Kim, K.-S., *Jpn. J. Appl. Phys.* **36**, 4989-96 (1997).
43. Childs, M. A., and Gallagher, A., *Bull. Amer. Phys. Soc.* **42**, No 8,1733 (1997).
44. Hugon, M. C., Delmotte, F., Agius, B., and Courant, J. L., *J. Vac. Sci. Technol. A.* **15**, 3143-53 (1997).
45. Delmotte, F., Hugon, M. C., Agius, B., and Courant, J. L., *J. Vac. Sci. Technol. B.* **15**, 1919-26 (1997).
46. Caquineau, H., Dupont, G., Despax, B., and Couderc, J. P., *J. Vac. Sci. Technol. A* **14**, 2071-82 (1996).
47. Meeks, E., Larson, R. S., Ho, P., Apblett, C., Han, S. M., Edelberg, E., Aydil, E. S., *J. Vac. Sci. Technol. A.* **16**, 544-63 (1998).
48. Gaillard, F., Brault, P., and Broquet, P., *J. Vac. Sci. Technol. A.* **15**, 2478-84 (1997).
49. Fracassi, F., d'Agostino, R., and Favia, P., *J. Electrochem. Soc.* **139**, 2636-44 (1992).
50. Basner, R., Foest, R., Schmidt, M., Hempel, F., and Becker, K., "Ions and Neutrals in the Ar-TEOS RF Discharge", in *Proceedings of the 23rd Int. Conf. on Phenomena in Ionized Gases, Vol IV,* Toulouse, July 1997, pp. 196-7.
51. Han, S. M. and Aydil, E. S., *J. Vac. Sci. Technol. A.* **15**, 2893-904 (1997).
52. Yoshimaru, M., Koizumi, S., and Shimokawa, K., *J. Vac. Sci. Technol. A.* **15**, 2908-14, 2915-22 (1997).

53. Tennant, D., Klemens, F., Sorsch, T., Baumann, F., Timp, G., Layadi, N., Kornblit, A., Sapjeta, B. J., Rosamilia, J., Boone, T., Weir, B., and Silverman, P., *J. Vac. Sci. Technol. B*. **15**, 2799-805 (1997).
54. Richter, H. H., Aminpur, M.-A., Erzgraber, H. B., Wolff, A., Kruger, D., Dehoff, A., and Reetz, M., *Jpn. J. Appl. Phys.* **36**, 4649-53 (1997).
55. Nishikawa, K., Ono, K., Tuda, M., Oomori, T., and Namba, K., *Jpn. J. Appl. Phys.* **34**, 3731-6 (1995).
56. Choe, J. Y. Herman, I. P., and Donnelly, V. M., *J. Vac. Sci. Technol. A* **15**, 3024-31 (1997).
57. Kota, G. P., Coburn, J. W., and Graves, D. B., *J. Vac. Sci. Technol. A*. **16**, 270-7 (1998).
58. Chang, J. P., Arnold, J. C., Zau, G. C. H., Shin, H.-S., and Sawin, H. H. *J. Vac. Sci. Technol. A* **15**, 1853-63 (1997).
59. Chang, J. P., Mahorowala, A. P., and Sawin, H. H., *J. Vac. Sci. Technol. A*. **16**, 217-24 (1998).
60. Levinson, J. A., Shaqfeh, E. S. G., Balooch, M., and Hamza, A. V., *J. Vac. Sci. Technol. A* **15**, 1902-12 (1997).
61. Ono, K., and Tuda, M., *Jpn. J. Appl. Phys.* **36** 4854-65 (1997).
62. Doshita, H., Ohtani, K., and Namiki, A., *J. Vac. Sci. Technol.A.* **16**, 265-9 (1998).
63. Layadi, N., Donnelly, V. M, and Lee, J. T. C., *J. Appl. Phys.* **81**, 6738-48 (1997).
64. Donnelly, V. M., *J. Vac. Sci. Technol. A* **14**, 1076-87 (1996).
65. Ono K., Tuda, M., Nishikawa, K., Oomori, T., and Namba, K., *Jpn. J. Appl. Phys.* **34**, 3731-6 (1995).
66. Lee, C., Graves, D. B., and Lieberman, M. A., *Plasma Chem. and Plasma Proc.* **16**, 99-120 (1996).
67. Hoekstra, R. J., Grapperhaus, M. J., and Kushner, M. J., *J. Vac. Sci. Technol. A* **15**, 1913-21 (1997).
68. Tuda, M., Nishikawa, K., and Ono, K., *J. Appl. Phys.* **81**, 960-7 (1997).
69. Banjo, T., Tsuchihashi, M., Hanazaki, M., Tuda, M., and Ono, K., *Jpn. J. Appl. Phys.* **36**, 4824-8 (1997).
70. Kazumi, H., Hamasaki, H., and Tago, K., *Jpn. J. Appl. Phys.* **36**, 4829-37 (1997).
71. Meeks, E., Ho, P., Ting, A., and R. J. Buss, submitted to *J. Vac. Sci. Technol. A*. (1998).
72. Fleddermann, C. B., and Hebner, G. A., *J. Vac. Sci. Technol. A* **15**, 1955-62 (1997).
73. Hebner, G. A., Fleddermann, C. B. and Miller, P. A., *J. Vac. Sci. Technol. A* **15**, 2698-708 (1997).
74. Tachi, S., Izawa, M., Tsujimoto, K., Kure, T., Kofuji, N., Suzuki, K., Hamasaki, R., and Kojima, M., *J. Vac. Sci. Technol.A.* **16**, 265-9 (1998).
75. Bell, F. H., Joubert, O., Oehrlein, G. S. Zhang, Y., and Vender, D., *J. Vac. Sci. Technol. A* **12**, 3095-101 (1997).
76. Marra, D. C., and Aydil, E. S., *J. Vac. Sci. Technol. A* **15**, 2508-17 (1997).

77. Den, S., Kuno, T., Ito, M., Hori, M., Goto, T., O'Keeffe, P. Hayashi, Y., and Sakamoto, Y., *J. Vac. Sci. Technol. A* **15**, 2880-4 (1997).
78. Rueger, N. R., Beulens, J. J., Schaepkens, M., Doemling, M. F., Mirza, J. M., Standaert, T. E. F. M., and Oehrlein, G. S. *J. Vac. Sci. Technol. A* **15**, 1881-9 (1997).
79. Bjorkman, C. H., Shan, H., Doan, K., Wang, J., Pu, B., and Welch, M., "Effect of hydrogen in high aspect ratio, small feature dielectric etch", presented at the American Vacuum Society 44th National Symposium, October 1997, *Book of Abstracts*, p. 149.
80. Standaert, T. E. F. M., Schaepkens, M., Rueger, N. R., Sebel, P. G. M., Oehrlein, G. S., and Cook, J. M., *J. Vac. Sci. Technol.A.* **16**, 239-49 (1998).
81. Stoffels, W. W., Stoffels, E., and Tachibana, K., *J. Vac. Sci. Technol.A.* **16**, 87-95 (1998).
82. Inayoshi, M., Ito, M., Hori, M., Goto, T., and Hiramatsu, M., *J. Vac. Sci. Technol.A.* **16**, 233-8 (1998).
83. Sugai, H., Mitsuoka, Y., and Toyoda, H., *J. Vac. Sci. Technol.A.* **16**, 290-3 (1998).
84. Meeks, E., Larson, R. S., Vosen, S. R., and Shon, J. W., *J. Electrochem. Soc.* **144**, 357-66 (1997).
85. Matsuo, P. J., Kastenmeier, B. E. E., Beulens, J. J., and Oehrlein, G. S., *J. Vac. Sci. Technol. A* **15**, 1801-13 (1997).
86. Brooks, C. B., Buie, M. J., and Vaidya, K. J., *J. Vac. Sci. Technol.A.* **16**, 260-4 (1998).
87. Graves, D. B., Kushner, M. J., Gallagher, J. W., Garscadden, A., Oehrlein, G. S., and Phelps, A. V., "Database Needs for Modeling and Simulation of Plasma Processing", National Academy Press, Washington, DC, 1996.
88. Christophorou, L. G., Olthoff, J. K., and Rao, M. V. V. S., *J. Phys. Chem. Ref. Data* **25**, 1341-88 (1996); **26**, 1-15 (1997); **26**, 1205-37 (1997); **28**, 1-29 (1998).
89. Morgan, W. L., *Kinema Research Web Site*, http://www.csn.net/kinema.
90. Oehrlein, *Surf. Sci.* **386**, 222-30 (1997).
91. Penetrante, B. M., Hsiao, M. C., Bardsley, J. N., Merritt, B. T., Vogtlin, G. E., Kuthi, A., Burkhart, C. P., and Bayless, J. R., *J. Adv. Oxid. Technol.* **2**, 299-305 (1997).
92. Penetrante, B. M., Hsiao, M. C., Bardsley J. N., Merritt, B. T., Vogtlin, G. E., Wallman, P. H., Kuthi, A., Burkhart C. P., and Bayless, J. R., "Power consumption and byproducts in electron beam and electrical discharge processing of volatile organic compounds", in *Proceedings of the 2nd Int. Symp. on Environmental Applications of Advanced Oxidation Technologies*, San Francisco, February 1996 (Electrical Power Research Institute), pp. 5-74 to 5-88.
93. Penetrante, B. M., Bardsley, J. N., and Hsiao, M. C., *Jpn. J. Appl. Phys.* **36**, 5007-17 (1997).
94. Itikawa, Y., Hayashi, M., Ichimura, A., Onda, K., Sakimoto, K., Takayanagi, K., Nakamura, M., Nishimura, H., and Takayanagi, T., *J. Phys. Chem. Ref. Data* **15**, 985-1010 (1986).

95. Majeed, T., and Strickland, D. J., *J. Phys. Chem. Ref. Data* **26**, 335-50 (1997).
96. Winters, H. F., *J. Chem. Phys.* **44**, 1472-6 (1966).
97. Cosby, P., *J. Chem. Phys.* **98**, 9544-53 (1993).
98. Schofield, *J. Phys. & Chem. Ref. Data,* **8**, 723-98 (1979).
99. Atkinson, R., Baulch, D. L., Cox, R. A., Hampson, R. F., Kerr, J. A., Rossi, M. J., and Troe, J., *J. Phys. Chem. Ref. Data* **26**, 521-1011 (1997).
100. Atkinson, R., Baulch, D. L., Cox, R. A., Hampson, R. F., Kerr, J. A., Rossi, M. J., and Troe, J., *J. Phys. Chem. Ref. Data* **26**, 1329-1499 (1997).
101. Guerra, V., Pinheiro, M. J., Gordiets, B. F., Loureira, J., and Ferreira, C. M., *Plasma Sources Sci. Technol.* **6**, 220-30 (1997).
102. Slavik, J. and Colonna, G., *Plasma Chem. Plasma Proc.* **17**, 305-14 (1997).
103. Vitello, P. A., Penetrante, B. M. and Bardsley, J. N., *Phys. Rev. E* **49**, 5574-98 (1994).
104. Morrow, R., and Lowke, J. J., *J.Phys. D.* **30**, 614-27 (1997).
105. Falkenstein, Z., and Coogan, J. J., *J.Phys. D.* **30**, 817-25 (1997).
106. Naidis, G. V., *J.Phys. D.* **30**, 1214-8 (1997).
107. Kulikovsky, A. A., *IEEE Trans. Plasma Sci.* **25**, 439-46 (1997).
108. Nilsson, J. O., and Eninger, J. O., *IEEE Trans. Plasma Sci.* **25**, 73-82 (1997).
109. Kitayama, J., and Kuzumoto, M., *J.Phys. D.* **30**, 2453-61 (1997).
110. Chang, J. S., Looy, P. C., Arquilla, M., Kamiya, I., and Sinjo, R., *J. Adv. Oxid. Technol.* **2**, 274-7 (1997).
111. Penetrante, B. M., Hsiao, M. C., Bardsley, J. N., Merritt, B. T., Vogtlin, G. E., Kuthi, A., Burkhart, C. P., and Bayless, J. R., *Plasma Sources Sci. Technol.* **6**, 251-9 (1997).
112. Matzing, H., *Adv. Chem. Phys.* **80**, 315-9 (1991).
113. Tas, M. A., van Veldhuizen, E. M., and Rutgers, W. R., *J.Phys. D.* **30**, 1636-45 (1997).
114. Penetrante, B. M., Hsiao, M. C., Merritt, B. T., Vogtlin, G. E., and Wallman, P. H., *IEEE Trans. Plasma Sci.* **23**, 679-87 (1995).
115. Penetrante, B. M., Hsiao, M. C., Merritt, B. T., Vogtlin, G. E., and Wallman, P. H., Neiger, M., Wolf, O., Hammer, T., and Broer, S., *Appl. Phys. Lett.* **68**, 3719-22 (1996).
116. Tas, M. A., van Hardeveld, R., and van Veldhuizen, E. M., *Plasma Chem. Plasma Proc.* **17**, 371-91 (1997).
117. Mizuno, A., Shimizu, K., Mastuoka, T and Furuta, S., *IEEE Trans. Ind. Applicat.*, **31**, 1463-7 (1995).
118. Dinelli, G., Civitano, L., and Rea, M., *IEEE Trans. Ind. Applicat.*, **26**, 535-41 (1995).
119. Song, Y. H., Shin, W. H., Choi, Y. S., and Kim, S. J., *J. Adv. Oxid. Technol.* **2**, 268-73 (1997).
120. Urashima, K., Chang, J. S., and Ito, T., *J. Adv. Oxid. Technol.* **2**, 286-92 (1997).
121. Vercammen, K. L. L., Berezin, A. A., Lox, F., and Chang, J.-S., *J. Adv. Oxid. Technol.* **2**, 312-329 (1997).

122. Penetrante, B. M., Hsiao, M. C., Bardsley, J. N., Merritt, B. T., Vogtlin, G. E., Kuthi, A., Burkhart, C. P., and Bayless, J. R., *Phys. Lett. A* **235**, 76-82 (1997).
123. Penetrante, B. M., Hsiao, M. C., Bardsley, J. N., Merritt, B. T., Vogtlin, G. E., Wallman, P. H., Kuthi, A., Burkhart, C. P., and Bayless, J. R., *Pure Appl. Chem.* **68**, 1083-7 (1997).
124. Hsiao, M. C., Merritt, B. T., Penetrante, B. M., Vogtlin, G. E., and Wallman, P. H., *J. Appl. Phys.* **78**, 3451-6 (1995).
125. Penetrante, B. M., Hsiao, M. C., Bardsley, J. N., Merritt, B. T., Vogtlin, G. E., Wallman, P. H., Kuthi, A., Burkhart, C. P., and Bayless, J. R., *Phys. Lett. A* **209**, 69-77 (1995).
126. Bromberg, L., Cohn, D. R., Koch, M., Patrick, R. M., and Thomas, P., *Phys. Lett. A* **173**, 293-9. (1993).
127. Koch, M., Cohn, D. R., Patrick, R. M., Schuetze, M. P., Bromberg, L., Reilly, D., and Thomas, P., *Phys. Letter A* **184**, 109-13 (1993).
128. Falkenstein, Z., *J. Adv. Oxid. Technol.* **2**, 223-38 (1997).
129. Vitale, S. A., Hadidi, K., Cohn, D. R., and Falkos, P., *Plasma Chem. Plasma Proc.* **17**, 59-91 (1997).
130. Chang, J. S., Yamamoto, T., Kohno, H., Tamura, M., Honda, S., Shibuya, A., and Berezin, A. A., *J. Adv. Oxid. Technol.* **2**, 346-52 (1997).
131. Zipf, E. C., *J. Geophys. Res.* **85**, 687-94 (1980).
132. Burch, R., *Pure Appl. Chem.* B68, 377-85 (1996).
133. Tonkyn, R., Barlow, S., Orlando, T., "Destruction of chlorinated hydrocarbons in a packed bed corona reactor" *Abstracts of the World Environmental Congress*, London, Ontario, September 1995, p. 214.
134. Prieto, G., Prieto, O., Gay, C. R., Mizuno, K., Tamori, I., and Yamamoto, T., *J. Adv. Oxid. Technol.* **2**, 330-6 (1997).
135. Kinoshita, K., Watanabe, S., Hayashi, N., Uchida, Y., Dykes, D., and Touchard, G., *J. Adv. Oxid. Technol.* **2**, 278-82 (1997).
136. Atkinson, R., *J. Phys. Chem. Ref. Data* **26**, 215-290 (1997).
137. Collison, W. Z., Ni, T. Q., and Barnes, M.S., *J. Vac. Sci. Technol.A.* **16**, 100-7 (1998).
138. Woodworth, J. R., Riley, M. E., Miller, P. A., Nichols, C. A., and Hamilton, T. W., *J. Vac. Sci. Technol. A.* **15**, 3015-23 (1997).
139. Hopwood, J., *Plasma Sources Sci. Technol.* **1**, 109 (1992).
140. Patrick. R., Schoenborn, R., and Toda, H., *J. Vac. Sci. Technol. A.* **11**, 1296 (1993).
141. Donnelly, V. M., *J. Appl. Phys.* **79**, 4597 (1992).
142. Gaddy, G. A., Webb., S. F., and Blumenthal, R., *Appl. Phys. Lett.* **71**, 3206-8 (1997).
143. Tokunaga, O., Suzuki, N., Nishimura, K, and Washino, M., *Int. J. Appl. Rad. Isotopes* **29**, 81 (1978).

Molecular Spectroscopy Data Needs

Jon T. Hougen

Optical Techology Division, National Institute of Standards and Technology, Gaithersburg, MD 20899

Abstract. At one time it was possible to envisage a molecular data center which critically evaluated the entire world's output of high-resolution spectroscopic studies for a given class of molecules within a certain frequency range. Such efforts are no longer possible due to the massive amounts of data produced each year. In order to design smaller, more focussed, and shorter term data efforts, a workshop was held at NIST, during which the most pressing immediate needs of various user communities were described in some detail. While these needs are now relatively clear, as indicated in the overview below, a good algorithm for prioritizing the needs of the very disparate user groups did not emerge from the workshop.

INTRODUCTION

This talk contains material from an internal report summarizing information presented on December 5-6, 1996 at a NIST Workshop on National Needs for Molecular Spectroscopy Data, as articulated by representatives of specific user communities. That report also summarized the activities of existing molecular spectroscopy data centers throughout the world. A previous report on the Workshop dealt with more global data questions concerning institutional responsibilities and information flow mechanisms (both human and electronic) among academic, industrial and government laboratories. It also contained reproductions of the overhead transparencies used by the eleven invited speakers at the Workshop.

The second report, like the first one, was primarily aimed at providing guidance for future activities of the Standard Reference Data Program (SRDP) at NIST. The presentation thus contains both information of general interest and information collected to answer specific questions posed by SRDP. For clarity, we begin by identifying various categories for the electromagnetic spectral regions, molecular species, uses of molecular data, data centers, and user communities.

Electromagnetic spectral regions. The table below gives a convenient terminology for the spectral regions of interest for molecular spectroscopy. Boundaries between regions are located where significant changes in light source, optical components, and/or detector technology are required.

Table I. Frequency vs Spectral Region

ν_{min}	ν_{max}	ν_{max}/ν_{min}	Region Name
1 MHz	3 GHz	3000	**Radiofrequency**
3 GHz	1 THz	300	**Microwave and Submillimeter**
30 cm^{-1}	300 cm^{-1}	10	**Far Infrared**
300 cm^{-1}	12000 cm^{-1}	40	**Infrared**
12000 cm^{-1}	25000 cm^{-1}	2	**Visible**
25000 cm^{-1}	50000 cm^{-1}	2	**Ultraviolet**

Molecular species. Six convenient categories for molecular species can be constructed by combining the words **diatomic** and **polyatomic** with the words **stables, unstables**, and **ions**. These six categories correspond to significant differences in species preparation and handling techniques, as well as to significant differences in spectral complications, theoretical formalisms, and intramolecular physical phenomena. For example, diatomic species have only one vibrational mode, which cannot change the linear geometry of the molecule. Polyatomic species with N atoms have 3N-5 (linear) or 3N-6 (nonlinear) vibrational modes which can interact with each other by Fermi and Coriolis resonance, and with electronic motion via the Renner-Teller (linear) and Jahn-Teller (nonlinear) effects. Stable species can usually be purchased and stored until needed. Unstable species are frequently present only at high temperatures, or only as chemically reactive free radicals, and must usually be generated *in situ* during the experiment. Ions have concentration limitations imposed by space-charge effects, but can be detected very selectively and efficiently using mass spectroscopy techniques. Since molecular species in the six different categories are essentially the subjects of six separate fields of study, users (and sometimes also producers and evaluators) of spectral data may experience confusion in crossing the boundaries separating these categories.

Data uses. The primary applications of molecular spectroscopic data are in fact quite unified, and can all be grouped under the heading of analytical chemistry. Qualitative analysis involves species identification. Quantitative analysis involves concentration determination. Slightly more generally, in gas phase applications one is usually trying to determine n_i/V (**concentration** of species i), T (**temperature**) or P (total **pressure**) in the equation of state

$$PV = (\Sigma_i n_i) RT$$

or molecular **velocities** (from Doppler shifts). Other applications of molecular spectroscopy measurements often involve determinations of molecular **structures** and **potential energy surfaces** or various thermodynamic quantities like **enthalpies** (from energy differences) or **entropies** (from energy levels and their degeneracies).

Data centers. We distinguish here between two types of data centers, namely those which produce annotated **bibliographies** of the research literature, and those which disseminate critically evaluated and conveniently packaged **data compilations** for users outside of the molecular spectroscopy community. Bibliographies are often extremely useful within the spectroscopic community itself, but only the need for data compilations will be discussed in this talk, since we wish to focus on the needs of users for whom the original research literature may appear both scattered and non-transparent.

High-resolution user communities.

1. Perhaps the most numerous and consistent users of high-resolution molecular spectroscopy data are the various **astrophysical and atmospheric** research communities (stellar envelopes, interstellar clouds, planetary atmospheres, earth's upper atmosphere, earth's lower atmosphere), for the obvious reason that four of these groups are almost entirely dependent on electromagnetic radiation for probing their objects of interest. As a consequence, NASA has consistently supported atmospheric data center work both in its own laboratories and elsewhere. On the other hand, stellar and interstellar research is funded by NSF, which does not have data center activity as part of its explicit mandate and which therefore relies on others to generate the data compilations needed for its own programs.
2. Another large and traditional user group consists of the gas-phase **chemical kineticists** interested in photolysis reactions, combustion modeling, reaction intermediates, etc., again because radiation provides a probe for their systems which is non-intrusive and often capable of good spatial and temporal resolution as well. Since kineticists have traditionally wanted most to be provided with line positions and intensities for new free radicals, and since the study of open shell molecules was also of great interest to practicing molecular spectroscopists, data needs of gas-phase kineticists have historically been satisfied "automatically," i.e., without direct funding connections between provider and user.
3. Closely related to kineticists are the **chemical thermodynamicists,** who often find spectroscopically determined enthalpies and entropies to be the best available for difficult to prepare or difficult to isolate chemical compounds (high temperature species, short lifetime species, etc.), and who, in fact, use spectroscopic data to determine all entropies, enthalpies and Gibbs energies of gaseous species. Thermodynamicists, working in their own data centers, normally

collect spectroscopic data from the research literature (again without direct funding connections between provider and user).

4. Users from all three branches of the **military** have explored (and financially supported) an almost endless variety of possible applications of high-resolution data and technology over the last half century (propellent chemistry, laser design, uranium isotope separation, chemical agent detection, etc.). But apart from the three decades of JANAF thermodynamic tables support, the military has not been a strong supporter of databases *per se*, presumably because their data needs change so rapidly and so unpredictably with each advance or attempted advance in weapons technology that long term commitments to data of a given type are normally not in their best interests. In most cases where the military does support its own data centers, they do so because the data is classified.

5. Users from **industry** have long relied on infrared and Raman spectral fingerprinting to identify and/or follow the generation, destruction or flow of various species in their products or process streams. The reference spectra required for such monitoring, such as those found in the Coblentz Society database described by Kathy Kalasinsky in her talk or those mentioned by Leslie May in her talk at the Workshop, are of modest resolution, since most industrial products and process streams are either liquids or solids, where high-resolution measurements are of little use. NIST has not been a traditional supplier of this type of data because of its historical laboratory focus on the development and application of very high precision measurement technology. Many of the newer industrial technologies, however (e.g., plasma etching, chemical vapor deposition, discharge lighting, etc), as well as gas-phase process streams and air-pollution or stack-gas monitors, are potential users of high resolution data, and we shall concentrate on their needs in this report. We note that industrial users, like the military users, have not been traditional supporters of data centers *per se*, presumably because their needs also change rapidly, and because they too must maintain reasonable secrecy concerning potentially profitable data applications.

Properties of substances versus properties of materials. This report is concerned with data centers providing information on the properties of pure substances or simple mixtures. Such data (e.g., the well known Iodine Atlas of Gerstenkorn and Luc for calibrating dye laser spectra) are immediately usable in any laboratory in the world with access to the substance and a suitable sample cell. In contrast, the properties of materials like glasses, ceramics, alloys, etc. depend greatly on the supplier and sometimes even on the particular batch from a given supplier. Rather than data *per se*, data centers for materials characterization must disseminate standard artifacts and standard measurement protocols. This important aspect of our national data needs will not be covered in this report.

HIGH RESOLUTION DATA NEEDS BY USER GROUP

Civilian Government Agencies and Academe

Astrophysics and Atmospheric Physics

High-temperature or high-photon-flux regions: **Visible and UV spectra.** Stars are normally too hot for molecule formation, but the cooler stellar envelopes often contain high-temperature diatomic species (and some small polyatomics) which are best studied in the visible or ultraviolet spectral regions. Also, as pointed out in the letter from Peter Smith and Kate Kirby of the Harvard-Smithsonian Center for Astrophysics (see first report on the Molecular Spectroscopy Workshop), electronic spectral data (particularly cross-sections and branching ratios for photo-destruction processes) are essential if we are to understand what is happening in photon-dominated regions of the interstellar medium. Electronic spectral line positions and intensities are similarly needed to interpret Hubble Space Telescope observations of comets and cometary impact events, the Io plasma torus, etc. Data needs for this community do not arise from the absence of information on five or six key molecules. Instead, as stated by Smith and Kirby, "there has been a 20-year lull in compilations of properties of electronic transitions in simple (2 to 4 atoms) molecules. ... There have been bibliographic activities ... but no substantial, critically evaluated compilations of electronic properties of molecules."

Interstellar clouds**: MW spectra of stable polyatomics and their ions** The precise composition of interstellar matter is still uncertain, particular with regard to larger organic molecules which might be implicated in extraterrestrial life or models for the origin of terrestrial life. As pointed out by Lew Snyder in his presentation at the Workshop, NSF and science funding analogs in other countries are investing considerable resources in large-radio-telescope-array hardware with better spatial resolution and in many cases also better sensitivity. Thus, an intense effort to identify new species and determine the total amount present can be anticipated in the next decade, particularly since such identifications have popular-press appeal when they can be connected to origin-of-life hypotheses. The radio astronomy community itself contains scientists capable of very sophisticated laboratory measurements, but that community alone cannot possibly measure, model and package all of the laboratory data it will need in the immediate future. Immediate priority species are stables, free radicals and ions containing four to ten atoms from the set H, C, N, O, F, Si, P, S, Cl, as well as interstellar candidates containing the more abundant metals like Na, Mg, Al, K, Ca, Fe. The packaging and dissemination of good spectral information on 50 to 200 carefully chosen new species from this list over the next decade should permit reasonable progress in this field.

Planetary atmospheres: **IR and Far IR spectra of stable polyatomics** Atmospheres of the hydrogen-rich larger planets and some of their moons contain relatively large quantities of H_2O, NH_3, CH_4, H_2S, PH_3, and presumably therefore smaller amounts of other species such as the hydrogen halides, silane, germane, etc. Many of these species have already been extensively studied, but in order to identify and determine the concentrations, temperature and pressure profiles, etc. of new minor species, it is essential to produce calculated atmospheric absorption profiles containing "all" lines of the major species. Such models rely heavily on IR databases containing a combination of (relatively few) precision measurements from high-resolution spectroscopic laboratories throughout the world and (relatively many) interpolated data calculated from these measurements using reliable quantum mechanical models. (Data calculated by extrapolation to quantum numbers and temperatures outside the region covered by measurements are usually not reliable.) Immediate priority species for this and the next category will not be stated explicitly here, since up-to-date needs are always available from the HITRAN data center described by Larry Rothman in his presentation at the Workshop.

Earth's upper atmosphere: **IR spectra of stable and unstable polyatomics** From a spectroscopic point of view, upper atmosphere research is directly concerned with heat and radiation budgets and with the chemistry that influences the storage and flow of these two forms of energy. Changes in the ultraviolet radiation reaching the ground caused by the ozone hole have prompted extreme interest in measuring the time and positionally-dependent concentrations of the hundreds of species involved in the ozone chemistry cycle. Fears concerning unwelcome changes in the average earth temperature have prompted similar interest in measurements of all species giving rise to significant infrared absorption in the atmosphere. The adverse effects of acid rain have prompted interest in measurements of species involved in the sulfur oxidation cycle. It can safely be anticipated that these and similar concerns will only increase as technological developments spread to more and more countries of the globe, and that reliable and automated concentration measurements of various stable and unstable molecular species by spectroscopic techniques will become more prevalent. Data needs of this community, as described by Mary Ann Smith in her talk at the Workshop, involve primarily high-resolution spectra because of the low pressures in the upper atmosphere. These data needs (dissemination of priorities, measurement and evaluation of reliable data, user-friendly packaging, etc.) are currently being served well by the HITRAN database.

Earth's lower atmosphere: **IR spectra of stable and unstable polyatomics** Atmospheric pressure at ground level leads to significant pressure broadening, so that unmodified high resolution spectra are in general not useful in open-path analytical applications such as fence-line monitoring, stack-gas monitoring, or real-time automobile exhaust monitoring. Thus, for *in situ* measurements, as described by George Russwurm in his presentation at the Workshop, some sort of resolution-degraded synthetic spectra are needed, together with suitable measurement algorithms to guarantee species specificity in the spectrally overlapped absorption profiles retrieved from field measurements. On the other hand, if extractive gas analysis techniques are used, high resolution spectra are required. Priority species can be determined from the 189 compounds given in the EPA 1990 Clean Air Act list, as augmented in an update by several hundred additional compounds.

Chemical Reaction Dynamics

Photochemistry: **Visible and UV spectra of unstable polyatomics** Up to the present, photochemists have used high resolution spectroscopy mainly to measure the concentration, temperature distribution, velocity distribution, etc. of diatomic fragments produced in the photolysis. (The photolysis act itself normally results from absorption into a relatively broad continuum region of the spectrum, so that low resolution spectral data are usually sufficient for experiment design and preliminary interpretation.) Good electronic data for triatomic fragments also exist, and the sensitivity of photochemical experiments has reached a point where the less favorable partition functions (compared to diatomics) can be tolerated. What is most needed here is a compendium of triatomic data similar to existing compendia for diatomic data. Unfortunately this will not be simple extension of the diatomic activity, since a whole host of new physical phenomena (Renner-Teller effect, quasilinearity, axis switching, Fermi resonances, Coriolis interactions, etc.) arise in triatomic molecules. Since many of the attendant theoretical concepts and notational complications are both unfamiliar and somewhat complicated, significant pedagogical material will have to be incorporated in any polyatomic data presentations aimed at non-spectroscopic user groups.

Combustion: **Spectra of unstable (visible and UV) and stable (IR and MW) polyatomics** As described in the talk of Bill Kirchhoff at the Workshop, combustion modeling requires the treatment of a large number of chemical reactions under non-equilibrium conditions, and is of interest both to various industries and to the Department of Energy. The desired role of high resolution spectroscopy here is to provide for as many species in the flame as possible a concentration and temperature probe which is thermodynamically and hydrodynamically non-intrusive and spatially and temporally resolved,. The end products of combustion (e.g., H_2O, CO_2, CO), and many of the smaller reaction intermediates (e.g., OH, CH, CH_2) can already be monitored. Most fuels and fuel-additives are stable polyatomic molecules (e.g., *iso*-octane, methyl *t*-butylether, etc.), and most reaction intermediates (pyrolysis products, partial oxidation products, etc.) are polyatomic free radicals. Present combustion models would greatly benefit from species-specific experimental measurements of moderate-sized (5 to 10 atoms) polyatomics at various points in the flame, but even the design of such experimental measurements requires detailed knowledge of the spectra of these polyatomics. Given the variety of combustion environments in actual use (internal combustion engine, diesel engine, waste incineration, electrical power plants, etc), a data project to fulfill all flame probe needs would be very large in scope indeed, but a good start has been made in the HITEMP data base described by Rothman at the Workshop. Immediate priority species for future activity will depend on which of the many possible applications is made the focus of the effort. In this context, the microgravity combustion data needs associated with space vehicle safety, as outlined by Nancy Piltch in her talk at the Workshop, represent a potential new area of NIST activity.

Thermodynamics. This is an example where one type of data center (thermodynamics) plays the role of user for another type of data center (high-resolution spectroscopy). In general spectroscopic data provide one of the basic tools for thermodynamic estimations and correlations. The latter can then be used by kineticists to calculate rate constants in one direction when they are known only in the other, etc. Needs at the present time are of two types. Uncertain entropy values for polyatomic molecules with one or more very anharmonic low-frequency vibrations could often be corrected if adequate spectroscopic data on these vibrational levels were available. (For such applications completeness of the energy level set is more desirable than extreme accuracy.) Good spectroscopic dissociation energies for the newer free radical species of importance are also needed, particularly those arising in non-petroleum-based chemistry (e.g., reactions of Si, As, P, B, etc.). Kerwin Dobb's talk at the Workshop described some of the ways that quantum chemists are attempting to calculate industrially important thermodynamic quantities when experimental numbers are completely lacking.

Molecular clusters. **MW and IR spectra of stable polyatomics** There has been an explosive growth of high-resolution molecular cluster studies, mainly using Fourier-transform microwave techniques, but also using molecular beam infrared techniques. These clusters permit "half-collision" kinetic studies, in the sense that one can photolyze one molecule in a two-molecule cluster in such a way that this molecule fires an atomic or molecular free radical projectile at the other molecule, leading to chemical reaction from controlled initial geometry. It is obvious that if this new field of half-collision reaction studies is to prosper (or if the potential new field of hydrogen bond kinetics is even to be born), good spectral data must be available for both geometrical structures and actual absorption lines of the known molecular clusters. Producing such a compilation is quite practical at the moment, since only several hundred complexes have been studied. It is not unreasonable to hope that these several hundred examples, when collected together and looked at globally, will indicate many of the dominant trends to be expected in all such complexes.

Department of Defense

Propellants: **Unstable polyatomics** The needs here are similar to those of the combustion and chemical kinetics community, except that the list of molecules should be restricted to those found in modern-day propellants (which are rich in C, N, O and H in such a way that they decompose rather exothermically into CO_2, H_2O and N_2). While the military will almost certainly not want to make a long term commitment to the evaluation and compilation of such data, it can safely be assumed that if such a compilation existed, they would make good use of it. Priority species would have to be determined partly by intuition and partly by examining presently funded propellant research (which is sometimes quite different in the three different branches of the military).

Friend or foe exhaust signatures: **IR spectra of stable polyatomics** A possible way of avoiding mistaken missile strikes is to seed the fuel mixtures of aircraft and missiles each day in such a way that infrared emission from their exhaust gases would carry a unique fingerprint for that day. Identifications at high altitudes, where the ambient pressure is low, would certainly benefit from the excellent species specificity of high-resolution spectral signatures.

Chemical agents: **IR and MW spectra of stable polyatomics** Almost instantaneous identification of chemical nerve agents and/or related decomposition products under battlefield conditions is highly desirable. Such agents are frequently rather large organic molecules relatively similar in composition and structure to a number of harmless chemical species. Rapid and accurate identification thus requires excellent real time species specificity, even in the presence of decoy additives, a capability which microwave spectroscopy is essentially unsurpassed at. One can reasonably anticipate that much of the military research in this area is classified, but also that military contractors trying to design field detectors for chemical agents could make good use of an appropriately designed MW database.

Reentry vehicle ablation: **Spectra of unstable diatomics and polyatomics** Some of the interest here is reminiscent of those of the thermodynamic community a decade or two ago, primarily because the nose cones of reentry vehicles are heated to a very high temperature. Aerodynamic modeling is thus complicated by unknown high-temperature chemistry and ablation effects. In addition, however, hot-target-seeking infrared detectors inside the nose cone could be blinded by infrared emission from heated gas layers at the nose-cone skin. The NIST Spectroscopy Applications Group is presently collaborating with staff at several wind-tunnel installations to develop various sensitive and reliable detection schemes. It seems likely that as long as rockets and spacecraft are of interest to NASA and/or the military, high-temperature reentry effects on the nose cones will also be studied. While no precise statement of needs exists at present, this is a potential focus area for new data activities.

Private Industry

Plasma etching: **Spectra of unstable polyatomics** This is obviously a multi-billion-dollar new technology area, where technological improvements can be expected. To the extent that some of these improvements could be suggested by better understanding and/or computer modeling of the plasma species, plasma temperatures (e.g., translational, vibrational, electron), plasma chemistry, etc. there is an immediate need for user-friendly and accessible high-resolution data on all potential free radical species present in the plasma. Some of these species have already been studied spectroscopically, so they are suitable for immediate data center activity. Others have not yet been studied; critical evaluation and packaging obviously cannot begin until such data has been measured in at least one laboratory. In any case, the number of chemical elements present in etching plasmas or present industrial interest is fairly limited, so that any data center activity focused on this user group would be of manageable size. User-friendly computer retrievability and computer manipulation software designed for the plasma modeling community would be important here. Michael Passow of IBM, in his presentation at the Workshop, listed the following plasma species of immediate interest: C_xF_y, C_xH_y, $C_xH_yF_z$, SF_x, Cl_2, HCl, HBr, BCl_3, NH_3, NF_3, SiF_x, $SiCl_x$, SiO, H_2, H_2O, OH, C, C_2, CO, CO_2, CN, NO, COF_2, HF, F, Si, H, and metal etch compounds containing W, Cu, Al, Ti, and Ta.

Chemical vapor deposition (CVD): **Spectra of unstable polyatomics** This discussion of CVD should logically precede that of plasma etching above, but the NIST Workshop did not actually have a speaker representing this user group. Remarks here would be almost identical to those above, however. In particular, the number of chemical elements (and their related compounds) of interest is relatively small. Immediate priority species include Si_xF_y, Si_xH_y, $Si_xH_yF_z$, Si_xO_y, as well as some from the list immediately above.

Discharge lighting: **Visible and UV spectra of diatomics** Coupling to the discharge lighting industry is particularly appealing to the atomic and molecular spectroscopy groups at NIST, since the product marketed by this industry is the radiation emitted from hot atoms and molecules. The high kinetic and electron temperatures involved in these lighting plasmas makes atomic species far more important than molecular, but when large concentrations of iodine or other halogen(s) are present, diatomic halogen-containing species are bound to play some role in either the thermodynamics, chemistry, or emitted radiation of the lamps. Industrial proprietary considerations prevent getting a precise wish-list of important species, but a combination of intelligence and intuition should suffice to set up a useful data activity for this user group. Graeme Lister, in his presentation at the Workshop, listed the following typical molecular species found in metal halide lamps: M_2, MI, MI_2, M_2I_2, where M = Hg, Na, Tl, Dy, Ho or Tm.

Medical

Breath analyses: **MW or IR spectra of stable polyatomics** There are extremely few applications of high-resolution spectroscopy to medical diagnostic procedures, but human breath is a gas-phase system, and therefore should be ideally suited for high resolution analysis. The high sensitivity and specificity of species discrimination should be useful for monitoring (or discovering) trace-molecule concentration changes correlated with various diseases (an obvious example being the acetone content of insulin-deficient diabetic breath). The real time capabilities should be useful for anesthesia monitoring etc. during surgical procedures. Since the technology needed for medical monitoring applications has not yet been developed or commercialized, the need for a data center is not driven by the existence of a large user group. Nevertheless, once such a data center (and a few demonstration measurements) existed, it could trigger the opening of a new high-profile field .

CONCLUSION

Given this large set of needs, the task facing NIST during the coming year is to develop a workable national strategy for choosing which ones to satisfy and how. At the lowest level of scientific involvement and cost, NIST could use its web site to act as a central registry of existing databases. At the next higher level, NIST could try to identify latent databases already existing in the spectroscopic community and encourage the extra work needed to make them truely useful and accessable. At the highest level of scientific involvement and cost, NIST could continue its historical role of collecting, evaluating, and disseminating molecular spectroscopy data to user groups outside the data generation community. Whatever mix of above activities is chosen, it is my personal opinion that NIST should offer the country some sort of leadership in molecular spectroscopy data activities.

Panel Discussions

International Coordination of A+M Data Efforts

Ratko Janev (Panel Chair)[1], Kurt Becker[2], Robert Clark[3], Graeme Lister[4], Kay Niemax[5]

[1] *International Atomic Energy Agency, Vienna, Austria*
[2] *Physics Department, Stevens Institute of Technology, Hoboken, USA*
[3] *Los Alamos National Laboratory, Los Alamos, USA*
[4] *OSRAM Sylvania Inc., Beverly, USA*
[5] *Institute of Spectroscopy and Applied Spectroscopy, Dortmund, Germany*

INTRODUCTION

A summary of the discussions at the ICAMDATA panel on international coordination of atomic and molecular (A+M) data efforts is presented. The status of databases and the needs for coordination of data generation, compilation and assessment efforts in the areas of spectrochemistry, low-temperature plasma applications, plasma processing of materials, lighting industry and fusion energy research are specifically addressed.

PANEL DISCUSSION

The progress in many scientific and technological areas in which atomic and molecular processes play an important role depends on the availability of accurate quantitative information on the collisional properties of these processes and spectroscopic characteristics of interacting species. The understanding of various astrophysical, atmospheric and biophysical phenomena, or the development and optimization of various technological processes and industrial devices relies on the knowledge of underlying physical and chemical mechanisms and on a detailed modeling of the corresponding atomic and molecular kinetics. The amount of required spectroscopic and collisional information for complete understanding and successful description of properties and behaviour of the gaseous and plasma media in natural or industrial systems may be extremely large. The generation of this information, either through compilation and critical assessment of the available literature data or through new experimental measurements and theoretical calculations, frequently appears to be an effort which is well beyond the capacity of a scientific or industrial laboratory, group of such laboratories, or even a nation. Astrophysics, fusion energy research, aeronomy and various low-temperature industrial plasma applications (such as plasma processing of materials, lighting devices, analytic spectrochemistry, etc) provide examples of research and technological development areas in which the establishment of required atomic and molecular databases needs coordination of the corresponding efforts on a transnational level. The international coordination of database establishment efforts in certain scientific or applied research fields accelerates the process of database establishment, ensures the required scientific expertise for critical data assessment (and, thereby, the high quality of the data), and considerably reduces the required resources. Such collaborative efforts are advantageous also from a technical point of view (e.g. unification of data presentation formats and data dissemination). Examples of successful international database establishment efforts can be found in astrophysics (the OPACITY and the IRON projects), in magnetic fusion research (the AMDIS/ALADDIN and the ADAS databases) and in atmospheric research (earth atmosphere photochemistry).

The ICAMDATA panel discussions on the international coordination of A+M data efforts focussed on the analysis of the A+M data status and needs for coordination in the fields of spectrochemistry, low-temperature plasma applications, plasma processing of materials, lighting devices and controlled thermonuclear fusion. The highlights and conclusions of these discussions are summarized below.

SPECTROCHEMICAL APPLICATIONS

Element analysis in spectrochemistry includes a wide range of measurement techniques, such as optical emission spectrometry, atomic absorption spectrometry, X-ray fluorescence, all of which involve a significant amount of spectroscopic data information. In recent years, the optical spectrometer in the techniques based on an inductively coupled plasma is more and more frequently replaced by a mass spectrometer in order to lower the element detection limits. Time-of-flight techniques have been also recently employed in this field because they allow for a simultaneous elements analysis. Further, laser induced breakdown spectrometry, where a sample material is ablated by a pulsed laser and the transient spectra of the laser produced plasma are measured with intensified diode arrays or CCD - cameras attached to the spectrometers, is currently considered as a very promising analytical technique.

The spectroscopic data information associated with the application of the above (and other, less standard) spectrochemical techniques is to a large extent already available (mainly due to the activity of NIST/NBS data centres) and includes: wavelengths and (source dependent) intensities of spectral lines, energy levels, transition probabilities and lifetimes of atomic states, hyperfine structure, isotope shifts and spectral line broadening and shift parameters. However, the spectroscopic data information for spectrochemical applications is far from being complete. Wavelength and spectral line intensity data are needed for metals and non-metals in the spectral range 150-800 nm (particularly for the weak lines which interfere with the strong ones). Atomic state energy and transition probability data are still needed for the species present in the analytical plasmas (for determining the plasma parameters) and for the laser spectrometry.

International collaboration in completing the spectroscopic data information required for advancing the spectrochemical analysis methods, as well as for other applications, would certainly be very useful for meeting the needs in a timely fashion. Such collaborations already exist on a bilateral level (e.g. between NIST and JAERI), but a wider coordination of the data efforts conducted in various national centres could be much more effective.

LOW-TEMPERATURE PLASMA TECHNOLOGIES

The use of non-equilibrium, low-temperature plasmas for materials processing is the key to the advancement of many rapidly developing technologies. The selective and highly anisotropic etching of materials and the controlled deposition of thin films in the fabrication of microelectronic structures are among the most important methods of plasma-assisted materials processing, plasma polymerization and plasma-assisted surface modification. Plasma-based processes are used in about 40 percent of the steps in the manufacture of semiconductor chips and the more sophisticated the chip, the larger the number of steps relying on plasma technology.

A low-temperature plasma is a system far from thermodynamic equilibrium. The electron temperature is much higher than the gas temperature and can drive "high-temperature" chemistry (without any adverse effects on the processed materials). The high-energy tail of the electron energy distribution function induces collisional break-up of the parent feedstock molecules which further leads to formation of chemically active neutral and ionic radicals and to a complex reaction kinetics. For modeling the behaviour and properties of such chemically active plasmas, detailed quantitative information is required for the most important collisional and radiative processes taking place both in the gas phase and on the surfaces. The available data information depends on the specific technological plasma, and in the case of plasma deposition applications (where Si-organic and metal-organic compounds are used) the existing databases are particularly scarce. The collisional and spectroscopic databases for most technological plasma applications are inadequate for a full understanding of the corresponding plasma chemistry dynamics and full exploitation of the optimization potential of these reactive plasmas. A coordinated international effort to improve the current status of the A+M databases relevant to low-temperature plasmas used in plasma processing applications would be extremely useful for advancement of the corresponding technological development. The coordination effort should include data generation (experimental and theoretical), data collection and assessment, data distribution, as well as stronger interaction with the relevant industrial partners and national funding agencies. Establishment of an international data collection, evaluation and dissemination centre, entrusted also with a coordinating role, would be an attractive organizational form for such an effort.

LOW-TEMPERATURE PLASMA APPLICATIONS IN THE LIGHTING INDUSTRY

Lighting devices is one of the traditional and still vigorously developing fields of low-temperature plasma applications. The gas discharges used in lighting devices (fluorescent lamps, HID, barrier discharge lamps, flat panel displays, etc) cover a wide range of plasma parameters, with electron temperatures in the range 0.1 - 2 eV, electron densities around 10^{11}-10^{13} cm^{-3} and gas pressures from about 3 torr (in the fluorescent lamps) up to a few atmospheres (in HID lamps). The understanding of the chemistry, dynamics and radiative properties of these plasmas, required to optimize the device parameters (the most important of which are the lamp efficiency and lifetime), relies on the knowledge of collisional and spectroscopic characteristics of all relevant plasma constituents. This understanding and the way to device optimization is achieved by detailed plasma behaviour modeling and, in particular, calculation of level populations of active gas atoms. Besides the knowledge of collisional and radiative characteristics of plasma constituents, the plasma modeling also requires knowledge of the electron energy distribution function, which in some cases may not be Maxwellian (e.g. in barrier discharge lamps). The existence of a high-energy tail in the electron energy distribution function requires inclusion of collisional and radiative processes involving highly excited states in the atomic database, which results in a drastic increase of the size of the database and the kinetic model calculations.

The recent developments in low pressure discharge technology for the lighting industry (electrodeless devices, small diameter cold cathode discharges) have led to new extensions of operating parameters of lighting devices with involvement of new important collisional and radiative processes and new working gas species (e.g. in the mercury free lamps). These developments have further expanded the diversity and quantity of the required collisional and spectroscopic information.

Despite the long and distinguished history of the lighting research community in establishing the required atomic databases, the present status is still unsatisfactory, particularly in view of new technology developments. A broad scale and well coordinated joint effort of the A+M and lighting research communities to establish the required databases would be highly desirable for the further development of light sources. The establishment of such joint efforts, however, meets certain difficulties in the area of the intellectual property rights.

FUSION ENERGY RESEARCH

Atomic and molecular collisional and radiative processes play an important role in magnetic fusion devices as they influence the energy balance of the confined plasma, the plasma transport and radiation properties and are used as the basis for many plasma diagnostic methods. The design and operation of some vital fusion device systems, such as neutral beam heating, impurity control and thermal power and particle exhaust systems, require large amounts of atomic and molecular collisional and radiative data. The overall A+M data information needed in fusion energy research is enormous and the need for coordination of the A+M data acquisition efforts on an international level already became evident at the beginning of seventies. In response to this need, the International Atomic Energy Agency (IAEA) has established (1975) an activity on compilation, evaluation and dissemination of A+M bibliographic and numerical data for fusion and was charged (by the International Fusion Research Council) to coordinate the national A+M data efforts for fusion. Soon thereafter, an international A+M Data Centre Network was established (with participating data centres from the USA, UK, Russia, Japan and France) which presently includes about 15 data centres. In addition to coordinating the activities of this Data Centre Network on the establishment of evaluated and/or recommended international A+M databases for fusion, the IAEA organizes and supports international coordinated research programmes for enhancing the A+M data generation for fusion. More than 40 laboratories and theoretical groups are currently involved in the data generation (and critical data assessment) effort. The results of the IAEA coordinated A+M data activities for fusion are contained in the IAEA Atomic and Molecular Data Information System (AMDIS), the databases of which are accessible online (via Internet and WWW).

The establishment of the international A+M database for fusion is an example (among many others) which illustrates the usefulness of the coordination of A+M data efforts for creating a comprehensive database for certain fields of scientific or technological research. This example also reveals the necessary conditions for a successful coordination of the A+M data efforts in establishing a database for a research field (or a specific research subject): (i) clear identification of the needs for such data (the users community), (ii) identification of the research groups, laboratories or data centres which jointly have the potential to provide (compile, evaluate and/or generate) the required data, (iii) establishment of an effective mechanism for work coordination, (iv) availability of adequate funds (normally provided by the data users or national funding agencies) and (v) a satisfactory agreement regarding the intellectual property rights.

CONCLUSIONS

Although limited in scope due to time constraints, the panel discussion has shown that the A+M collisional and radiative processes play an essential role in many modern science and technology areas and that quantitative information about these processes is instrumental for further progress in these areas. The research and development fields analysed by the panel (spectrochemistry, industrial and technology plasma applications, fusion energy research) have demonstrated that there are urgent needs for establishing comprehensive and reliable A+M databases for the advancement of these fields. Coordination of world-wide efforts in establishing such databases is found to be advantageous for rapid and efficient achievement of the development goals (as has been demonstrated in the case of fusion energy research). The coordination of database establishment efforts in the areas of industrial and technology plasma applications requires fulfillment of certain conditions (see the last part of the previous section), the most critical of which are those related to the funding and intellectual properties rights. The producer-user interactions and constructive consideration of all factors involved in a database establishment process may help in removing these difficulties.

Increasing the Visibility and Publicity for Data Activities and Assuring the Open Exchange of Data

David Schultz (Panel Chair)[1], R. Stephen Berry[2], Claudio Mendoza[3], Stephen Younger[4]

[1] Oak Ridge National Laboratory, Oak Ridge, USA
[2] Department of Chemistry, University of Chicago, Chicago, USA
[3] Venezuelan Institute for Scientific Research, Caracas, Venezuela
[4] Los Alamos National Laboratory, Los Alamos, USA

INTRODUCTION

This panel was charged with leading and stimulating discussion regarding two principal issues: (1) the need to increase the visibility of atomic data production and collection activities in recognition of their role as vitally important resources for diverse applications, and (2) the need to assure the open exchange of this data. Comments by the panelists, supplemented by interaction with the audience, are summarized here along with the principal conclusions.

PANEL DISCUSSION

Atomic, molecular, and optical (AMO) data are indispensable for such diverse applications as commercial and residential lighting, astrophysics, the development of fusion energy, semiconductor manufacturing, flat panel display technology, detection and remediation of pollutants, *etc*. In addition, the collection and organization of such data is of great aid to these endeavors as well as to the advancement of the AMO physics field itself. However, it is felt that the justifications for the generation and collection of AMO data are not sufficiently well recognized within scientific funding agencies nor within the applications communities themselves. Thus, the goal of this panel session, as well as a crucial theme running throughout the entire ICAMDATA meeting, was to discuss ways in which greater visibility, and ultimately greater funding, for these activities could be obtained, and to highlight the need to assure free exchange of these vital data.

The panel concluded that
- efforts must be made to increase the connections and interactions between producers of data, collectors/evaluators of data, and users of data,
- input to governmental bodies (e.g. the U.S. Congress) and to professional societies must be given to urge improvement of the Internet and to register the community's opinion on the freedom of data exchange in advance of legislation which would limit access,
- communities and funders should assure access to data and implement standards of exchange, preferably through multi-national data centers, and should adopt the World Wide Web as the standard for communication, and
- it should be recognized that AMO data are of increasing importance for applications, but that the community must make a strong case to document and emphasize this to potential customers and funders.

The panel consisted of comments from Stephen Berry [1], Claudio Mendoza [2], and Stephen Younger [3], followed by discussion and questions from the audience. Summarized below are the panelists' comments.

PANELISTS' REMARKS

In his comments, Dr. Berry emphasized issues of access to data rather than generation of data and stressed particularly the vehicles through which the community can call for action to influence policy decisions. For example, citing the congestion currently present on the Internet, especially for transoceanic connections, caused by the swelling commercial and entertainment sectors crowding out the science and education users, he called on the community to speak out within our professional organizations and especially directly to the U.S. Congress.

Dr. Berry also pointed out the recent furious debate over the degree of protection to be applied to the Internet, reflecting attempts to balance public interest with private intellectual property rights. This translates in the scientific and applications arenas to issues of whether data produced using public funds should be freely exchanged or restricted. This issue is particularly thorny regarding the transborder flow of information, and was a principal focus of the National Research Council's U.S. National Committee for CODATA report entitled "Bits of Power" [4]. This committee's report opened the debate and tried to reflect the diversity of all the concerned parties' points of view. Dr. Berry concluded by urging the AMO community to watch American Physical Society publications for calls to provide input to this debate, and to directly reach out to members of Congress who will be considering legislation pertaining to the freedom of access to information over the Internet.

Dr. Mendoza's comments reflected his experience in the theoretical production of large amounts of AMO data and in computer-based methodologies for distributing these data. For example, he cited the shear size of the AMO data files produced by such enterprises as the Opacity Project as necessitating electronic distribution of the information through databases. Even so, scientists involved in producing and making available such data are not in business, but rather obtain funding from traditional science funding agencies. Thus, he emphasized the need for institutionally based data resources such as the Centre de Donnes Astronomiques de Strasbourg (CDS) which provides a reliable, persisting source on the World Wide Web through which Opacity Project data, and other resources, can be distributed. Such centers allow the user to not question the authenticity and reliability of the electronic resource.

Moreover, he stressed the role of data centers in providing standards for data exchange, assuring the timely updating of information, and guaranteeing quick and reliable access. Dr. Mendoza called for data centers to seek to minimize the differences that users see from center to center, and to actively promote data collection and dissemination, a job which they are better suited to than the individual producers of data. Further, he emphasized that data centers would best cover the wide scope of AMO data and best assure universal access if they were multi-national in constitution and use the present *de facto* standard of Web-based data exchange.

Recognizing the need of users to have access to evaluated data, since potential users may not be expert in all relevant subfields, Dr. Mendoza pointed out the difficulty, or, in fact, near impossibility, of assigning individual evaluations to each datum in enormous sets of data generated by theoretical models that is nonetheless crucial for applications. He concluded his remarks by giving his perspective of the accessibility of data in developing countries. He noted that third world countries have to be motivated by economic reasons to provide good connections to the World Wide Web since the cost of maintaining traditional libraries is so high. However, to this time developing nations lack the required technical expertise and sufficient support of infrastructure to allow the much dreamed-of seamless access. Dr. Mendoza also pointed out that many resources which are free to U.S. and European users are only available in the third world for a fee.

Dr. Younger pointed out from the outset of his comments that the defense community does indeed have important needs for AMO data, and that these needs should be even greater in the future. He described the broad present and potential requirements for such data by tracing a typical battlefield scenario. For example, command, control, communication, and intelligence (C^3I) functions involve the use of sensors which must look through the atmosphere, and therefore, AMO data is needed which provides the foundation of analysis of atmospheric phenomena. The ordinance must then reach the target, and the physics of ionized air and of the interactions of solids (such as missiles) with plasmas is important in predicting how weapon systems will perform. Explosives involve molecular processes, and recent interest is focussed on making high explosives safer during storage and handling without degrading performance when delivered onto a target.

The simulation of nuclear weapons explosions, which encompasses the description of matter over a very wide range of temperature and density, has been a major application of a broad class of AMO data. Dr. Younger pointed out that the Comprehensive Test Ban Treaty which eliminates the ability to test nuclear weapons, could drive the need for AMO data further, due to the exclusive reliance on simulation to assure the safety and reliability of the nation's nuclear stockpile. In the U.S. the Accelerated Strategic Computing Initiative has been instituted in order to increase by a factor of one hundred thousand the computational power available by the year 2004. This would result in a 100 TeraFLOPS machine capable of aiding in such weapons simulation through complex hydrodynamics and radiation transport calculations.

Thus, the AMO community should expect these efforts to drive new needs for data, and to present a new opportunity to use facilities to generate AMO data. In a broad sense, these new computers could revolutionize how machines are used to generate information. Dr. Younger concluded by emphasizing the opportunity that the AMO community has to provide data and utilize these computational resources, but that it must also clearly demonstrate the relevance of AMO physics to defense to have its requests heard.

REFERENCES

1. R. Stephen Berry is the James Franck Distinguished Service Professor, Department of Chemistry, at the University of Chicago. He is a member of the National Academy of Science, and Chair of the National Research Council Committee on Issues in the Transborder Flow of Scientific Data.
2. Claudio Mendoza is the Head of the Physics Center at the Venezuelan Institute for Scientific Research, Caracas, Venezuela. He is a member of the Opacity Project and the IRON Project, and has been involved in the development and maintenance of atomic databases for the astrophysical community at the Centre de Donnes Astronomiques de Strasbourg, France, and Goddard Space Flight Center, U.S.A.
3. Stephen Younger is the Program Director, Nuclear Weapons Technologies Programs, at Los Alamos National Laboratory. He has been an active member of the atomic physics community and producer of atomic data.
4. "Bits of Power: Issues in Global Access to Scientific Data" (National Academy Press, 1997) is accessible at
http://www.nap.edu/readingroom/enter2.cgi?RI.html.
See also the report in this volume by R.S. Berry.

Standardization of Databases; Data Assessment; and Uncertainty Statements

Peter Smith (Panel Chair)[1], Linda Brown[2], John Rumble[3], Hiroyuki Tawara[4]

[1] *Harvard-Smithsonian CfA, Cambridge, USA*
[2] *Jet Propulsion Laboratory, Pasadena, USA*
[3] *National Institute of Standards and Technology, Gaithersburg, USA*
[4] *National Institute for Fusion Science, Nagoya, Japan*

The usefulness of databases comprising atomic and molecular parameters depends upon their reliability, completeness, and ease of use. The classic database, a book with limited availability containing data critically evaluated by experts, is gradually becoming obsolete. Databases today can be (and are being) prepared, distributed, and utilized by anyone through the World Wide Web (WWW). Producers and users of databases can be expert or naive. Therefore standards for databases are required in order to ensure quality.

Today, databases are, for historical reasons, often discipline-specific. With standardization of database content, format, and assessment, expert software will be able to merge small, separately created and maintained, datasets to create large "virtual" databases capable of dealing with disparate users with multiparameter requests.

Without standards, users get flawed, incomplete, unreliable, and/or out-of-date data. There is no configuration control; data can change from day to day without notice. Data in different formats cannot be merged.

Standards – a broadly agreed upon, public set of rules for data format, database content, data assessment and uncertainty statements, and documentation – ensure that database users get reliable information. With standards, producers can concentrate on content not form.

Documentation is an important part of database standards. Documentation must include information for the most naive legitimate user about the sources (*i.e.*, references), accuracies, and completeness of the data, as well as information about the evaluation process. Documentation should also explain how related databases differ.

Standards, however, cost money, and the effort required for production of quality databases is often not appreciated by scientific users, in part because the standards, the compilation and assessment procedure, and the data themselves are often not considered to be a very interesting, exciting scientific endeavor.

Assessment is a particularly time consuming and thus expensive part of database production. It is not peer review, but an independent evaluation by experts who assess the methods used to generate the data, and who also intercompare results obtained using different techniques. Standards are required to ensure that assessments of different data sets are consistent. Since standards need to be based on broad agreements within the community, the ICAM-DATA conference series may be a good vehicle to advance the standardization process.

Utilization of the World Wide Web

Peter Mohr (Panel Chair)[1], Gary Mallard[1],
Uri Ralchenko[2], David Schultz[3]

[1] National Institute of Standards and Technology, Gaithersburg, USA
[2] Weizmann Institute of Science, Rehovot, Israel
[3] Oak Ridge National Laboratory, Oak Ridge, USA

INTRODUCTION

This panel session examined two aspects of utilization of the World Wide Web. One aspect is communication of technical data through web sites that provide repositories of atomic and molecular data accessible through searchable databases. The other aspect is use of the World Wide Web as a means of communication about issues of mutual concern among data producers, data compilers and evaluators, and data users. The latter aspect includes communication of information relevant to future meetings of the ICAMDATA series. Conclusions and recommendations based on remarks at the session are given below.

PANEL DISCUSSION

The main points made at this panel session are briefly given in this section.

The World Wide Web is the medium of choice for communicating data to a wide audience. It is widely used for dissemination of data by data centers world wide and its use promises to increase in the future. At the same time, the very fact that data may easily be distributed by anyone means that critical evaluation by experts is crucial in order for users to be able distinguish reliable information from all that can be found on the Web.

A Web site that gives a comprehensive listing of atomic data Web sites is maintained at *http://plasma-gate.weizmann.ac.il/*. Users are requested to provide information on sites that might be added to this listing so that it will be as complete as possible.

A database that gives a wide variety of properties of chemical compounds in an integrated searchable format is maintained at *http://webbook.nist.gov/chemistry/*.

Data compiled and distributed on the Web should not be hindered by copyright of the original source literature. The quotation of published results on the Web is equivalent to quoting results in publications provided the original source is credited. In fact, it is critical to give full credit to the original source of data in order for data producers to be recognized. The related issue of international copyright laws directed at databases is discussed in one of the accompanying articles in this volume by R. S. Berry.

CONCLUSIONS AND RECOMMENDATIONS

Some general conclusions may be drawn concerning utilization of the World Wide Web for communication of and about data.

- The World Wide Web is an important means of data exchange now and for the future. It is already the preferred way of distributing data by data centers; its use is increasing, and the techniques of making data accessible are becoming more sophisticated.

- There is a large quantity of information available on the Web, and its quality and integrity is not necessarily known. It is important that data be available on the Web that has been critically evaluated by experts. Support and continuation of data compilation and dissemination coupled with critical evaluation is essential.

- Web sites that distribute atomic and molecular data should provide means for users to communicate their needs for data or comments and suggestions for improvement of the databases.

- Web databases should provide information to users about changes made to the data over time. One way to do this is to provide access to earlier versions to the database so that the user can obtain information about earlier contents of the database.

- The ICAMDATA Web site, at *http://physics.nist.gov/icamdata* should maintained indefinitely and distributed over mirror sites in order to provide efficient access to the international community.

- An ad hoc committee will provide advice on use of the Web to publicize ICAMDATA issues and to help prepare for the second meeting. The committee consists of:
  - James Babb, Harvard-Smithsonian Center for Astrophysics, USA
  - Peter Mohr, National Institute of Standards and Technology, USA
  - Izumi Murakami, National Institute for Fusion Science, Japan
  - David Schultz (Chair), Oak Ridge National Laboratory, USA
  - Yuri Ralchenko, Weizmann Institute of Science, Israel.

Author Index

A

Adler, H. G., 315
Anderson, H., 37, 259

B

Badnell, N. R., 259
Bardsley, J. N., 333
Becker, K., 391
Berry, R. S., 7, 397
Bliek, F. W., 37, 259
Brown, L., 401
Brown, L. R., 159
Butler, K., 23

C

Clark, R., 391
Crandall, D. H., 287

D

Dalgarno, A., 193
Dere, K., 213
Deslattes, R. D., 89
Dunn, G. H., 57

F

Flaud, J.-M., 221

G

Griffin, D. C., 259

H

Hoekstra, R., 37, 259
Horton, L. D., 259
Hougen, J. T., 377

H (cont.)

Howman, A., 259

I

Indelicato, P., 89

J

Janev, R., 391

K

Kallman, T. R., 203
Kelleher, D. E., 105
Kessler, Jr., E. G., 89
Konig, R., 259

L

Lane, N. F., 3
Lindroth, E., 89
Lister, G., 391
Lubinski, G., 37

M

Maggi, C. F., 259
Mallard, G., 403
Mason, H., 213
McCracken, G. M., 259
Mendoza, C., 397
Mohr, P., 403
Morgenstern, R., 37

N

Niemax, K., 391

O

Olson, R. E., 37, 259
O'Mullane, M. G., 259

P

Pindzola, M. S., 259
Post, D. E., 233
Presnyakov, L. P., 147

R

Ralchenko, U., 403
Rumble, J., 401

S

Schmoranzer, H., 67
Schultz, D. R., 119, 397, 403
Smith, P., 401

Sommerer, T. J., 295
Stamp, M. F., 259
Summers, H. P., 37, 259

T

Tawara, H., 131, 401

V

Volz, U., 67
von Hellermann, M., 259

W

Wiese, W. L., 105

Y

Younger, S., 397